AF606514

Adrenal Cortex

Butterworths International Medical Reviews

Clinical Endocrinology

Published titles

1 **The Pituitary**
Colin Beardwell and Gary L. Robertson

2 **Calcium Disorders**
David A. Heath and Stephen J. Marx

3 **Essential Hypertension as an Endocrine Disease**
Christopher R. W. Edwards and Robert M. Carey

Adrenal Cortex

Edited by

David C. Anderson, MD, MSc, MRCPath, FRCP
Reader in Medicine (Endocrinology), University of Manchester, School of Medicine, Hope Hospital, Salford, UK

and

Jeremy S. D. Winter, MD, FRCP(C)
Professor, Department of Paediatrics, University of Manitoba; Director, Section of Endocrinology–Metabolism, Children's Hospital of Winnipeg, Winnipeg, Manitoba, Canada

Butterworths
London Boston Durban Singapore Sydney Toronto Wellington

First published 1985

British Library Cataloguing in Publication Data

Adrenal cortex.–(Clinical endocrinology,
ISSN 0260-0072; 4)–(Butterworths international medical reviews)
1. Adrenal cortex–Diseases
I. Anderson, David C. II. Winter, Jeremy, S. D.
III. Series
616.4′5 RC659

ISBN 0-407-02275-9

Photoset by Butterworths Litho Preparation Department
Printed and bound in England by Robert Hartnoll Ltd., Bodmin, Cornwall

Preface

As Neena Schwartz (1984) so elegantly described in a recent presidential address to the Endocrine Society, intellectual progress in endocrinology has occurred through sequential scientific revolutions, each resulting in the formulation of a paradigm to explain in a coherent framework hitherto paradoxical observations. These revolutions are followed by stable periods during which new methodologies and data merely reinforce accepted truths, but eventually dissonant clinical and laboratory observations accumulate which raise questions about facets of the operative model, and force the development of a new one.

Many endocrinologists would argue that our current model of adrenocortical function and regulation has stood the test of time, and can still serve both the clinician and the investigator. They might point out that all the important steroid hormones have probably been isolated, and their biosynthetic pathways elucidated – it is interesting that Lieberman, Greenfield and Wolfson. (1984), in a recent review, have taken issue with these traditional two-dimensional pathways of steroidogenesis and the thought patterns which they have imposed. The discovery of parallel ACTH–cortisol and renin–angiotensin–aldosterone systems, together with the apparent structural and functional zonation of the cortex, seemed to resolve the dual physiological roles of the adrenals – but what factors impose and maintain this separation of glomerulosa and fasciculata? Furthermore, what is the significance of the considerable secretion of C_{19} 'androgens' during both fetal and adult life, and how is this regulated, if at all? The demonstration of high-affinity receptors for steroid hormones and their interaction with the target cell genome greatly clarified mechanisms of steroid action; but, as Funder discusses, there are disquieting anomalies in the specificity and distribution of what have been viewed as unique glucocorticoid or mineralocorticoid receptors.

In clinical medicine the past decade has seen the advent of new investigative and therapeutic techniques for the management of adrenal disease. However, as discussed by Ratcliffe and by Faiman, these modalities have not resolved the basic questions regarding the aetiology and pathogenesis of hypercortisolism that some thought had been settled by Harvey Cushing. A role for the adrenals in the pathogenesis of hypertension has been suspected and discussed for several decades, but, as Valloton points out, practical results in this area have been disappointing. Even the inborn errors of steroidogenesis, definitively described in the days of Lawson Wilkins, now betray a disturbing trend to phenotypic variability. As New discusses, one can categorize these variant forms of congenital adrenal hyperplasia by their clinical presentation; but certainly in the next decade the new techniques of molecular biology will uncover an even greater range of genetic heterogeneity in these disorders, similar to that seen in the thalassaemias.

A common theme running through several of the essays in this volume, including those of Hornsby, Waterman and Simpson, Weinkove and Anderson, Winter, and Dewis and Anderson, is the logical relationship which must exist in the adrenal cortex between structure and function. In particular, the unique centripetal vascular system of the adrenal imposes a significant transcortical gradient in intracellular steroid concentrations and exposes the adrenal medulla to high levels of cortisol and other steroids. Certainly this is one factor which contributes to the functional and structural zonation of the cortex, and may be a major determinant of both adrenal androgen secretion and medullary catecholamine biosynthesis.

Finally, Gustavson and Benet have prepared a comprehensive review of the pharmacokinetics of administered glucocorticoids. Such information has not been made readily available to clinicians, even though these agents are widely used for both endocrine and non-endocrine conditions.

It should be clear that we have not set out to prepare another textbook, but rather to gather together the opinions of seasoned investigators regarding the current state of adrenal physiology and disease, with particular emphasis upon those observations which question the validity of accepted truths. The volume is intended to be provocative and possibly even seditious, to the extent that it may promote controversy and even an intellectual revolution in this field. Each essay poses some unsettling questions, and the discerning reader will be made aware that not all the authors share the same vision of reality. This is as it should be, since we are concerned only with the construction of a more effective paradigm, not with the imposition of a new authority.

David C. Anderson
Jeremy S. D. Winter

References

SCHWARTZ, N. B. (1984) Endocrinology as paradigm, endocrinology as authority. *Endocrinology*, **114,** 308–313

LIEBERMAN, S., GREENFIELD, N. J. AND WOLFSON, A. (1984) A heuristic proposal for understanding steroidogenic processes. *Endocrine Reviews*, **5,** 128–148

Contributors

David C. Anderson, MD, MSc, FRCP
Reader in Medicine (Endocrinology) and Consultant Physician, University of Manchester, Department of Medicine, Hope Hospital, Salford, UK

Leslie Z. Benet, PhD
Professor and Chairman, Department of Pharmacy, School of Pharmacy, University of California, San Francisco, California, USA

P. Dewis, BSc, MB, ChB, MRCP
Tutor in Medicine, University of Manchester, Department of Medicine, Hope Hospital, Salford, UK

Charles Faiman, MD, MSc, FRCP(C)
Professor, Departments of Medicine and Physiology, University of Manitoba; Head, Section of Endocrinology and Metabolism, Health Sciences Centre, Winnipeg, Manitoba, Canada

L. Favre, MD
Division of Endocrinology, Department of Medicine, Faculty of Medicine, Geneva, Switzerland

John W. Funder, MD, PhD, FRACP
Senior Principal Research Fellow, NH & MRC; Deputy Director, Medical Research Centre, Prince Henry's Hospital, Melbourne, Victoria, Australia

Linda E. Gustavson, PhD
Graduate Fellow, Department of Pharmacy, School of Pharmacy, University of California, USA

Peter J. Hornsby, PhD
Department of Medicine, University of California, San Diego, La Jolla, California, USA

Maria I. New, MD
Professor and Chairman, Department of Pediatrics; Chief, Pediatric Endocrinology; Associate Program Director, Pediatric Clinical Research Center; and Harold and Percy Uris Professor of Pediatric Endocrinology and Metabolism, Division of Pediatric Endocrinology, The New York Hospital–Cornell Medical Center, New York, New York, USA

J. G. Ratcliffe, DM, MRCPath, FRCP (Glas)
Professor of Chemical Pathology, University of Manchester, Hope Hospital, Salford, UK

Evan R. Simpson, PhD
Cecil H. and Ida Green Center for Reproductive Biology Sciences, and the Departments of Obstetrics-Gynecology and Biochemistry, University of Texas Health Science Center, Dallas, Texas, USA

M. B. Vallotton
Professor, Division of Endocrinology, Department of Medicine, Faculty of Medicine, Geneva, Switzerland

Michael R. Waterman, PhD
Cecil H. and Ida Green Center for Reproductive Biology Sciences, and the Departments of Obstetrics-Gynecology and Biochemistry, University of Texas Health Science Center, Dallas, Texas, USA

C. Weinkove, BSc, MB ChB, FCP(SA), PhD
Senior Lecturer in Chemical Pathology, University of Manchester, Hope Hospital, Salford, UK

Jeremy S. D. Winter, MD, FRCP(C)
Professor, Department of Paediatrics, University of Manitoba, and Head, Section of Endocrinology and Metabolism, Children's Hospital of Winnipeg, Winnipeg, Manitoba, Canada

Contents

1
The regulation of adrenocortical function by control of growth and structure

Peter J. Hornsby

Overview

The major proposal to be presented here is that the regulation of the structure of the mammalian adrenal cortex is an intrinsic component of the regulation of the synthesis of its steroid hormones. The two important structural factors to be considered are (1) adrenocortical mass and (2) zonation. Adrenocortical structure is involved in the regulation of the synthesis of the mineralocorticoid, glucocorticoid and androgenic steroids secreted by the adrenal under both normal and pathological conditions.

THE STRUCTURE OF THE ADRENAL CORTEX

The adrenal glands vary in shape from almost spherical, in fetal life and in the adult of some species (e.g. the rat, mouse, rabbit and guinea pig), through somewhat flattened to more or less leaf-like, e.g. in humans and cattle (Bachmann, 1954). Although the size of the gland varies with the size of the animal, the thickness of the cortex remains remarkably constant. In large, flattened glands it may become highly convoluted (Neville and O'Hare, 1982). This apparent requirement for a constant cortical width almost certainly results from the arrangement of the blood supply (*see below*). A diagrammatic representation of the vascular arrangement is illustrated in *Figure 1.1,* together with a scanning electron microscopic image of a cast of the capillary structure (Ohtani *et al.*, 1983). The arterial supply derives from several vessels to the capsule, where a network of arterioles is formed. A few arteries penetrate the cortex and extend directly to the medulla. However, most of the capsular arterioles give rise to capillaries that cross the cortex, supplying the adrenocortical cells and finally uniting in a plexus at the corticomedullary boundary. The significance of this centripetal arrangement is a major focus of this review. Apparently, the cortex cannot increase beyond a certain thickness because this would cause the venous end of the capillary bed to be too far from the arterial supply.

Traditionally, since the description of the histology of the mammalian adrenal gland by Arnold in 1866, the cortex has been subdivided into three concentric

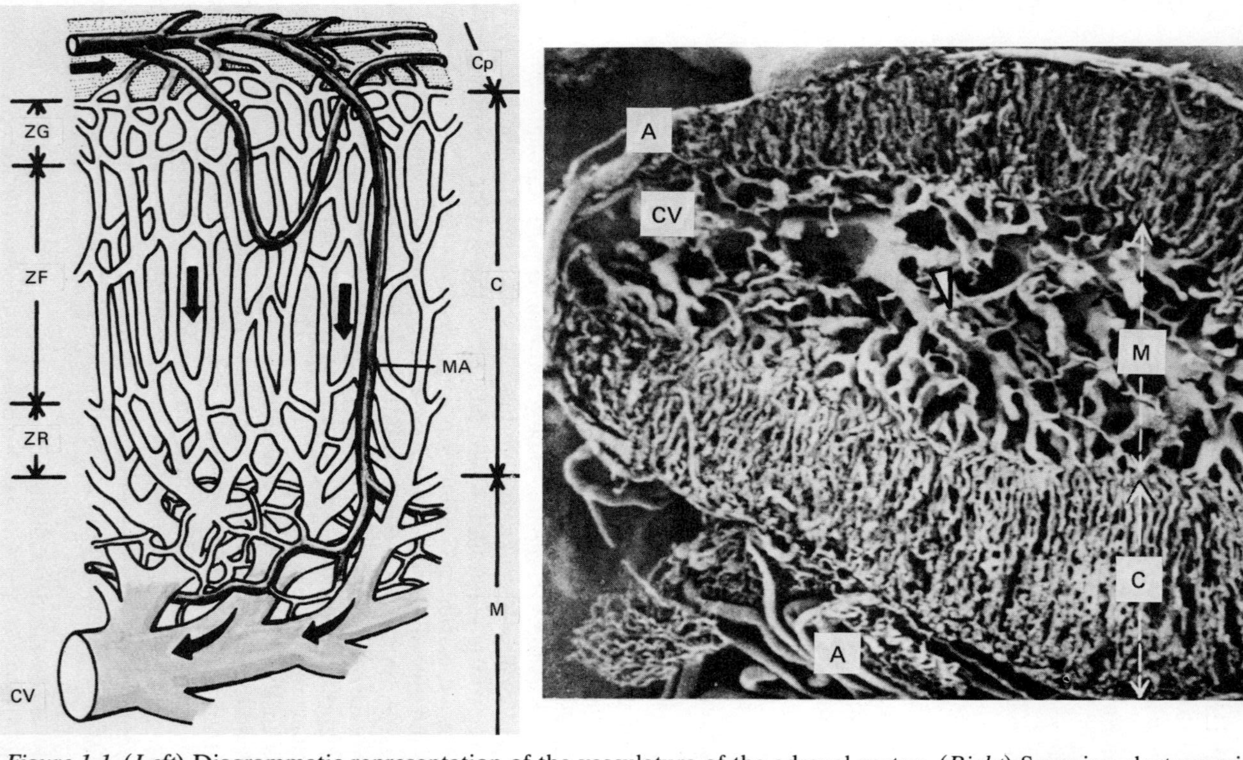

Figure 1.1 (*Left*) Diagrammatic representation of the vasculature of the adrenal cortex. (*Right*) Scanning electron micrograph of a cast of the injected vasculature of the rat adrenal cortex. A = adrenal artery; C = cortex; M = medulla; ZG = zona glomerulosa; ZF = zona fasciculata; ZR = zona reticularis; Cp = capsule; CV = central vein; MA = medullary artery. (From Ohtani *et al.*, 1983, courtesy of the Editor and Publishers, *Archivum Histologicum Japonicum*)

zones, the outer zona glomerulosa, middle zona fasciculata, and inner zona reticularis, named for the different cellular arrangements in the different regions of the gland (*glomus*, ball; *fascis*, bundle; and *rete*, net). The current definition of the zones still rests on recognition of these different cellular arrangements. The traditional three zones may not always be present. The reticularis may be absent or it may be replaced by zones of different morphology, and sometimes function, such as the X-zone in the female mouse (Dunn, 1970; Shire, 1979) and the fetal zone in the human and some other primates (Lanman, 1953; Idelman, 1970; Villee, 1972). These zones generally exhibit sufficient similarity to the zona reticularis for them to be considered together in this review. Concerning zonation, the following important points must be considered and are discussed in this chapter:

(1) The definition of the different zones is morphological rather than functional, even though it is the difference in function between the zones that is biologically important.
(2) Functional zonation should not be expected always to follow morphological zonation.
(3) The evidence available from both *in vivo* and cell culture observations strongly supports the concept that all cells of the different zones of the adrenal cortex are of the same cell type, exhibiting temporary functional and morphological differences according to their location in the cortex.
(4) There is no evidence for separate mitogenic regulation of zonal mass. The mass of the zones is controlled by the combination of the regulation of adrenocortical mass as a whole and the regulation of zonal width.

The morphological and biochemical zonation of the adrenal cortex can be clearly illustrated with the guinea-pig gland. Inside the capsule is the zona glomerulosa, the cells of which are arranged in ball-like clusters (some 20–30 cells per cluster, approximately 13 μm in diameter), usually enclosed in capsular-like tissue. These cells characteristically have a high nuclear/cytoplasmic ratio. Zona fasciculata cells (21 μm) are usually vacuolated due to the loss of lipid which occurs during some histological procedures (e.g. toluidine blue, and haematoxylin and eosin staining) and have been classically described as 'clear cells'. They are arranged in columns extending from the zona glomerulosa to the zona reticularis. The cells of this innermost zone of the cortex (approximately 23 μm in diameter) have a relatively lipid-sparse cytoplasm and are commonly described as 'compact'. The above morphological zonation can also be seen macroscopically if the gland is sliced in the mid-line along the major axis (*see Plate 1.1a opposite page 6*). Immediately abutting the thin reddish/brown border surrounding the gland is a thick yellow zone (zona fasciculata) which contains lipid-rich cells (*Plate 1.1b*) and occupies about 30–40% of the cortical surface in this plane. This zone can sometimes be seen to interdigitate into the dark red/brown gelatinous zone (zona reticularis) which extends to the medulla in the centre of the gland.

The reason for the different morphological arrangement of the cells in the zones is not known. It may be that the primary cause is a variation in the contacts between cells, or in the type and amount of extracellular matrix laid down. Another possibility is that the vasculature may dictate the parenchymal cell arrangements. Furthermore, lipid content, although readily observable, provides an example of a distribution pattern that does not always follow morphological zonation. The zonation of lipid droplets is subject to change under some conditions and this has often led to erroneous identification of the zones (Neville and O'Hare, 1982). For example, when there is an excessive demand for adrenal steroid synthesis, as during the reaction to stress, depletion of lipid from the cortex is observed, which may appear as a narrowing of the zona fasciculata and concomitant widening of the zona glomerulosa and zona reticularis. Examination of the cortex by other criteria shows that morphological zonation has not changed (Neville and O'Hare, 1982). As will be discussed later, changes in both morphological and functional zonation also occur during chronic ACTH stimulation, with the glomerulosa width being diminished rather than increased.

The biochemical zonation of the gland can be well illustrated using cytochemical techniques. The distribution of 3β-hydroxysteroid dehydrogenase (HSD) reveals a lower level of activity in the inner zone, indicating an important biochemical difference between the zona reticularis and the outer zones (*Plate 1.1c*). This photomicrograph was taken under conditions of dark-ground illumination, which has the effect of reversing the colour intensities seen under normal light, but highlights the zones beautifully. The capsule and medulla are without reaction product and show up as a translucent blue colour under these conditions. The lower activity of 3β-hydroxysteroid dehydrogenase in the innermost zone may be important in the regulation of the synthesis of the adrenal androgens, principally dehydroepiandrosterone (DHA) and its sulfate, in the adult reticularis and in the fetal zone (*see below*).

The distribution of the enzymic sequence responsible for NADPH reoxidation (the so called 'diaphorase') is shown in *Plate 1.1d*. This reaction clearly delineates the zona glomerulosa from the zona fasciculata. In contrast, alkaline phosphatase activity is localized in the zona fasciculata, being very low in the zona glomerulosa and entirely absent in the zona reticularis (*Plate 1.1e*).

REGULATION OF NORMAL STEROIDOGENESIS BY ADRENOCORTICAL STRUCTURE

The rate of secretion of any particular adrenocortical steroid is determined by the product of its rate of synthesis (per unit mass of adrenocortical tissue) and the mass of adrenocortical tissue involved in its synthesis; this may not be the entire cortex but may consist of one or more zones or parts of zones. In turn, the rate of production of a given adrenocortical steroid is determined by the product of (i) the total rate of synthesis of all steroids in the part of the cortex involved in its synthesis and (ii) the proportion of total steroid output that the steroid represents within this part.

(1) The rate of total steroidogenesis is determined by the rate of supply of cholesterol to the cytochrome P-450 that cleaves it to yield pregnenolone. The control of this level of adrenocortical steroidogenesis is reviewed in Chapter 3. The rate of flux through the rate-limiting step determines the rate of synthesis of the sum of the steroid products, but does not determine the rate of synthesis of any individual steroid.
(2) The relative activities of the enzymes of the steroidogenic pathway beyond the formation of pregnenolone determine the pattern of steroidogenesis, i.e. which steroids are produced and in what ratio.

A failure to recognize that the rate of synthesis of a given steroid is regulated at these two different levels may lead to the mistaken search for individual hormones that may specifically stimulate its output. Examples of this will become apparent later in this survey.

Thus, an important means of regulating the synthesis of a steroid is the regulation of the mass of tissue involved in its synthesis. For the mineralocorticoid aldosterone this mass corresponds approximately to the zona glomerulosa, since synthesis of this steroid appears to be confined to this zone. For species in which the adrenal cortex secretes large quantities of androgenic steroids, there is evidence that most of the biosynthesis of these steroids is localized to the inner zones, whose mass is therefore a regulatory factor in adrenal androgen production. However, since adrenal androgens may also be synthesized by the zona fasciculata, this is not as unequivocal as the regulation of aldosterone production by the mass of the zona glomerulosa. The glucocorticoid cortisol is synthesized by both the zona fasciculata and the zona reticularis, but not by the zona glomerulosa; in addition, in species that secrete large amounts of adrenal androgens, proportionally more cortisol is synthesized by the zona fasciculata than by the zona reticularis. Thus, a factor in the regulation of cortisol synthesis is the mass of the zona fasciculata. Similarly, the mass of the zona fasciculata is a regulatory factor for corticosterone synthesis, although a small amount of corticosterone is synthesized by the zona glomerulosa. For both glucocorticoids, however, the major structural determinant of secretion rate is the mass of the adrenal cortex as a whole, both because glucocorticoid secretion is not confined to a single zone and because the zona fasciculata is the predominant zone of the cortex.

Thus, the following topics are appropriate to this subject:

(1) Control of the mass of the adrenal cortex;
(2) Evidence for the localization of the synthesis of particular steroids to particular zones;
(3) Control of the mass of the zona glomerulosa in proportion to the rest of the cortex;
(4) Control of the mass of the zona reticularis, or other specialized inner zones, in proportion to the rest of the cortex.

Control of the mass of the adrenal cortex

An appreciation of the mode of regulation of the mass of the adrenal cortex requires first an understanding of the life history of the adrenocortical cell. Adrenocortical mass is controlled by the balance of increases and decreases of growth stimuli and by the rate of cell death, occurring both normally in turnover and in involution in more specialized circumstances.

The life history of the adrenocortical cell

The adrenal cortex is a mesodermal tissue; its embryological development has been reviewed elsewhere (Jirasek, 1968; Neville and O'Hare, 1982). The basic pattern of growth of the adrenal gland is similar in all mammalian species, but differs in detail regarding (a) the overall shape of the gland and (b) the transient development and regression of specialized inner zones such as the fetal zone and the X-zone or, conversely, the absence of a separate zone in this position. The regulation of the fetal adrenal cortex in the human is reviewed in Chapter 2.

In its simplest form, the adrenal cortex can be visualized as a ball of cells, more or less spherical in shape. Evidently, therefore, an increase in substance is possible only if that increase occurs on the periphery of the gland, unless considerable internal reorganization is to occur. During both embryonic and postnatal growth, most cell division takes place in the outer adrenal cortex, the zona glomerulosa and outer zona fasciculata (Ford and Young, 1963; Wright, 1971; Dhom, 1973; Wright, Voncina and Morley, 1973; Belloni *et al.*, 1978), or in the definitive zone in the fetal adrenal cortex (Johannisson, 1979). The actual mechanism restricting cell division to the outer region of the cortex is not known, but is probably related to the pattern of the blood supply described earlier. Probably the outer region of the cortex is better supplied with some nutrient essential to proliferation because it is closer to the arterial blood supply. As will be discussed later, adrenocortical differentiated functions may also be regulated by the proximity of the cells to the arterial or venous sides of the capillary bed.

Once the adrenal cortex has reached its mature size, the rate of cell division in the outer cortex decreases to that required to balance the rate of loss of cells due to death, with temporary alterations in rate when more or less adrenocortical tissue is required by the organism. In the regulation of adrenocortical mass in the mature cortex, most cell division still occurs in the outer cortex (Wright, 1971; Pappritz, Trieb and Dohm, 1972; Stocker and Schmid, 1973; Wright, Voncina and Morley, 1973; Payet, Lehoux and Isler, 1980). It has sometimes been assumed that each zone of the adrenal cortex is self-maintaining, i.e. that cells in each zone are derived from other cells in that zone only. This concept became popular when it became clear that there were distinct functional differences between the zones ('zonal theory'), and was followed logically by a tendency to regard the adrenal cortex as 'two (or three) glands in one'. However, the observed distribution of mitoses in the cortex makes this concept of independence of the zones unlikely. Direct measurements show that the rate of production of cells in the zona glomerulosa exceeds the requirements of this zone for self-maintenance (Wright, 1981). Conversely, the rate of cell division in the zona reticularis is insufficient to maintain this zone (Wright, 1981). The zona reticularis was early observed to be the 'senescent' zone of the adrenal cortex (Bachmann, 1954). Again, the concept of functional independence of the zones made this view unpopular. Evidence for the localization of most cell death in the zona reticularis has come from the finding that most apoptotic bodies are found in this zone (Wyllie, Kerr and Currie, 1973; Wyllie *et al.*, 1973) and also from deposits of large amounts of age pigment (lipofuscin) in this zone (Sahinen and Soderwall, 1965; Weglicki, Reichel and Nair, 1968; Dunn, 1970; Akazawa *et al.*, 1980; Walton, 1982). The likely origin of these lipofuscin deposits has been discussed in detail elsewhere (Hornsby and Crivello, 1983); they probably result from lipid peroxidation originating from the interaction of steroids and cytochrome P-450s. This interaction may contribute to the ageing process of the adrenocortical cell (Hornsby and Crivello, 1983).

These observations clearly support the concept that cells in the inner zona glomerulosa are pushed inwards into the cortex by the pressure of cell division and become fasciculata cells; further in, they become reticularis cells, and may end their life span in this zone. This is an old concept, originally known as the 'escalator' or 'cell migration' theory (Bachmann, 1954). These terms, however, are somewhat misleading. Parenchymal cells in a tissue *in vivo* are not likely to move relative to one another, although adrenocortical cells isolated in culture, like all mesodermal cell types, are capable of migration on a suitable substratum (Gospodarowicz *et al.*,

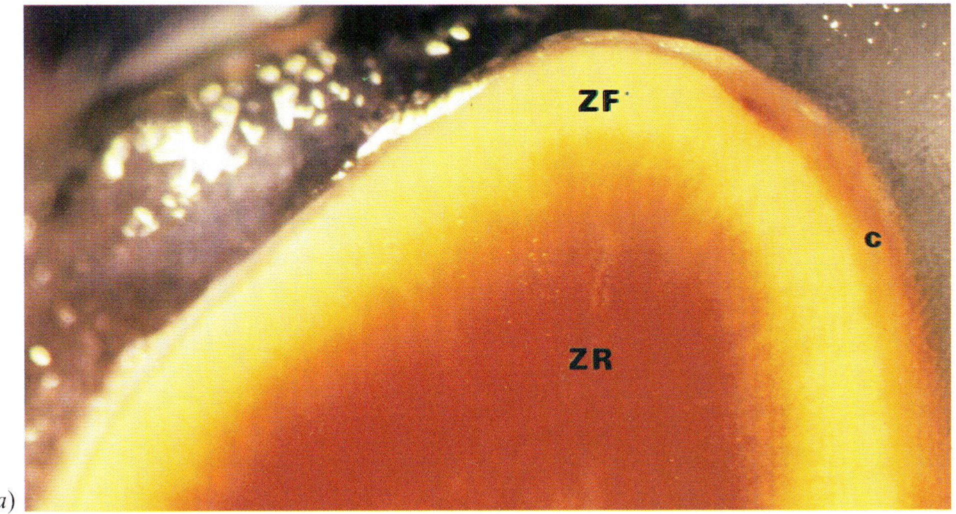

Plate 1.1 Morphological and biochemical zonation of the adrenal cortex. (*a*) Mid-line slice of the guinea-pig adrenal gland, unstained, viewed along the major axis (×10). ZF = zona fasciculata; ZR = zona reticularis; c = capsule

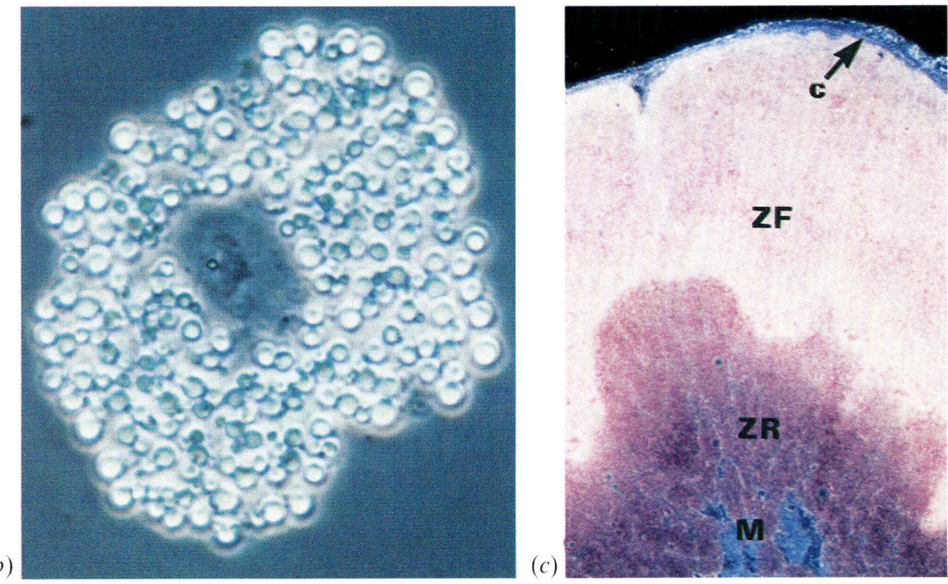

Plate 1.1(*b*) Individual cell from the zona fasciculata (×1000 phase contrast) showing lipid droplets. (*c*) Distribution of 3β-hydroxysteroid dehydrogenase (×20, dark-ground illumination) in a fresh-frozen section (10 μm) of the guinea-pig adrenal gland (*see text for details*). This enzyme assay employed glycyl-glycine (0.1 mol/l, pH 7.2) as the buffer, epiandrosterone (0.4 mmol/l) as the substrate, NAD^+ (3.0 mmol/l) as the co-factor and nitro-blue tetrazolium (2.5 mmol/l) as the hydrogen acceptor. The reaction was for 60 min at 37 °C. M = medulla

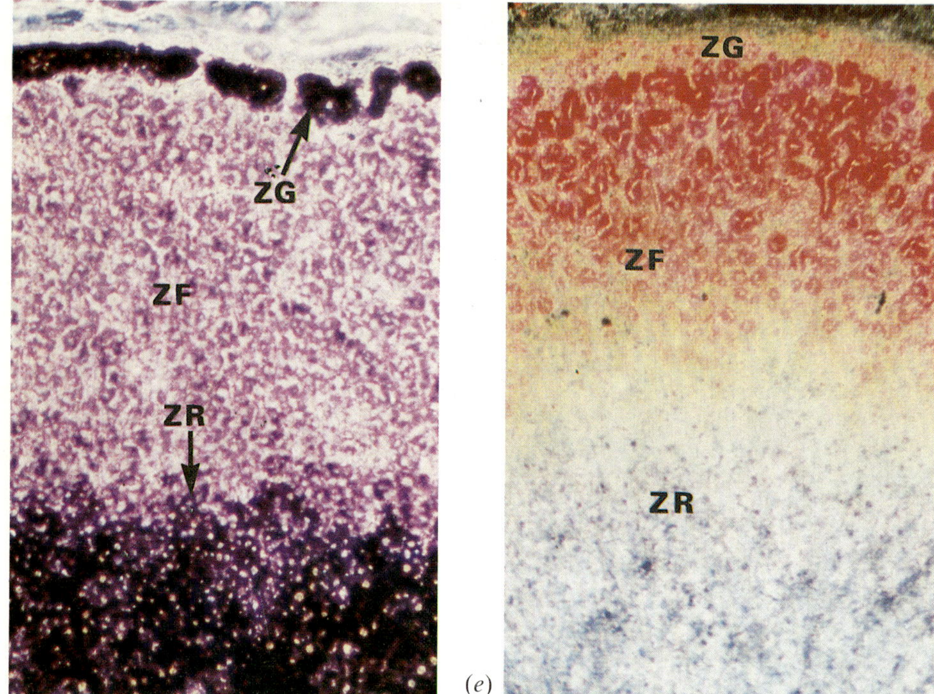

Plate 1.1(*d*) Distribution of NADPH diaphorase (×50). This enzyme sequence was assayed for 20 min in the presence of NADPH (3 mmol/l) and nitro-blue tetrazolium (2.5 mmol/l). All other details are as for (*c*). (*e*) Distribution of alkaline phosphatase (×50). The activity of this enzyme was assayed for 40 min at 37 °C. The buffer was glycyl-glycine (pH 9.0) and the substrate naphthol AS-B1-phosphate (1 mmol/l). The end-product was visualized by a post-coupling technique using Fast Garnet GBC (1 mg/ml) in ice-cold acetate buffer (0.1 mol/l, pH 6.2) for 10 min. ZG = zona glomerulosa, ZF = zona fasciculata; ZR = zona reticularis. (Photomicrographs reproduced by kind permission of Dr W. R. Robertson and Dr M. Gordon, Department of Chemical Pathology, University of Manchester, Hope Hospital, Salford, UK)

1977; Hornsby and Gill, 1977; Gill, Hornsby and Simonian, 1979). This appears to develop as a reaction to the cell culture environment, involving development of an extensive cytoskeleton (Rainey, Hornsby and Shay, 1983). *In vivo* the cells may be visualized as being pushed *en masse* toward the medulla by cell division in the outer cortex, balanced by cell death in the inner cortex. Cell culture experiments have provided direct evidence of such interconversion of zonal cell types (Hornsby, O'Hare and Neville, 1974; Crivello, Hornsby and Gill, 1982; Crivello and Gill, 1983; *see later*).

The cellular ageing process of the adrenocortical cell *in vivo* is unique in that it encompasses different phases of steroid secretion (mineralocorticoid, glucocorticoid, and adrenal androgen) and so represents a maturational as well as a senescent process. While the effects of the steroid gradient on adrenocortical functions may be investigated in isolated cells, a senescence process different from that *in vivo* occurs in adrenocortical cells in culture. Cultured adrenal cells have a finite replicative life span similar to other cell types, such as human fibroblasts (Hornsby, Simonian and Gill, 1979). The cellular senescence process resulting in a finite replicative life span and the cellular ageing process that appears to occur *in vivo* are not mutually exclusive processes.

The adrenal cortex is also capable of dramatic regenerative growth when required. A functional, zoned adrenal cortex will regenerate from fragments of adrenocortical tissue, either remaining *in situ* after 'enucleation' (a process of extruding the medulla and most of the cortex through a slit in the capsule) or when transplanted elsewhere in the body (Greep and Deane, 1949; Plzak, 1960; Voitkevich, 1970; Holzwarth, Shinsako and Dallman, 1980; Srougi, Gittes and Underwood, 1980; Wilkinson, Shinsako and Dallman, 1981), and even from a trypsinized cell suspension inoculated under the skin (Usadel *et al.*, 1970; Schwedes *et al.*, 1974). In regeneration from small fragments, the pattern of growth is similar to that occurring during normal development, with formation of a morphologically unorganized cell mass, followed by a period of growth with mitoses in the outer region of the regenerating tissue, and finally re-establishment of functional zones. Apparently, regeneration can take place from adrenocortical fragments from any zone; fragments of medullary tissue with adherent cortex transplanted to the eye regenerate and are apparently also functional (Coupland, 1957).

It therefore appears that, as in some other tissues, such as the liver, all of the parenchymal cells are capable of division when required even though they may usually be mitotically quiescent. This is also demonstrable in bovine adrenocortical cells in primary culture; all of the cells attaching to the culture dish from the primary suspension enter DNA synthesis and form part of the proliferating population (Hornsby, Aldern and Harris, 1984). Thus there are no stem cells, although mitoses are normally confined to the outer regions of the cortex. The process of movement from the zona glomerulosa to the zona reticularis involves predominantly a simplification of enzyme content rather than acquisition of new differentiated features, as will be discussed.

Stimulation of adrenocortical growth

It has long been observed that the size of the adrenal cortex is closely regulated *in vivo* and that adrenocortical volume is closely associated with the functional activity of the gland. Diseases of excessive adrenocortical activity are associated with a large adrenal gland and diseases of deficiency with adrenal atrophy. It has been known for some time that removal of one adrenal gland stimulates compensatory growth of the other (Engeland and Dallman, 1975; Dallman, Engeland and Shinsako, 1976). However, the mechanisms for regulation of adrenocortical growth have only recently begun to be elucidated. Whereas some other polypeptide hormones are direct mitogens for the adrenal cortex, ACTH is probably an indirect mitogen, acting to increase the delivery of growth factors to adrenocortical tissue, e.g. by effects on the adrenal vascular system (O'Hare, Ellison and Neville, 1978; Neville and O'Hare, 1979, 1982; Gill *et al.*, 1982; Hornsby and Crivello, 1983). In addition, adrenal growth is at least partially regulated by neural mechanisms (Engeland and Dallman, 1975; Dallman, Engeland and Shinsako, 1976). An important distinction will also be made between control of adrenocortical growth by regulation of adrenocortical cell size, and regulation by adrenocortical cell number.

ACTH

Despite these cautions, ACTH is clearly the major hormonal regulator of adrenocortical growth *in vivo.* Excess pituitary secretion of ACTH in diseases such as Cushing's syndrome or the various forms of congenital hyperplasia is associated with enlargement of the adrenal glands (Neville and O'Hare, 1982). When purified pituitary ACTH or synthetic ACTH is administered *in vivo*, a substantial increase in adrenocortical mass is seen 24 hours after treatment (Studzinski, Hay and Symington, 1963; Bransome, 1968). The growth stimulation by ACTH is usually accounted for predominantly by an increase in cell size rather than an increase in cell number or DNA content (Nickerson, 1975; Bransome, 1968). Blocking ACTH secretion from the pituitary, e.g. by administration of a glucocorticoid, results in an early cessation of adrenocortical growth, including DNA synthesis and cell division, and later in atrophy with loss of protein, RNA and DNA content (Bransome, 1968, Wright, Appleton and Morley, 1974). In the absence of the pituitary, in anencephalic fetuses, adrenal growth is poor, and is restored on administration of ACTH (Lanman, 1961; Johannisson, 1979; Reynolds, 1981; Parker *et al.*, 1983). However, ACTH is not a direct mitogen for the adrenocortical cell, as demonstrated in a number of cell culture systems from several species, including rat, cow and man (O'Hare and Neville, 1973; Ramachandran and Suyama, 1975; Hornsby and Gill, 1977, 1978; Rainey, Hornsby and Shay, 1983). With very few exceptions (Morera and Saez, 1980), natural pituitary ACTH as well as synthetic ACTH has been observed to inhibit cell proliferation in these systems, often powerfully so. ACTH does not inhibit increased synthesis of RNA and protein (Gill, Hornsby and Simonian, 1979, 1980; Gill *et al.*, 1982). In culture, the effect of combined treatment with a mitogen and ACTH is cellular hypertrophy, resulting from a restriction on DNA replication without restriction of other aspects of growth. The increase in adrenocortical cell size observed *in vivo* on ACTH treatment is therefore consistent with the combined action of a growth stimulator and a substance that directly inhibits replication. Furthermore, cultured adrenocortical cells become densensitized to the inhibitory effects of ACTH on replication and then proliferate at varying rates, up to that seen in the absence of ACTH (Hornsby and Gill, 1977). It is not clear whether this happens *in vivo*, but it has been observed that in some cases most stimulation of proliferation by ACTH occurs 5 to 9 days after commencing treatment (Farese and Reddy, 1963; Imrie *et al.*, 1965), possibly due to the combined effects of desensitization to its inhibitory effects on replication and its indirect growth-promoting properties. The adrenocortical cell itself does appear to be the target for ACTH acting as an *indirect* mitogen. In congenital unresponsiveness to ACTH, a syndrome caused probably by an absence of the ACTH receptor on adrenocortical cells, there is a lack of adrenal growth, despite very high levels of circulating ACTH (Migeon *et al.*, 1968). In contrast, in congenital adrenal hyperplasia, there is enhanced adrenal growth in the presence of high circulating ACTH levels, despite the lack of secretion of the appropriate steroids (Neville and O'Hare, 1982). This indicates that the indirect mitogenic effect of ACTH is not mediated by secreted adrenal steroids. On the other hand, the indirect mitogenic effect does appear to be mediated by cyclic adenosine 3′,5′-monophosphate (cAMP), since both cholera toxin and dibutyryl cAMP stimulate growth and proliferation of the adrenal cortex in the hypophysectomized animal (Marton, Palfreyman and Schulster, 1976; Lewinski and Szkudlinski, 1981).

MITOGENIC PEPTIDES

Fibroblast and epidermal growth factors

Several polypeptides are directly mitogenic to cultured adrenocortical cells. The most potent so far identified are the substances termed fibroblast growth factors (FGFs) (Gospodarowicz, Moran and Mescher, 1978; Gospodarowicz, Moran and Bialecki, 1976; Gospodarowicz *et al.*, 1977), which act on a wide variety of cells, principally those of mesodermal origin (Gospodarowicz, Moran and Bialecki, 1976; Gospodarowicz, Moran and Mescher, 1978). They comprise two structurally unrelated groups of peptides, from the pituitary and the brain (Gospodarowicz *et al.*, 1977; Gospodarowicz, Lui and Cheng, 1982). Their physiological importance is still uncertain, since the conditions under which they may be secreted *in vivo*, either locally or for systemic circulation, have not been identified.

Epidermal growth factor (EGF) is mitogenic for human but not bovine adrenocortical cells in culture (Gospodarowicz *et al.*, 1977; Crickard, Ill and Jaffe, 1981; Hornsby *et al.*, 1983), a species difference which correlates with a difference in content of receptors for EGF (Crickard, 1981). EGF also fails to bind to rat adrenal glands *in vivo* (Chabot *et al.*, 1982). Although EGF is clearly of physiological importance in development, its role in the maintenance of adrenocortical size in the adult is unclear.

Angiotensin II

Angiotensin II differs from the previously mentioned growth factors in that it is specific for adrenocortical cells, at least in culture (Gill, Ill and Simonian, 1977; Simonian and Gill, 1979). The receptor responsible for the mitogenic response to angiotensin is probably the same as that mediating steroidogenic effects of this hormone (Gill, Ill and Simonian, 1977; Simonian and Gill, 1979). Interestingly, under certain circumstances angiotensin is also capable of exhibiting replication-inhibitory effects, e.g. in the presence of the more powerful growth stimulator FGF (Simonian and Gill, 1979). There has been only limited investigation of a possible mitogenic action of angiotensin II in the adrenal cortex *in vivo*. Chronic administration of angiotensin causes an increase in adrenocortical mass (Marx *et al.*, 1963), and sodium deprivation, known to cause increased secretion of angiotensin, results in an increase in the adrenocortical mitotic index (Race and Green, 1955). The interpretation of these results is complex. Firstly, *in vivo* there may be complex interactions between angiotensin and ACTH, both through direct effects on peptide hormone secretion and indirectly through stimulation and feedback of adrenocortical steroids (Mason *et al.,* 1979). This could result in either an over- or underestimation of the direct mitogenic effect of angiotensin. Secondly, although both inner and outer zones of the adrenal cortex have angiotensin receptors and show steroidogenic and mitogenic effects (Gill, Ill and Simonian, 1977; McKenna *et al.,* 1978; Hyatt, Bhatt and Tait, 1983), there are more receptors in the zona glomerulosa (Brecher *et al.,* 1973; Cantin *et al.,* 1982) and, at least in some species, the steroidogenic effect of angiotensin is greater in this zone (Muller, 1970; Hyatt, Bhatt and Tait, 1983). This zonation of angiotensin receptors might be expected to result in a greater mitogenic effect on the glomerulosa than on the zona fasciculata. An increase in mitoses in this zone does indeed occur during sodium depletion (Race and Green, 1955), but because this zone normally has the highest rate of cell proliferation and because ACTH also increases mitoses mainly in the outer cortex, including the zona glomerulosa (Machemer and Oehlert, 1964; Payet, Lehoux and Isler, 1980), this is not necessarily either a specific or a direct action of angiotensin. Sodium depletion also results in an increase in the mitotic index of the zona fasciculata (Race and Green, 1955). Additionally, an increase in the mass of the glomerulosa might erroneously be taken as evidence of a specific mitogenic action of angiotensin on this zone when what is really occurring is a movement inward of the boundary between the glomerulosa and the fasciculata or an increase in cell size in this zone (*see below*). Thus, the role of angiotensin as a direct mitogen in adrenocortical cell growth *in vivo* remains unknown. If it is a direct mitogen, it is not likely that, being a steroidogenic agent, it would mediate the indirect mitogenic effects of ACTH. The limited evidence available does, however, suggest that it is at least an indirect mitogen *in vivo*. It may be significant that aldosterone administration appears to inhibit compensatory adrenal growth (Dunlap and Grizzle, 1984; Grizzle and Dunlap, 1984), perhaps implying a role for angiotensin in this process (*see below*).

Insulin and the insulin-like growth factors

Insulin, somatomedins, multiplication-stimulating activity (MSA), and the insulin-like growth factors (IGFs) are mitogenic for bovine adrenocortical cells in culture (Gill, Ill and Simonian, 1977; Simonian and Gill, 1979; Simonian, White and Gill, 1982; Ill and Gospodarowicz, 1982; Li *et al.,* 1983). The insulin group of peptides synergize strongly with either angiotensin or FGF (Gill, Ill and Simonian, 1977; Simonian and Gill, 1979). Although specific effects of insulin on the function of the adrenal cortex *in vivo* have been demonstrated (Piras, Bindstein and Piras, 1973) and the adrenal cortex has receptors for insulin (Bergeron *et al.,* 1980), it is likely that its mitogenic effects here are mediated by IGF receptors. Synthetic IGF is a potent growth factor for adrenocortical cells in culture (Li *et al.,* 1983). Data on a possible role for insulin or the IGFs *in vivo* in adrenocortical cell proliferation are lacking.

Other pituitary factors

There are probably other pituitary factors besides ACTH that stimulate adrenocortical cell growth. Early work on the effects of crude pituitary preparations administered *in vivo* showed that their mitogenic effect decreased as ACTH content increased (Studzinski, Hay and Symington, 1963) and that preparations partially purified with respect to growth hormone were mitogenic, synergizing with ACTH (Cater and Stack-Dunne, 1955). Some experimental transplantable pituitary tumors secrete factors that result in adrenocortical cell proliferation *in vivo* (Nickerson, Brownie and Molleni, 1970; Nickerson, 1975;

Cathiard and Saez, 1981). Both tumors that secrete ACTH and those that do not may stimulate growth. In the cases where the tumor secretes ACTH, the adrenocortical cells appear to be hypertrophic (Nickerson, Brownie and Molleni, 1970; Nickerson, 1975). Tumors that secrete prolactin also appear to stimulate growth, and hyperprolactinemia has been noted to be accompanied by growth of the adrenal cortex (Boyar and Hellman, 1974; Bartke *et al.*, 1977; Negro-Villar, Saad and McCann, 1977; Degenhart, 1979). It has been suggested that prolactin may be a growth factor for the adrenal during fetal development (Jaffe *et al.*, 1981), but this view has been challenged by Neville and O'Hare (1982) and by Winter in Chapter 2 of this volume. Receptors for prolactin and other lactogenic hormones are abundant in the adrenal cortex, and their regulation has been examined by Bolander *et al.* (1976), Marshall, Bruni and Meites (1979), Calvo *et al.* (1981) and Katikineni, Davies and Catt (1981). Both growth hormone and prolactin, in common with other purified pituitary hormones, consistently fail to show direct mitogenic activity in cultured adrenocortical cells (Gill, Hornsby and Simonian, 1979; Cathiard and Saez, 1981; Gill *et al.*, 1982; Hornsby *et al.*, 1983), but prolactin, like ACTH, could be an indirect mitogen *in vivo*. The mitogenic effects of luteinizing hormone (LH) in culture were found to result from the presence of an impurity, either FGF or an FGF-like substance (Gospodarowicz, Moran and Bialecki, 1976). It is possible, therefore, that some such substance was responsible for the observed growth-promoting effects of crude pituitary preparations. Other observations support the concept of non-ACTH adrenocortical growth factors in the pituitary. Pituitary cells in culture synthesize factors mitogenic for adrenocortical cells, other than FGF, that stimulate the proliferation of the Y-1 adrenocortical tumor cell line (Kudlow and Gerrie, 1983). The plasma of patients with Cushing's disease contains circulating factors, distinct from ACTH but so far unidentified, that stimulate adrenocortical growth when administered to rats (Segal and Christy, 1968; Segal *et al.*, 1970). It is therefore possible that pituitary growth factors are released into the circulation *in vivo* under circumstances of excess ACTH secretion, and that these may enhance the indirect growth-stimulatory effects of ACTH.

Many other hormones and factors have been tested for their growth-promoting effects on bovine and human adrenocortical cells in culture, but with negative results (Gill *et al.*, 1978, 1982; Gill, Hornsby and Simonian, 1979, 1980; Hornsby *et al.*, 1983).

Mitogens: summary

In summary, ACTH stimulates growth *in vivo*, but directly inhibits replication; it is therefore an indirect mitogen. Insulin and some pituitary factors, but not pure hormones, stimulate growth in culture, but their role in control of adrenocortical size *in vivo* is unknown. Angiotensin is mitogenic in culture and perhaps also *in vivo* under circumstances where a greater adrenocortical mass is required for an increase in steroidogenic capacity, e.g. in sodium deficiency.

NEURAL MECHANISMS OF GROWTH REGULATION

Within the last ten years it has become clear that a major mechanism for regulation of adrenocortical growth comprises a neural reflex that signals the mass of one adrenal cortex to the cortex of the contralateral gland. The simplest demonstration of this reflex consists of manipulation of the adrenal gland, which results in a dramatic stimulation of proliferation of the cells of the contralateral adrenal cortex (Dallman, Engeland and Shinsako, 1976). The stimulation of DNA synthesis is much greater than that normally observed on administration of ACTH, although the overall increase in mass is comparable, due to the cellular hypertrophy occurring with ACTH. Significantly, ACTH administered at the same time as activation of the reflex inhibits the increase in DNA synthesis (Dallman *et al.*, 1980), demonstrating that ACTH inhibits replication *in vivo*, as it does in culture. The pathway of the reflex is via the hypothalamus, and is abolished by surgically created hypothalamic hemi-islands (Engeland and Dallman, 1975; Holzwarth, Shinsako and Dallman, 1980).

The mechanism whereby nerve activation is capable of stimulating proliferation is unknown, but it has long been known that regeneration of limbs in amphibians requires the presence of nerves in the regenerating tissue. Traditionally, the adrenal cortex was thought to be poorly supplied with nerves and its nerve supply therefore unimportant (Bachmann, 1954). Catecholamine fluorescence and electron microscopy have shown that there is an autonomic nerve supply to the adrenal cortex, but it is not yet clear whether innervation is to the parenchymal cells themselves or to blood vessels (Unsicker and Groschel-Stewart, 1978). Some adrenocortical autonomic nerve fibers do not degenerate after sectioning of the splanchnic nerve, and therefore some at least may arise from cell bodies in the cortex or in the adventitia of the adrenal arteries (Robinson *et al.,* 1977). Similarly, a local nerve supply from medulla to cortex has been demonstrated by immunohistochemical localization of vasoactive intestinal peptide (VIP) (Hokfelt *et al.,* 1981), which has been reported to have steroidogenic effects on adrenocortical cells (Birnbaum, Alfonzo and Kowal, 1980). (Aspects of interactions between the cortex and medulla are reviewed in Chapter 10.) Several reports indicate that the nerve supply to the cortex is involved in the regulation of steroidogenesis, by mechanisms yet to be fully elucidated (Henry *et al.,* 1976; Henry and Stephens, 1977; Kolta and Soliman, 1981; Hadjian, Guidicelli and Chambaz, 1982; Karteszi *et al.,* 1982; Ottenweller and Meier, 1982; Wilkinson, Shinsako and Dallman, 1982; Lilly, Engeland and Gann, 1983; Wilson *et al.,* 1983). Cholinergic stimulation of steroidogenesis may be mediated by muscarinic receptors in the adrenal cortex (Hadjian, Ventre and Chambaz, 1981; Hadjian, Guidicelli and Chambaz, 1982). The inhibition of compensatory adrenal growth by aldosterone (Dunlap and Grizzle, 1984; Grizzle and Dunlap, 1984) could indicate a role for angiotensin in this process. Additionally, it may be significant that powerful growth factors, such as brain FGF, are found in neural tissue (Gospodarowicz, Moran and Bialecki, 1976; Gospodarowicz, Lui and Cheng, 1976). There is also evidence for neurally mediated proteolytic cleavage of circulating pro-γ-melanocyte stimulating hormone (MSH) to a growth-stimulatory form (Lowry *et al.*, 1983).

The reflex may be triggered most simply by the manipulation of the nerve bundle, but it was originally discovered as a reflex mediating compensatory adrenal growth after unilateral adrenalectomy (Engeland and Dallman, 1975). Compensatory adrenal growth was formerly believed to be mediated by ACTH, but careful work has established that the increase in ACTH occurring on removal of one gland is insufficient to account for the increase in size of the contralateral gland (Engeland, Shinsako and Dallman, 1975). This powerful neural reflex must be important in the control of growth of the adrenal cortex *in vivo*, but it is not essential for this process. Regeneration of the adrenal cortex may occur both in an ectopic site, from transplanted adrenocortical tissue or a suspension of adrenocortical cells reintroduced into the animal, or from adrenocortical tissue remaining *in situ* after enucleation; in the latter case, the nerve and blood supplies to the gland remain intact, but denervation does not affect regeneration (Holzwarth, Shinsako and Dallman, 1980). Additionally, such regeneration appears to be dependent on the secretion of ACTH and is inhibited by the presence of a normal or otherwise adequate mass of adrenocortical tissue in the body (Plzak, 1960; Wilkinson, Shinsako and Dallman, 1981). Even though regeneration does not depend on the nerve supply, there is some evidence for re-innervation of transplanted adrenocortical tissue (Engeland, Feeley and Wilke, 1982).

POSSIBLE ROLE OF THE VASCULAR SYSTEM

The currently available data on growth factor and neural stimulation of adrenocortical mitogenesis are apparently at variance with the role of ACTH as the chief factor involved in the regulation of adrenocortical growth *in vivo*. I suggest that its growth-promoting properties may be connected with its action on the adrenal vascular system, an immediate effect of ACTH being to increase the blood flow through the gland (Neville and O'Hare, 1982). The effect of indomethacin and other inhibitors on ACTH-induced hyperemia suggests that this effect may be mediated by prostaglandins (Varga, Stark and Folly, 1979; Banks and Beilin, 1981). Both ACTH and angiotensin stimulate prostaglandin synthesis in the adrenal cortex (Brown and Smith, 1982); also, in other tissues, prolactin stimulates prostaglandin synthesis (Manku *et al.*, 1979). This short-term increase in blood flow may itself stimulate adrenocortical proliferation by increased delivery of critical nutrients to the tissue, but it may also trigger the neural reflex mentioned above and thus stimulate growth in the contralateral cortex. This would happen simultaneously in both glands. The hyperemia could trigger the nerve reflex by stimulating baroreceptors, which are probably present in the adrenal cortex (Niijima and Winter, 1968). Such an ACTH-mediated neural reflex would perhaps explain some discrepancies in the reported extent of the early DNA synthetic response to administration of ACTH. In some circumstances, a large increase in DNA synthesis is observed 24 hours after administration of ACTH (Masui and Garren, 1970, 1971). The variability may be caused by the level of prior stimulation of the cortex; dexamethasone-suppressed animals show a greater DNA synthetic response to ACTH than intact animals (Bransome, 1968). In this case, the stimulation of blood flow and activation of the neural reflex may overcome any direct antimitotic effects. In the longer term, if the change in blood flow is connected with the increase in adrenocortical proliferation, it may be that the enhancement of growth arises from a secondary permanent increase in the vascular supply resulting from growth of capillaries in the tissue. Capillary growth might be stimulated by prostaglandins, or perhaps by other angiogenic factors. Ovarian tissue transplanted into the anterior chamber of the eye will not grow until it has been revascularized (Gospodarowicz and Thakral, 1978), a process that is stimulated by LH. Similar, less detailed, experiments with transplanted adrenocortical tissue have given results consistent with the conclusions reached in the experiments on ovarian tissue (Coupland, 1957). Capillary growth could cause parenchymal tissue growth by either enhanced supply of nutrients or growth factors, or provision of a greater area of extracellular matrix. In culture, coating of the substratum with corneal endothelial cell-produced extracellular matrix, or its major component, fibronectin, has a powerfully growth-stimulating effect on bovine and human adrenocortical cells (Gospodarowicz, Delgado and Vlodavsky, 1980; Crickard, Ill and Jaffe, 1981; Ill and Gospodarowicz, 1982; Simonian, White and Gill, 1982; Hornsby *et al.*, 1983). Under some circumstances the degree of stimulation is comparable to that produced by the most powerful growth factors, such as FGF, and it has been hypothesized that the mechanism of growth stimulation by FGF may involve stimulation of production of extracellular matrix (Gospodarowicz, Delgado and Vlodavsky, 1980).

STIMULATION OF ADRENOCORTICAL GROWTH *IN VIVO*: HYPOTHESIS

The following hypothesis is proposed. The principal regulation of adrenocortical growth is through the indirect growth-stimulatory effects of ACTH. ACTH may act by stimulation of adrenocortical blood flow and increased vascularization. At the same time as indirectly increasing adrenocortical cell proliferation, ACTH inhibits DNA synthesis by its direct action. The net result is cellular hypertrophy rather than hyperplasia. This basic process operates whether the adrenocortical tissue is in its normal location or has been transplated to a remote site. A secondary mechanism, operating only in the former circumstance, is via a neural reflex which stimulates growth in the contralateral adrenal cortex when blood flow increases in the gland, presumably due to the action of ACTH. This would act in concert with the basic ACTH-stimulation process and speed the increase in adrenocortical growth that is required.

HYPERTROPHY VERSUS HYPERPLASIA

It has also been noted that the fact that ACTH causes predominantly an increase in adrenocortical cell size, rather than proliferation, may be explained teleologically by the need for the increase in adrenocortical mass to be readily reversible (Gill, Hornsby and Simonian, 1979). Cellular hypertrophy is easily reversed by a temporary decrease in synthesis of cellular components or by increased degradation, while rapid reversal of hyperplasia would necessitate a marked increase in the rate of cell death. Although increased cell death is observed on withdrawal of ACTH, it may be advantageous to the organism, where possible, to avoid the tissue remodeling that this entails.

Degeneration of adrenocortical tissue

Adrenocortical mass is determined by the balance of growth and death of adrenocortical cells. Cell death is not normally a major factor in the control of adrenal size, but under certain circumstances the mass of the innermost zone appears to be regulated by mechanisms other than those for the rest of the cortex. At certain points during life such zones may undergo rapid involution or atrophy (Lanman, 1953; Neville and O'Hare, 1982). The usual explanation is that these zones are 'maintained' by specific hormones (e.g. placental hormones for the human fetal zone and estrogens for the mouse X-zone) and that degeneration occurs on their withdrawal (Lanman, 1953, Neville and O'Hare, 1982). The actual mechanisms by which such specialized inner zones develop and involute, and the relationship between these rapid degenerative processes and the usual cell death occurring continuously in the reticularis, are unknown.

Another example of rapid cell death in the cortex is that occurring during chronic excessive stimulation with ACTH – the Waterhouse–Friderichsen syndrome (Neville and O'Hare, 1982). We have hypothesized that degeneration following such overstimulation may be mediated by oxidative damage arising from the interaction of steroids with adrenocortical cytochrome P-450s (Hornsby and Crivello, 1983). In degeneration of the fetal zone after birth, necrosis could be triggered by such a mechanism. ACTH stimulation is high in the antenatal period and this may lead to damage of the inner cortex. Necrosis could occur when ACTH stimulation and blood flow decrease after birth. An additional factor may be the higher tissue oxygen tensions after transfer from the intrauterine environment.

Some insight into the mechanism of necrosis is provided by the action of adrenolytic toxic chemicals. These include mitotane (*o,p'*-DDD), dimethylbenzanthracene, carbon tetrachloride and acrylonitrile (Bergenstal *et al.*, 1960; Huggins and Morii, 1961; Szabo *et al.*, 1980; Malendowicz and Colby, 1982). Some of these substances appear to act by oxidative damage to adrenocortical tissue (Hornsby and Crivello, 1983). The adrenolytic action of acrylonitrile and similar compounds was found to depend on the presence of the medulla; acrylonitrile was not adrenolytic in glands after enucleation and regeneration (Szabo *et al.*, 1981): examination of the cortex revealed fragments of the medulla in the cortex, suggesting that retrograde medullary-cell embolism had occurred. Catecholamines from medullary fragments were hypothesized to mediate the toxic effects. These results suggest a general mechanism for necrosis of the inner cortex during degeneration. Alterations of blood flow, either increases or decreases, together with damage of the adrenocortical cells, lead to the introduction of medullary fragments into the cortex, and subsequent degeneration.

The zonation of adrenocortical function

The functional differences between the zona glomerulosa and the zona fasciculata, and between the zona fasciculata and the zona reticularis and its equivalents, will be reviewed below. This is followed by a discussion of the possible origin of such differences.

Aldosterone synthesis and the zona glomerulosa

The realization in the 1940s that the width of the zona glomerulosa varied with the state of the salt and water balance of the animal led to the conclusion that this zone was responsible for the secretion of a hormone involved in regulation of this balance (Tait and Tait, 1979). Deoxycorticosterone (DOC), which had been isolated earlier, was believed to be the salt-retaining steroid hormone until the 1950s, when aldosterone was isolated and characterized (Tait and Tait, 1979) and *in vitro* experiments conclusively established that the mechanically separated zona glomerulosa was the only site of synthesis of this steroid in the cortex (Giroud, Stachenko and Venning, 1956). So far as can be established with microdissection techniques there is a correspondence between the zona glomerulosa and the aldosterone-secreting zone (Giroud, Stachenko and Venning, 1956; Stachenko and Giroud, 1964; Miao and Black, 1982; Crivello, Hornsby and Gill, 1983). Aldosterone secretion results from the presence of a cytochrome P-450 species with corticosterone methyl oxidase (CMO) activity which is confined to the zona glomerulosa (Ulick, 1976). The zona fasciculata can catalyze all the steps leading to aldosterone synthesis except the final one, the conversion of corticosterone to aldosterone. Synthesis of 18-hydroxycorticosterone in the fasciculata probably results from the activity of cytochrome P-$450_{11\beta}$, which can both 11β-hydroxylate and 18-hydroxylate DOC (Finkelstein and Shaefer, 1979; Kramer, Gallant and Brownie, 1979).

The other significant difference in cytochrome P-450 content between the zones is the absence of 17-hydroxylase in the glomerulosa and its presence in the fasciculata, except in those species (principally the rat and some other rodents) that lack 17-hydroxylase throughout the adrenal cortex (Stachenko and Giroud, 1964; Harkins *et al.*, 1974; Miao and Black, 1982; Crivello and Gill, 1983). $C_{17,20}$-Lyase is also present only in the inner zones; this enzymatic activity is probably catalyzed by the same cytochrome P-450 (Kominami, Shinzawa and Takemori, 1982; Nakajin, Shinoda and Hall, 1983). Although ACTH-stimulated cAMP production in membrane particles from the glomerulosa and fasciculata is similar (Shima, Kawashima and Hirai, 1979), cyclic AMP levels may normally be lower in the glomerulosa. This may be due to (1) differences between the zones in the activity of phosphodiesterase (Hornsby and O'Hare, 1977), which is considerably higher in the glomerulosa than in the fasciculata (Gallant, Kauffman and Brownie, 1974; Koletsky, Brown and Williams, 1983), and (2) higher levels in the glomerulosa of receptors for angiotensin and somatostatin (Brecher *et al.*, 1973; Aguilera, Parker and Catt, 1982; Cantin *et al.*, 1982), both of which have been shown to inhibit cyclic AMP synthesis (Correa and Saavedra, 1983; Marie and Jard, 1983). 17-Hydroxylase is a strongly cAMP-dependent enzyme, both *in vivo* and in culture. In several species, 17-hydroxylase levels *in vivo* are low, except during above-normal ACTH stimulation; this causes such species to alternate between being cortisol-secretors and corticosterone-secretors (Badr and Spickett, 1965; Varon, Touchstone and Christian, 1966; Ogunsua *et al.*, 1971; Fevold and Drummond, 1973; Slaga and Krum, 1973; Slanina and Fevold, 1982). In human and bovine adrenocortical cells in culture, 17-hydroxylase is more dependent on ACTH or cAMP, and more strongly induced by these stimuli, than any other steroidogenic enzyme (Neville and O'Hare, 1978; Simonian *et al.*, 1979; O'Hare, Nice and Neville, 1980; Crivello and Gill, 1983; Kramer *et al.*, 1983; McCarthy *et al.*, 1983; Hornsby and Aldern, 1984). On the other hand, the glomerulosa-specific cytochrome $P\text{-}450_{CMO}$ is induced by potassium in a process which does not depend on cAMP (Hornsby and O'Hare, 1977). Thus, the glomerulosa may normally maintain a lower cAMP level than the rest of the cortex, thereby preventing induction of 17-hydroxylase.

There are thus at least two primary differences between the glomerulosa and the fasciculata that result in the zonation of aldosterone and cortisol synthesis: (1) restriction of cytochrome $P\text{-}450_{CMO}$ to the glomerulosa and (2) differences in cyclic AMP levels. There are other differences, principally in content of various receptors, which may be important. Other receptors that may possibly differ in content between the zona glomerulosa and the fasciculata include those for β-lipotropin, MSH, serotonin and dopamine (Muller, 1970; Matsuoka, Mulrow and Li, 1980; Dunn and Bosmann, 1981; Bevilacqua *et al.*, 1982; Dell *et al.*, 1982; Whitehouse, Vinson and Thody, 1982).

The zona reticularis and adrenal androgen secretion

Although the zona reticularis has been linked to adrenal androgen synthesis for some time, it is not clear that there is a simple correspondence between the region of the adrenal cortex that secretes adrenal androgens and the morphological reticularis. For example, a zona reticularis also occurs in species that do not secrete appreciable quantities of adrenal androgens, such as the rat, cow and many others. However, it is clear that in the human, other primates, and some other species there is an association between the development of large glands with distinct inner zones and adrenal androgen synthesis (Schiebinger *et al.*, 1981a; Neville and O'Hare, 1982). The major adrenal androgen is dehydroepiandrosterone sulfate

(DHAS), with lesser amounts of free dehydroepiandrosterone and androstenedione (Neville and O'Hare, 1982). Two conflicting hypotheses have been presented: one postulates that adrenal androgen synthesis results from stimulation by a specific hormone, as yet undescribed (Parker and Odell, 1980), and the other that the synthesis of adrenal androgens may be accounted for by relative enzyme activities of 3β-HSD, $C_{17,20}$-lyase, and DHA sulfotransferase, with no hormones other than ACTH being involved (Anderson, 1980; Hornsby and Aldern, 1984). In species which secrete large amounts of adrenal androgens the balance of these enzymes favors adrenal androgen synthesis regardless of the zonal origin of the cells (Hornsby and Aldern, 1984). The initial product derived from pregnenolone in human cells is 17-hydroxypregnenolone rather than progesterone; 17-hydroxypregnenolone is then converted about equally to cortisol (via 17-hydroxyprogesterone and 11-deoxycortisol) and dehydroepiandrosterone sulfate (via dehydroepiandrosterone). Thus, the low 3β-HSD activity results in conversion of 17-hydroxypregnenolone being the rate-limiting step, allowing $C_{17,20}$-lyase to compete with 3β-HSD for this substrate. Experiments on cultured human adrenocortical cells under serum-free conditions, with no exposure to hormones other than ACTH, show that the high DHA secretion in this species does not require a special hormone, but results from the ratio of the activities of 3β-HSD, 17-hydroxylase and $C_{17,20}$-lyase, which is an intrinsic property of the human adrenocortical cell (Hornsby and Aldern, 1984).

In the human, the synthesis of adrenal androgens is associated with the development of inner zones which are present in fetal life and in maturity but not in childhood. Between birth and adrenarche DHAS levels are very low, coinciding with the lesser development of the zona reticularis in this period (Neville and O'Hare, 1982). Adrenocortical changes during adrenarche are reviewed in Chapter 5. 17-Hydroxylase and $C_{17,20}$-lyase activities are also low before adrenarche, and then increase (Anderson, 1980; Rich *et al.*, Schiebinger *et al.*, 1981b; Kelnar and Brook, 1983). As described above, these enzyme activities are strongly ACTH-dependent and their increase may therefore simply reflect a higher chronic level of stimulation with ACTH. However, the large glands of both fetal and mature adult life in the human are also characterized by inner zones (fetal zone and zona reticularis) that have lower 3β-HSD activity, higher DHA sulfotransferase activity, and possibly also higher $C_{17,20}$-lyase and 17-hydroxylase activities than the outer zones (Dawson, Pryse-Davies and Snape, 1961; Cavallero and Chiappino, 1962; Cameron *et al.*, 1969; Jones, Groom and Griffiths, 1970; Cooke and Taylor, 1971; Sakhatskaya and Altukhova, 1973; Anderson, 1980; Kennerson, McDonald and Adams, 1983). In the guinea pig there also appears to be zonation of $C_{17,20}$-lyase and DHA sulfotransferase (Jones and Griffiths, 1968; Miao and Black, 1982; Davison *et al.*, 1983; Hyatt *et al.*, 1983). These zones secrete adrenal androgens because they are low in 3β-HSD and high in $C_{17,20}$-lyase and DHA sulfotransferase. Apart from 3β-HSD, $C_{17\text{-}20}$-lyase and DHA sulfotransferase, 11β-hydroxylase is lower in the zona reticularis of the rat and possibly other species (Bell *et al.*, 1978). There may be differences in some receptor levels between the reticularis and the fasciculata; in the rat, ACTH-stimulated cAMP production rates differ in the two zones (Hyatt *et al.*, 1980).

As far as can be ascertained, in these cases the morphological and functional boundaries do coincide. Apparently, therefore, while there is no all-or-nothing zonation, as in the case of the zona glomerulosa and aldosterone synthesis, there are distinct and important enzyme differences between the zona reticularis and the zona fasciculata.

The different enzyme levels in the inner zones result in a greater rate of adrenal androgen synthesis relative to that of the zona fasciculata, or that of the definitive zone in the fetal cortex. This has been shown directly by separate culture of the fasciculata and the reticularis, and the definitive and fetal zones (O'Hare, Nice and Neville, 1980; Simonian and Gill, 1981; Simonian and Capp, 1984). Cells from the zona reticularis or the fetal zone in culture initially produced greater quantities of DHAS, and Δ^5-3β-hydroxy- and sulfated steroids generally than the zona fasciculata or the definitive zone. Zona fasciculata cells do secrete adrenal androgens but not at the same rate as reticularis cells (O'Hare, Nice and Neville, 1980). In common with other separate zonal cell types in culture, the distinction between the zones was lost within a few days. This supports the hypothesis, presented above, that cells of the different zones are all of the same basic cell type, exhibiting temporary functional and morphological differences.

The gradient hypothesis of regulation of adrenocortical zonation

A hypothesis of functional and morphological zonation of the adrenal cortex assigns a central role to the unusual adrenocortical vascular system in the maintenance of structure and function (Anderson, 1980; Neville and O'Hare, 1982; Hornsby and Crivello, 1983). In its essence this concept is not new, having been first proposed by Greep and Deane in 1949 (Greep and Deane, 1949). They speculated that the division of the cortex into the mineralocorticoid and glucocorticoid secreting zones may be the result of a requirement for different environments, as provided by the centripetal capillary system, for the synthesis of the different steroid hormones. This concept is strongly supported by cell culture experiments which show that zone glomerulosa, zona fasciculata, zona reticularis, fetal zone and definitive zone cells all become very similar in their properties when removed from their natural environment, dominated presumably by the vascular system, and placed in culture (Hornsby and Crivello, 1983). This hypothesis has been expanded to suggest that the role of the vasculature in zonation is to create a gradient of steroids which may regulate differentiated functions of adrenocortical cells (Hornsby, O'Hare and Neville, 1974; Hornsby and O'Hare, 1977; O'Hare, Ellison and Neville, 1978; Neville and O'Hare, 1979, 1982; Kahri, Voutilainen and Salmenpera, 1979; Hornsby, 1980, 1982; Anderson, 1980; Crivello, Hornsby and Gill, 1982, 1983; Gill *et al.*, 1982; Hornsby and Crivello, 1983). The postulate of steroids as the constituent of the gradient, and of the intermediacy of lipid peroxidation, concerns details of the more general idea that zonation results from the action of a substance secreted by adrenocortical cells creating a gradient across the capillary bed. The hypothesis may stand apart from these detailed considerations. However, several aspects of the hypothesis, as will be discussed below, do depend on the gradient substance being stimulated by ACTH.

ZONATION HYPOTHESIS; ALDOSTERONE SYNTHESIS AND THE ZONA GLOMERULOSA

The aldosterone-synthesizing cytochrome P-450, corticosterone methyloxidase, may be confined to the zona glomerulosa because its existence is compatible only with conditions on the arterial side of the capillary bed (Neville and O'Hare, 1982; Hornsby and Crivello, 1983). The zona glomerulosa is always present only on the extreme arterial side of the capillary bed. This is true in several unusual situations other than the normal gland.

(1) In the human adrenal gland a cuff of cortex surrounding the central vein penetrates the medulla. The blood supply in the cuff is from a plexus that surrounds and supplies the wall of the central vein. Capillaries radiate outwards from this plexus supplying the cortical cuff. A zona glomerulosa occurs on the arterial side of the capillary bed (Neville and O'Hare, 1982).
(2) Accessory adrenal glands are miniature glands consisting only of adrenal cortex, no medulla being present, and are frequently seen as developmental abnormalities in some strains of animals (Bachmann, 1954). Although normally small, they are capable of dramatic growth when the animal is adrenalectomized; indeed, when accessory glands are present, the usually lethal process of adrenalectomy is survived. When the accessory adrenal glands are large enough to show an internal structure, the capillary bed is seen to have its normal structure, with the zona glomerulosa on the arterial side (Gruenwald, 1946).
(3) Adrenocortical tissue that has regenerated from tissue fragments also shows a glomerulosa on the arterial side of the capillary bed (Ingle and Higgins, 1938). This zone reforms after most of the regenerative growth is completed. During regenerative growth, when the blood supply is disorganized, zonation of the adrenocortical tissue is not apparent.
(4) When adrenocortical tissue is transplanted into the anterior eye chamber the grafts may revascularize from the iris and thus receive their blood on one surface of the graft only. A zona glomerulosa is seen to form along this surface (Coupland, 1957).

In these unusual circumstances functional, as opposed to morphological zonation, has not been investigated, although in certain cases it may be inferred that aldosterone secretion is intact.

The restriction of the zona glomerulosa and aldosterone secretion to the arterial side of the capillary bed implies the existence of a gradient of a substance or substances presumably produced, or consumed, by the adrenocortical cells. On the basis of experiments on isolated adrenocortical cells in culture I have proposed that steroids, synthesized by the adrenocortical cells themselves, form the gradient substance(s). In cultured rat and bovine adrenocortical cells, corticosterone methyloxidase (cytochrome P-450_{CMO}) appears to be subject to inactivation by steroids added to the cultures at concentrations likely to exist in the adrenal cortex *in vivo* (Hornsby and O'Hare, 1977; Neville and O'Hare, 1982; Crivello, Hornsby and Gill, 1983). In rat glomerulosa cell cultures, added cortisol and other steroids cause loss of P-450_{CMO} activity, as evidenced by loss of aldosterone and 18-hydroxycorticosterone production, without affecting P-$450_{11\beta}$ activity, as evidenced by continued formation of corticosterone and 18-hydroxy-deoxycorticosterone (Hornsby and O'Hare, 1977); similar findings have also been reported in explant cultures of fetal rat and fetal human adrenals (Kahri, Voutilainen and Salmenpera, 1979). In bovine glomerulosa cell cultures, cortisol and other steroids caused loss of both cytochromes P-$450_{11\beta}$ and P-450_{CMO} activities (Crivello, Hornsby and Gill, 1983). In this case, P-450_{CMO} appeared to be so unstable in the presence of endogenous steroids and oxygen that basal activity could only be preserved in the presence of the inhibitor metyrapone (Crivello, Hornsby and Gill, 1982; Hornsby and Crivello, 1983). These experiments suggest that the steroids responsible could include cortisol, corticosterone and androstenedione, but not aldosterone (Hornsby and O'Hare, 1977; Crivello, Hornsby and Gill, 1983). We have discussed in detail elsewhere possible mechanisms of loss of cytochromes P-450$_{CMO}$ and P-450$_{11\beta}$ in the presence of added steroids (Hornsby and Crivello, 1983). Some steroids may act as pseudosubstrates for these enzymes stimulating release of oxygen radicals, initiation of lipid peroxidation and loss of enzymatic activity. It is possible that the level of antioxidants in the adrenal cortex is low enough to permit destruction of P-450$_{CMO}$ at a certain steroid concentration but high enough to protect P-450$_{11\beta}$, at least at the glomerulosa/fasciculata interface. Steroid-induced loss of P-450$_{11\beta}$ may occur further into the cortex, however, as discussed later.

The other differences between glomerulosa and fasciculata are of unknown origin. Possibly, in each case, the initial step is the steroid–cytochrome P-450 interaction leading to lipid peroxidation, with the other glomerulosa/fasciculata differences as secondary consequences. The origin of the difference in phosphodiesterase between the glomerulosa and the fasciculata, which was noted earlier as being a critical difference between the zones, has not been investigated. However, steroids have been observed to depress phosphodiesterase activity (Schmidtke *et al.*, 1976). This may be significant in the process of the transition from glomerulosa cell to fasciculata cell (Hornsby and O'Hare, 1977).

The early observation that ACTH, by causing hypertrophy of the adrenal cortex, brought about a diminution in the width of the zona glomerulosa (Bachmann, 1954) led to the 'transformation field' hypothesis, which suggested that the borders of the glomerulosa and fasciculata (and also of the fasciculata and reticularis) were changeable, with the width of the fasciculata increasing at the expense of the other zones to supply an increased demand for adrenocortical steroids (Bachmann, 1954). This theory is similar to the concept, still sometimes encountered, that the zona glomerulosa is an 'undifferentiated' reserve zone for the 'differentiated' fasciculata. This concept, however, is untenable in the light of the highly specialized functions of the zona glomerulosa and its unique cytochrome P-450 species. Nevertheless, it does appear that the movement of the border between the glomerulosa and the fasciculata under the action of ACTH is an important regulatory mechanism, not so much for regulation of glucocorticoid output but for homeostasis of mineralocorticoid output. When the width of the adrenal cortex as a whole increases during long-term growth under the action of ACTH, the width of the glomerulosa must decrease in order to maintain a constant mass of the zone. ACTH stimulates corticosterone synthesis in the glomerulosa but does not increase cytochrome P-450$_{CMO}$ levels; thus the local concentration of corticosterone in the glomerulosa will rise, causing a change in the gradient of steroids across the capillary bed. The distance from the capsule at which the concentration of steroids becomes sufficient to cause loss of cytochrome P-450$_{CMO}$ will decrease. Thus the width of the glomerulosa will decrease. The process of the transition may be promoted by depression of phosphodiesterase (which, as indicated earlier, marks an important difference between the glomerulosa and the fasciculata), since this will cause increased cAMP, increased steroidogenesis (but no stimulus to increased synthesis of cytochrome P-450$_{CMO}$) and thus increased local corticosterone. Two points should be noted with regard to the determination of the glomerulosa/fasciculata boundary by the steroid gradient:

(1) Zonation is probably determined by the steroid concentration integrated over a relatively long period (hours and days). Short-term variations in steroidogenesis are probably unimportant since the processes of enzyme depression and induction are long-term.

(2) The gradient at the glomerulosa/fasciculata boundary is likely to be somewhat steeper than in the remainder of the cortex. Glomerulosa cells synthesize less corticosterone and cortisol than fasciculata cells, and therefore there may be an increase in the slope of the gradient at the zone boundary. This may serve to keep the boundary between the zones sharp, since each zone is subject to a positive feedback; a low steroid concentration in the glomerulosa maintains it as glomerulosa, and a higher concentration in the fasciculata maintains it as fasciculata. Cells may therefore undergo a rather sudden transition from glomerulosa to fasciculata on reaching the critical point in the gradient when moving inward in the cortex.

During growth of the cortex these mechanisms may serve to decrease the width of the glomerulosa, but keep its mass constant and thus maintain a constant potential capacity for aldosterone secretion. The opposite process may be envisaged as occurring when adrenocortical mass decreases.

The above discussion assumes maintenance of a constant requirement for aldosterone by the organism. The concept also applies, with some modification, when this requirement changes. The demand for aldosterone decreases, for example, under chronic excessive ACTH stimulation, as occurs in some animals with implanted pituitary tumors, when the amount of DOC secreted by the zona fasciculata increases sufficiently to satisfy the mineralocorticoid requirement of the animal (Nickerson, Brownie and Molten, 1970; Nickerson, 1975). The glomerulosa may become limited to foci under the capsule, no longer forming a continuous layer (Neville and O'Hare, 1982; Payet and Lehoux, 1982). This indicates that under the combined effect of chronic overstimulation with ACTH and lack of glomerulosa stimulation the critical point in the gradient has moved outward to reach the capsule itself. When more aldosterone secretion is required, e.g. under sodium deficiency, volume and width of the glomerulosa increase, as discussed earlier in the context of the possible mitogenic effect of angiotensin. Whereas the mitogenic effect may well be important, an adjustment of the boundary between the zones may also occur as follows: the stimulus to the glomerulosa cell (angiotensin or potassium ion concentration) increases cytochrome P-450$_{CMO}$ activity (Hornsby and O'Hare, 1977), and possibly also phosphodiesterase, both of which will tend to limit corticosterone secretion while permitting continued aldosterone synthesis. Thus, the critical point in the gradient will move inward, allowing expansion of the zona glomerulosa.

An extension of this concept to account for the mixed steroidogenic pattern of tumors secreting aldosterone has been discussed elsewhere (Neville and O'Hare, 1982; Symington, 1982). Briefly, disorganized growth of the cortex may give rise to nodules or benign tumors. When such a tumor has a blood supply relatively close to the arterial side of the capillary bed, it has the potential for secreting aldosterone in that part of the tumor. Cells in regions of the tumor that are not close to the arterial side of the bed will act as fasciculata cells and secrete cortisol. If no part of the tumor is close to the arterial supply, only cortisol is produced. This may result in a so-called 'non-functioning' nodule; such nodules do in fact secrete some cortisol, but the amount secreted is insufficient to alter the glucocorticoid balance of the body as a whole. On the other hand, aldosterone secretion by such tumours is likely to cause symptoms of aldosterone excess, because of the normally very low levels of secretion of this steroid.

ZONATION HYPOTHESIS: ADRENAL ANDROGEN SYNTHESIS AND INNER ZONES (ZONA RETICULARIS OR FETAL ZONE)

A further development of the zonation hypothesis suggests that similar gradient effects may be responsible for the distinction between the zona fasciculata and zona reticularis, or the fetal zone and the definitive zone. I have proposed that cytochromes P-$450_{11\beta}$ and P-450_{21} may act as sensitizers for the steroid gradient farther into the cortex (Hornsby and Crivello, 1983), and it is possible that the same mechanisms of a steroid gradient and lipid peroxidation initiated by steroid/cytochrome P-450 interactions are involved. The important enzymatic features of functional zonation in the inner cortex are low 3β-HSD, and high DHA sulfotransferase and $C_{17,20}$-lyase activities. It is possible that 3β-HSD is also regulated through a primary sensing of the steroid gradient by P-$450_{11\beta}$ or P-450_{21}, and subsequent lipid peroxidation. The activity of 3β-HSD is known to be modulated by lipids of the endoplasmic reticulum (Gallay *et al.,* 1981). Direct suppression by steroids is also suggested by experiments on cultured human fetal adrenocortical cells where 3β-HSD activity is lower in the presence of estrogens (Winter *et al.,* 1980). The other enzymes involved in adrenal androgen synthesis appear to be stable in the presence of high steroid concentrations (Hornsby, 1982; Crivello and Gill, 1983) and so may tolerate high steroid levels in the inner cortex.

A detailed hypothesis of the regulation of adrenal androgen synthesis by control of reticularis width is not currently feasible. The mechanism by which such a steroid gradient might result in a fasciculata/reticularis transition remains speculative (Anderson, 1980; Hornsby and Crivello, 1983). However, it may be assumed that, as for the glomerulosa/fasciculata boundary, there is a critical concentration in the gradient that results in the transition. In the human, the presence of inner zones (reticularis or fetal cortex) is associated with large adrenal glands. Presumably, when the cortex reaches a critical size, the concentration of the gradient substance reaches a level in the inner cortex that is sufficient to cause the development of the reticularis. This development could become self-sustaining in the following way: ACTH may stimulate an increase in adrenocortical size; stimulate an increase in the highly ACTH-dependent cytochrome P-450_{17} and $C_{17,20}$-lyase; and cause a decrease in the activity of 3β-HSD in the inner cortex through stimulation of the gradient substance. Thus, ACTH will stimulate less glucocorticoid secretion from the reticularis. This will tend to increase ACTH secretion by the pituitary and so complete a positive feedback cycle. Consequently it may be proposed that quite small initial changes in ACTH secretion may result in large changes in the reticularis and adrenal androgen synthesis.

SUMMARY: THE CONTROL OF STEROIDOGENESIS BY REGULATION OF ADRENOCORTICAL ZONATION

It is proposed here that a major mechanism for the long-term regulation of adrenocortical steroidogenesis is the regulation of the mass of the various zones. This mode of regulation of steroidogenesis has frequently been overlooked. The mass of the zona glomerulosa is a major factor in the long-term regulation of aldosterone synthesis, and that of the inner zone (zona reticularis or fetal zone) is of major importance for the regulation of adrenal androgen synthesis. To some extent the mass of the zona fasciculata is also important for the regulation of glucocorticoid synthesis, but, as discussed earlier, here the mass of the cortex as a whole is the critical morphological factor. Whereas it is zonal mass which is the determining morphological factor in control of steroidogenesis, I propose that it is not mass as such that is under primary regulation. Rather, if the gradient model of zonation is correct, zonal width is likely to be the primary regulated factor, since boundaries between the zones will be determined by critical concentrations of the gradient substance, reached at certain distances from the capsule. Zonal mass would be regulated indirectly by feedback mechanisms. Although I have proposed that the gradient substance comprises one or more adrenocortical steroids, this is not essential to the general concept. It is simpler to construct a model of zonation if the gradient substance is assumed to be a steroid or steroids, particularly one secreted by all the zones, such as corticosterone. The general concept is that at certain distances from the capillary bed, and therefore at certain critical levels of the gradient, the concentration of the gradient substance is high enough to cause the adrenocortical cell to lose its glomerulosa characteristics and acquire fasciculata characteristics, or to lose fasciculata characteristics and acquire reticularis characteristics. Since cell division is higher in the outer cortex and cell death higher in the inner cortex, cells will pass through stages of being glomerulosa cells, then fasciculata cells, and finally reticularis cells during their life history. The adrenal cortex exhibits a unique system of physiological regulation, incorporating three separately regulated systems of hormone synthesis – mineralocorticoid, glucocorticoid and androgen – into the life history of a single cell type.

Acknowledgments

I am very grateful to Drs W. R. Robertson and M. Gordon, Department of Chemical Pathology, University of Manchester, for supplying the photomicrographs presented in *Plate 1.1*.

Work in original publications from this laboratory cited in this review was supported by Research Grants AG00936 and CA32468 from the National Institutes of Health.

References

AGUILERA, G., PARKER, D. S. and CATT, K. J. (1982) Characterization of somatostatin receptors in the rat adrenal glomerulosa zone. *Endocrinology,* **111,** 1376–1384

AKAZAWA, N., TAKANOHASHI, T., MIKAMI, S., SANO, M. and YAMADA, S. (1980) Histological and electron microscopical changes of the adrenal cortex in the rats fed vitamin E deficient and linoleic acid supplemented diet. *Bitamin,* **54,** 563–572

ANDERSON, D. C. (1980) Hypothesis: the adrenal androgen-stimulating hormone does not exist. *Lancet,* **2,** 454–456

BACHMANN, R. (1954) Die Nebenniere. In *Handbuch der mikroskopischen Anatomie des Menschen,* edited by W. Bargman, Vol. **6,** *Blutgefäss- und Lymphgefässapparat. Innersekretorische Drüsen,* Part 5, pp. 1–952. Berlin: Springer-Verlag

BADR, F. M. and SPICKETT, S. G. (1965) Genetic variation in the biosynthesis of corticosteroids in *Mus musculus. Nature,* **205,** 1088–1090

BANKS, R. A. and BEILIN, L. J. (1981) The effects of meclofenamate, captopril and phentolamine on organ blood flow in the conscious rabbit. *Clinical Science,* **61,** 97–105

BARTKE, A., SMITH, M. S., MICHAEL, S. D., PERON, F. G. and DALTERIO, S. (1977) Effects of experimentally induced chronic hyperprolactinemia on testosterone and gonadotropin levels in male rats and mice. *Endocrinology,* **100,** 182–186

BELL, J. B. G., GOULD, R. P., HYATT, P. J., TAIT, J. F. and TAIT, S. A. S. (1978) Properties of rat adrenal zona reticularis cells: preparation by gravitational sedimentation. *Journal of Endocrinology,* **77,** 25–41

BELLONI, A. S., MAZZOCCHI, G., MENEGHELLI, V. and NUSSDORFER, G. G. (1978) Cytogenesis in the rat adrenal cortex: evidence for an ACTH-induced centripetal cell migration from the zona glomerulosa. *Arch. Anat. Hist. Embr. Norm. Exp.*, **61**, 195–206

BERGENSTAL, D. M., HERTZ, R., LIPSETT, M. and MOY, R. H. (1960) Chemotherapy of adrenocortical cancer with *o,p'*-DDD. *Annals of Internal Medicine*, **53**, 672–680

BERGERON, J. J. M., RACHUBINSKI, R., SEARLE, N., BORTS, D., SIKSTROM, R. and POSNER, B. I. (1980) Polypeptide hormone receptors *in vivo*: demonstration of insulin binding to adrenal gland and gastrointestinal epithelium by quantitative radioautography. *Journal of Histochemistry and Cytochemistry*, **28**, 824–835

BEVILACQUA, M., VAGO, T., SCORZA, D. and NORBIATO, G. (1982) Characterization of dopamine receptors by 3H-ADTN binding in calf adrenal zone glomerulosa. *Biochemical and Biophysical Research Communication*, **108**, 1661–1669

BIRNBAUM, R. S., ALFONZO, M. and KOWAL, J. (1980) Vasoactive intestinal peptide- and adrenocorticotropin-stimulated adenyl cyclase in cultured adrenal tumor cells: evidence for a specific vasoactive intestinal peptide receptor. *Endocrinology*, **106**, 1270–1275

BOLANDER, F. F. JR, HURLEY, T. W., HANDWERGER, S. and FELLOWS, R. E. (1976) Localization and specificity of binding of subprimate placental lactogen in rabbit tissues. *Proceedings of the National Academy of Sciences USA*, **72**, 2932–2935

BOYAR, R. M. and HELLMAN, L. (1974) Syndrome of benign nodular adrenal hyperplasia associated with feminization and hyperprolactinemia. *Annals of Internal Medicine*, **80**, 389–394

BRANSOME, E. D. JR (1968) Regulation of adrenal growth. Differences in the effects of ACTH in normal and dexamethasone-suppressed guinea-pigs. *Endocrinology*, **83**, 956–960

BRECHER, P., TABACCHI, M., PYUN, H. Y. and CHOBANIAN, A. V. (1973) Angiotensin binding to rat adrenal capsular cell suspensions. *Biochemical and Biophysical Research Communications*, **54**, 1511–1515

BROWN, B. E. and SMITH, M. J. JR (1982) Adrenal venous prostaglandin response to ACTH or angiotensin. *Federation Proceedings*, **41**, 1242

CALVO, J. C., FINOCCHIARO, L., LUTHY, I. *et al.*, (1981) Specific prolactin binding in the rat adrenal gland: its characterization and hormonal regulation. *Journal of Endocrinology*, **89**, 317–325

CAMERON, E. H. D., JONES, T., JONES, D., ANDERSON, A. B. M. and GRIFFITHS, K. (1969) Further studies on the relationship between C19- and C21-steroid synthesis in the human adrenal gland. *Journal of Endocrinology*, **45**, 215–230

CANTIN, M., GUTKOWSKA, Y., ANAND-SRIVASTAVA, M. B., LEDOUX, S., BIANCHI, C., CARRIERE, P. and GENEST, J. (1982) Binding and internalization of ^{125}I-angiotensin II in the rat adrenal. An ultrastructural radioautographic study. *Journal of Cell Biology*, **95**, 411a

CATER, D. B. and STACK-DUNNE, M. P. (1955) The effects of growth hormone and corticotrophin upon the adrenal weight and adrenocortical mitotic activity in the hypophysectomized rat. *Journal of Endocrinology*, **12**, 174–184

CATHIARD, A.-M. and SAEZ, J. M. (1981) Adrenal growth factors in the rat mammotropic pituitary tumor (MtT-F4). *Biochemical and Biophysical Research Communications*, **99**, 196–204

CAVALLERO, C. and CHIAPPINO, G. (1962) Histochemistry of steroid-3β-ol dehydrogenase in the human adrenal cortex. *Experientia*, **18**, 119–120

CHABOT, J.-G., GUY, J., ST-ARNAUD, R., WALKER, P. and PELLETIER, G. (1982) Autoradiographic localization of epidermal growth factor (EGF) receptors in different tissues of the rat. *Journal of Cell Biology*, **95**, 179a

COOKE, B. A. and TAYLOR, P. D. (1971) Site of dehydroepiandrosterone sulphate biosynthesis in the adrenal gland of the previable foetus. *Journal of Endocrinology*, **51**, 547–556

CORREA, F. M. A. and SAAVEDRA, J. M. (1983) Somatostatin inhibits the isoproterenol-stimulated adenylate cyclase in the intermediate lobe of the male rat pituitary gland. *Neuroendocrinology*, **37**, 284–287

COUPLAND, R. E. (1957) Factors affecting the survival of the adrenal medulla and associated cortical cells in the anterior chamber of the rabbit's eye. *Journal of Endocrinology*, **15**, 162–170

CRICKARD, K., ILL, C. R. and JAFFE, R. B. (1981) Control of proliferation of human fetal adrenal cells *in vitro*. *Journal of Clinical Endocrinology and Metabolism*, **53**, 790–796

CRIVELLO, J. F. and GILL, G. N. (1983) Induction of cultured bovine adrenocortical zona glomerulosa cell 17-hydroxylase activity by ACTH. *Molecular and Cellular Endocrinology*, **30**, 97–107

CRIVELLO, J. F., HORNSBY, P. J. and GILL, G. N. (1982) Metyrapone and antioxidants are required to maintain aldosterone synthesis by cultured bovine adrenocortical cells. *Endocrinology*, **111**, 469–479

CRIVELLO, J. F., HORNSBY, P. J. and GILL, G. N. (1983) Suppression of cultured bovine adrenocortical zona glomerulosa cell aldosterone synthesis by steroids and its prevention by antioxidants. *Endocrinology*, **113**, 235–242

DALLMAN, M. F., ENGELAND, W. C. and SHINSAKO, J. (1976) Compensatory adrenal growth: a neurally mediated reflex. *American Journal of Physiology*, **231**, 408–415

DALLMAN, M. F., ENGELAND, W. C., HOLZWARTH, M. A. and SCHOLZ, P. M. (1980) Adrenocorticotropin inhibits compensatory adrenal growth after unilateral adrenalectomy. *Endocrinology,* **107,** 1397–1404

DAVISON, B., LARGE, D. M., ANDERSON, D. C. and ROBERTSON, W. R. (1983) Basal steroid production by the zona reticularis of the guinea-pig adrenal cortex. *Journal of Steroid Biochemistry,* **18,** 285–290

DAWSON, I. M. P., PRYSE-DAVIES, J. and SNAPE, I. M. (1961) The distribution of six enzyme systems and of lipid in the human and rat adrenal cortex before and after administration of steroid and ACTH, with comments on the distribution in human foetuses and in some natural disease conditions. *Journal of Pathology and Bacteriology,* **81,** 181–190

DEGENHART, H. G. (1979) Normal and abnormal steroidogenesis in man. In *Genetic Variation in Hormone Systems,* Vol. **1,** edited by J. G. M. Shire, pp. 11–42. Boca Raton: CRC Press

DELL, A., ETIENNE, T., PANICO, M., MORRIS, H. R., VINSON, G. P., WHITEHOUSE, B. J., BARBER, M., BORDOLI, R. S., SEDGWICK, R. D. and TYLER, A. N. (1982) Characterization of an adrenal zona glomerulosa-stimulating component of posterior pituitary extract as bisacetyl-ser$_1$-α-MSH. *Neuropeptides,* **2,** 233–240

DHOM, G. (1973) The prepubertal and pubertal growth of the adrenal (adrenarche). *Beiträge zur Pathologie,* **150,** 357–377

DUNLAP, N. E. and GRIZZLE, W. E. (1984) Golden Syrian hamsters: a new experimental model for compensatory adrenal hypertrophy. *Endocrinology,* **114,** 1490–1495

DUNN, T. B. (1970) Normal and pathologic anatomy of the adrenal gland of the mouse, including neoplasms. *Journal of the National Cancer Institute,* **44,** 1323–1380

DUNN, M. G. and BOSMAN, H. B. (1981) Peripheral dopamine receptor identification: properties of a specific dopamine receptor in the rat adrenal zona glomerulosa. *Biochemical and Biophysical Research Communications,* **99,** 1081–1087

ENGELAND, W. C. and DALLMAN, M. F. (1975) Compensatory adrenal growth is neurally mediated. *Neuroendocrinology,* **19,** 352–362

ENGELAND, W. C., SHINSAKO, J. and DALLMAN, M. F. (1975) Corticosteroids and ACTH are not required for compensatory adrenal growth. *American Journal of Physiology,* **229,** 1461–1470

ENGELAND, W. C., FEELEY, S. and WILKE, C. (1982) Cholinergic reinnervation of adrenals transplanted to the anterior chamber. *Federation Proceedings,* **41,** 1111

FARESE, R. V. and REDDY, W. J. (1963) Observations on the interrelations between adrenal protein, RNA and DNA during prolonged ACTH administration. *Biochimica et Biophysica Acta,* **76,** 145–148

FEVOLD, H. R. and DRUMMOND, H. B. (1973) Factors affecting the adrenocorticotropic hormone stimulation of rabbit adrenal 17α-hydroxylase activity. *Biochimica et Biophysica Acta,* **313,** 211–220

FINKELSTEIN, M. and SHAEFER, J. M. (1979) Inborn errors of steroid biosynthesis. *Physiological Reviews,* **59,** 353–380

FORD, J. K. and YOUNG, R. W. (1963) Cell proliferation and displacement in the adrenal cortex of young rats injected with tritiated thymidine. *Anatomical Record,* **146,** 125–133

GALLANT, S., KAUFFMAN, F. C. and BROWNIE, A. C. (1974) Cyclic nucleotide phosphodiesterase activity in rat adrenal gland zones. *Life Sciences,* **14,** 937–944

GALLAY, J., VINCENT, M., DE PAILLERETS, C., ROGARD, M. and ALFSEN, A. (1981) Relationship between the activity of the 3β-hydroxysteroid dehydrogenase from bovine adrenal cortex microsomes and membrane structure. Influence of proteins and steroid substrates on lipid microviscosity. *Journal of Biological Chemistry,* **256,** 1235–1241

GILL, G. N., ILL, C. R. and SIMONIAN, M. H. (1977) Angiotensin stimulation of bovine adrenocortical cell growth. *Proceedings of the National Academy of Sciences of the United States of America,* **74,** 5569–5573

GILL, G. N., HORNSBY, P. J., ILL, C. R., SIMONIAN, M. H. and WEIDMAN, E. R. (1978) Regulation of adrenocortical cell growth. In *The Endocrine Function of the Human Adrenal Cortex,* edited by V. H. T. James, M. Serio, G. Giusti and L. Martini, pp. 207–228. New York: Academic Press

GILL, G. N., HORNSBY, P. J. and SIMONIAN, M. H. (1979) Regulation of growth and differentiated function of cultured bovine adrenocortical cells. *Cold Spring Harbor Conferences on Cell Proliferation,* **6,** 701–715

GILL, G. N., HORNSBY, P. J. and SIMONIAN, M. H. (1980) Hormonal regulation of the adrenocortical cell. *Journal of Supramolecular Structure,* **14,** 353–369

GILL, G. N., CRIVELLO, J. F., HORNSBY, P. J. and SIMONIAN, M. H. (1982) Growth, function, and development of the adrenal cortex: insights from cell culture. *Cold Spring Harbor Conferences on Cell Proliferation,* **9,** 461–482

GIROUD, C. J., STACHENKO, J. and VENNING, E. H. (1956) Secretion of aldosterone by the zona glomerulosa of rat adrenal glands incubated *in vitro. Proceedings of the Society for Experimental Biology and Medicine,* **92,** 154–157

GOSPODAROWICZ, D. and THAKRAL, K. K. (1978) Production of a corpus luteum angiogenic factor responsible for proliferation of capillaries and neovascularization of the corpus luteum. *Proceedings of the National Academy of Sciences of the United States of America,* **75,** 847–851

GOSPODAROWICZ, D., MORAN, J. S. and BIALECKI, H. (1976) Mitogenic factors from the brain and the pituitary: physiological significance. In *Growth Hormone and Related Peptides,* edited by A. Pecile and E. E. Muller, pp. 141–155. Amsterdam: Excerpta Medica

GOSPODAROWICZ, D., ILL, C. R., HORNSBY, P. J. and GILL, G. N. (1977) Control of bovine adrenal cortical cell proliferation by fibroblast growth factor. Lack of effect of epidermal growth factor. *Endocrinology,* **100,** 1080–1089

GOSPODAROWICZ, D., MORAN, J. S. and MESCHER, A. L. (1978) Cellular specificities of fibroblast growth factor and epidermal growth factor. In *Proliferation and Differentiation,* edited by J. Papaconstantinou and W. J. Rutter, pp. 33–63. New York: Academic Press

GOSPODAROWICZ, D., DELGADO, D. and VLODAVSKY, I. (1980) Permissive effect of the extracellular matrix on cell proliferation *in vitro. Proceedings of the National Academy of Sciences of the United States of America,* **77,** 4094–4098

GOSPODAROWICZ, D., LUI, G.-M. and CHENG, J. (1982) Purification in high yield of brain fibroblast growth factor by preparative isoelectric focusing at pH 9.6. *Journal of Biological Chemistry,* **257,** 12266–12276

GREEP, R. O. and DEANE, H. W. (1949) Histological, cytochemical and physiological observations on the regeneration of the rat's adrenal gland following enucleation. *Endocrinology,* **45,** 42–60

GRIZZLE, W. E. and DUNLAP, N. E. (1984) Aldosterone blocks compensatory adrenal hypertrophy in the rat. *American Journal of Physiology,* **246,** E306–E310

GRUENWALD, P. (1946) Embryonic and postnatal development of the adrenal cortex, particularly the zona glomerulosa and accessory nodules. *Anatomical Record,* **95,** 391–400

HADJIAN, A. J., VENTRE, R. and CHAMBAZ, E. M. (1981) Cholinergic muscarinic receptors in bovine adrenal cortex. *Biochemical and Biophysical Research Communications,* **98,** 892–900

HADJIAN, A. J., GUIDICELLI, C. and CHAMBAZ, E. M. (1982) Cholinergic muscarinic stimulation of steroidogenesis in bovine adrenal cortex fasciculata cell suspensions. *Biochimica et Biophysica Acta,* **714,** 157–163

HARKINS, J. B., NELSON, E. B., MASTERS, B. S. S. and BRYANT, G. T. (1974) Preparation and properties of microsomal membranes from isozonal cells of beef adrenal cortex. *Endocrinology,* **94,** 897–905

HENRY, J. P. and STEPHENS, P. M. (1977) The social environment and essential hypertension in mice: possible role of the innervation of the adrenal cortex. In *Hypertension and Brain Mechanisms,* edited by W. De Jong and A. P. Provoost, pp. 263–276. Amsterdam: Elsevier

HENRY, J. P., KROSS, M. E., STEPHENS, P. M. and WATSON, F. M. C. (1976) Evidence that differing psychosocial stimuli lead to adrenal cortical stimulation by autonomic or endocrine pathways. In *Catecholamines and Stress,* edited by E. Usdin, R. Kvetnanski and I. J. Topin, pp. 457–468. Oxford: Pergamon Press

HOKFELT, T., LUNDBERG, J. M., SCHULTZBERG, M. and FAHRENKRUG, J. (1981) Immunohistochemical evidence for a local VIP-ergic neuron system in the adrenal gland of the rat. *Acta Physiologica Scandinavica,* **113,** 575–576

HOLZWARTH, M. A., SHINSAKO, J. and DALLMAN, M. F. (1980) Adrenal regeneration. Time course, effect of hypothalamic hemi-islands and response to unilateral adrenalectomy. *Neuroendocrinology,* **31,** 168–176

HORNSBY, P. J. (1980) Regulation of cytochrome P-450-supported 11β-hydroxylation of deoxycortisol by steroids, oxygen, and antioxidants in adrenocortical cell cultures. *Journal of Biological Chemistry,* **255,** 4020–4027

HORNSBY, P. J. (1982) Regulation of 21-hydroxylase activity by steroids in cultured bovine adrenocortical cells: possible significance for adrenocortical androgen synthesis. *Endocrinology,* **111,** 1092–1101

HORNSBY, P. J. and ALDERN, K. A. (1984) Steroidogenic enzyme activities in cultured human definitive zone adrenocortical cells: comparison with bovine adrenocortical cells and resultant differences in adrenal androgen synthesis. *Journal of Clinical Endocrinology and Metabolism,* **58,** 121–130

HORNSBY, P. J., ALDERN, K. A. and HARRIS, S. E. (1984) Adrenocortical cultures as model systems for investigating cellular aging. In *Aging,* edited by D. Armstrong, R. S. Sohal, R. G. Cutler and T. F. Slater, Vol 27, pp. 203–222. New York: Raven Press

HORNSBY, P. J. and CRIVELLO, J. F. (1983) The role of lipid peroxidation and biological antioxidants in the function of the adrenal cortex. Part 2. *Molecular and Cellular Endocrinology,* **30,** 123–147

HORNSBY, P. J. and GILL, G. N. (1977) Hormonal control of adrenocortical cell proliferation. Desensitization to ACTH and interaction between ACTH and fibroblast growth factor in bovine adrenocortical cell cultures. *Journal of Clinical Investigation,* **60,** 342–352

HORNSBY, P. J. and GILL, G. N. (1978) Characterization of adult bovine adrenocortical cells throughout their life span in tissue culture. *Endocrinology,* **102,** 926–936

HORNSBY, P. J. and O'HARE, M. J. (1977) The roles of potassium and corticosteroids in determining the pattern of metabolism of [3H]deoxycorticosterone by monolayer cultures of rat adrenal zona glomerulosa cells. *Endocrinology,* **101,** 997–1005

HORNSBY, P. J., O'HARE, M. J. and NEVILLE, A. M. (1974) Functional and morphological observations on rat adrenal zona glomerulosa cells in monolayer culture. *Endocrinology,* **95,** 1240–1251

HORNSBY, P. J., SIMONIAN, M. H. and GILL, G. N. (1979) Aging of adrenocortical cells in culture. *International Review of Cytology,* Suppl. 10, 131–162

HORNSBY, P. J., STUREK, M., HARRIS, S. E. and SIMONIAN, M. H. (1983) Serum and growth factor requirements for proliferation of human adrenocortical cells in culture: comparisons with bovine adrenocortical cells. *In Vitro,* **19,** 863–869

HUGGINS, C. and MORII, S. (1961) Selective adrenal necrosis and apoplexy induced by 7,12-dimethylbenz(a)anthracene. *Journal of Experimental Medicine,* **114,** 741–750

HYATT, P. J., WALE, L. W., BELL, J. B., TAIT, J. F. and TAIT, S. A. (1980) Cyclic AMP levels in purified rat adrenal zonae fasciculata and reticularis cells and the effect of adrenocorticotrophic hormone. *Journal of Endocrinology,* **85,** 435–442

HYATT, P. J., BHATT, K. and TAIT, J. F. (1983) Steroid biosynthesis by zona fasciculata and zona reticularis cells purified from the mammalian adrenal cortex. *Journal of Steroid Biochemistry,* **19,** 953–960

HYATT, P. J., BELL, J. B. G., BHATT, K. and TAIT, J. F. (1983) Preparation and steroidogenic properties of purified zona fasciculata and zona reticularis cells from the guinea-pig adrenal gland. *Journal of Endocrinology,* **96,** 1–14

IDELMAN, S. (1970) Ultrastructure of the mammalian adrenal cortex. *International Review of Cytology,* **27,** 181–281

ILL, C. R. and GOSPODAROWICZ, D. (1982) Factors involved in supporting the growth and steroidogenic functions of bovine adrenal cortical cells maintained on extracellular matrix and exposed to a serum-free medium. *Journal of Cellular Physiology,* **113,** 373–384

IMRIE, R. C., RAMAIAH, T. R., ANTONI, F. and HUTCHINSON, W. C. (1965) The effect of adrenocorticotrophin on the nucleic acid metabolism of the rat adrenal gland. *Journal of Endocrinology,* **32,** 303–312

INGLE, D. J. and HIGGINS, G. M. (1938) Autotransplantation and regeneration of the adrenal gland. *Endocrinology,* **22,** 458–470

JAFFE,E R. B., SERON-FERRE, M., CRICKARD, K., KORITNIK, D., MITCHELL, B. F. and HUHTANIEMI, I. T. (1981) Regulation and function of the primate fetal adrenal gland and gonad. *Recent Progress in Hormone Research,* **37,** 41–103

JIRASEK, J. E. (1969) Morphological and histochemical analysis of the development of adrenals and gonads in man. In *Progress in Endocrinology,* edited by C. Gual and F. J. G. Ebling, pp. 1100–1107. Amsterdam: Excerpta Medica

JOHANNISSON, E. (1979) Aspects of the ultrastructure and function of the human fetal adrenal cortex. *Contributions to Gynecology and Obstetrics,* **5,** 109–130

JONES, T. and GRIFFITHS, K. (1968) Ultramicrochemical studies on the site of formation of dehydroepiandrosterone sulphate in the adrenal cortex of the guinea-pig. *Journal of Endocrinology,* **42,** 559–565

JONES, T., GROOM, M. and GRIFFITHS, K. (1970) Steroid biosynthesis by cultures of normal human adrenal tissue. *Biochemical and Biophysical Research Communications,* **38,** 355–361

KAHRI, A. I., VOUTILAINEN, R. and SALMENPERA, M. (1979) Different biological action of corticosteroids, corticosterone and cortisol, as a base of zonal function of adrenal cortex. *Acta Endocrinologica,* **91,** 329–337

KARTESZI, M., DALLMAN, M. F., MAKARA, G. B. and STARK, E. (1982) Regulation of the adrenocortical response to insulin-induced hypoglycemia. *Endocrinology,* **111,** 535–541

KATIKINENI, M., DAVIES, T. F. and CATT, K. J. (1981) Regulation of adrenal and testicular prolactin receptors by adrenocorticotropin and luteinizing hormone. *Endocrinology,* **108,** 2367–2375

KELNAR, C. J. H. and BROOK, C. G. D. (1983) A mixed longitudinal study of adrenal steroid excretion in childhood and the mechanism of adrenarche. *Clinical Endocrinology,* **19,** 117–129

KENNERSON, A. R., MCDONALD, D. A. and ADAMS, J. B. (1983) Dehydroepiandrosterone sulfotransferase localization in human adrenal glands: a light and electron microscopic study. *Journal of Clinical Endocrinology and Metabolism,* **56,** 786–795

KOLETSKY, R. J., BROWN, E. M. and WILLIAMS, G. H. (1983) Calmodulin-like activity and calcium-dependent phosphodiesterase in purified cells of the rat zona glomerulosa and zona fasciculata. *Endocrinology,* **113,** 485–490

KOLTA, M. G. and SOLIMAN, K. F. A. (1981) Effect of peripheral cholinergic activation on the adrenal cortex function. *Endocrine Research Communications,* **8,** 239–246

KOMINAMI, S., SHINZAWA, K. and TAKEMORI, S. (1982) Purification and some properties of cytochrome P-450 specific for steroid 17α-hydroxylation and C17–C20 bond cleavage from guinea pig adrenal microsomes. *Biochemical and Biophysical Research Communications,* **109,** 916–921

KRAMER, R. E., GALLANT, S. and BROWNIE, A. C. (1979) The role of cytochrome P-450 in the action of sodium depletion on aldosterone biosynthesis in rats. *Journal of Biological Chemistry,* **254,** 3953–3958

KRAMER, R. E., MCCARTHY, J. L., SIMPSON, E. R. and WATERMAN, M. R. (1983) Effects of ACTH on steroidogenesis in bovine adrenocortical cells in primary culture – increased secretion of 17α-hydroxylated steroids associated with a refractoriness in total steroid output. *Journal of Steroid Biochemistry,* **18,** 715–723

KUDLOW, J. E. and GERRIE, B. M. (1983) Production of growth factor activity by cultured bovine calf anterior pituitary cells. *Endocrinology,* **113,** 104–110

LANMAN, J. T. (1953) The fetal zone of the adrenal gland. Its development course, comparative anatomy, and possible physiologic functions. *Medicine,* **32,** 389–920

LANMAN, J. T. (1961) The adrenal gland in the human fetus. An interpretation of its physiology and unusual developmental pattern. *Pediatrics,* **27,** 140–190

LEWINSKI, A. and SZKUDLINSKI, M. (1981) Effect of dibutyryl cyclic adenosine 3′5′-monophosphate on the adrenal cortex mitotic activity in hypophysectomized rats. *Endokrinologie,* **77,** 371–374

LI, C. H., YAMASHIRO, D., GOSPODAROWICZ, D., KAPLAN, S. L. and VAN VLIET, G. (1983) Total synthesis of insulin-like growth factor I (somatomedin C). *Proceedings of the National Academy of Sciences of the United States of America,* **80,** 2216–2220

LILLY, M. P., ENGELAND, W. C. and GANN, D. S. (1983) Responses of cortisol secretion to repeated hemorrhage in the anesthetized dog. *Endocrinology,* **112,** 681–688

LOWRY, P. J., SILAS, L., MCLEAN, C., LINTON, E. A. and ESTIVARIZ, F. E. (1983) Pro-γ-melanocyte-stimulating hormone cleavage in adrenal gland undergoing compensatory growth. *Nature,* **306,** 70–73

MACHEMER, R. and OEHLERT, W. (1964) Autoradiographische Untersuchungen über den physiologischen Zellumsatz und die gesteigerte Zellneubildung der Nebenniere der ausgewachsenen Ratte nach Behandlung mit ACTH. *Endokrinologie,* **46,** 77–91

MALENDOWICZ, L. K. and COLBY, H. D. (1982) Effects of carbon tetrachloride on adrenocortical function in rats. *Toxicology and Applied Pharmacology,* **65,** 32–37

MANKU, M. S., HORROBIN, D. F., KARMAZYN, M. and CUNNANE, S. C. (1979) Prolactin and zinc effects on rat vascular reactivity: possible relationship to dihomo-γ-linolenic acid and to prostaglandin synthesis. *Endocrinology,* **104,** 774–779

MARIE, J. and JARD, S. (1983) Angiotensin II inhibits adenylate cyclase from adrenal cortex glomerulosa zone. *FEBS Letters,* **159,** 97–101

MARSHALL, S., BRUNI, J. F. and MEITES, J. (1979) Effects of hypophysectomy, thyroidectomy, and thyroxine on specific prolactin receptor sites in kidneys and adrenals of male rats. *Endocrinology,* **104,** 390–400

MARTON, J., PALFREYMAN, J. W. and SCHULSTER, D. (1976) The effect of cholera toxin on the adrenal weight in hypophysectomized rats. *Molecular and Cellular Endocrinology,* **5,** 147–149

MARX, A. J., DEANE, H. W., MOWLES, T. F. and SHEPPARD, H. (1963) Chronic administration of angiotensin in rats: changes in blood pressure, renal and adrenal histophysiology and aldosterone production. *Endocrinology,* **73,** 329–340

MASON, P. A., FRASER, R., SEMPLE, P. F. and MORTON, J. J. (1979) The interaction of ACTH and angiotensin II in the control of corticosteroid plasma concentration in man. *Journal of Steroid Biochemistry,* **10,** 235–239

MASUI, H. and GARREN, L. D. (1970) On the mechanism of action of adrenocorticotropic hormone. Stimulation of deoxyribonucleic acid polymerase and thymidine kinaseo activities in adrenal glands. *Journal of Biological Chemistry,* **245,** 2627–2632

MASUI, H. and GARREN, L. D. (1971) On the mechanism of action of adrenocorticotropic hormone. The stimulation of thymidine kinase activity with altered properties and changed subcellular distribution. *Journal of Biological Chemistry,* **246,** 5407–5413

MATSUOKA, H., MULROW, P. J. and LI, C. H. (1980) β-Lipotropin: a new aldosterone-stimulating factor. *Science,* **209,** 307–308

MCCARTHY, J. L., KRAMER, R. E., FUNKENSTEIN, B., SIMPSON, E. R. and WATERMAN, M. R. (1983) Induction of 17α-hydroxylase (cytochrome $P\text{-}450_{17\alpha}$) activity by adrenocorticotropin in bovine adrenocortical cells maintained in monolayer culture. *Archives of Biochemistry and Biophysics,* **222,** 590–598

MCKENNA, T. J., ISLAND, D. P., NICHOLSON, W. E. and LIDDLE, G. W. (1978) Angiotensin stimulates cortisol biosynthesis in human adrenal cells *in vitro. Steroids,* **32,** 2315–2320

MIAO, P. and BLACK, V. H. (1982) Guinea pig adrenocortical cells: *in vitro* characterization of separated zonal cell types. *Journal of Cell Biology,* **94,** 241–252

MIGEON, C. J., KENNY, F. M., KOWARSKI, A., SNIPES, C. A., SPAULDING, J. S., FINKELSTEIN, J. W. and BLIZZARD, R. M. (1968) The syndrome of congenital adrenocortical unresponsiveness to ACTH. Report of six cases. *Pediatric Research,* **2,** 501–513

MORERA, A. M. and SAEZ, J. M. (1980) *In vitro* mitogenic and steroidogenic effects of ACTH analogues on an adrenal tumor cell line (Y-1). *Experimental Cell Research,* **127,** 446–451

MULLER, J. (1970) Steroidogenic effect of stimulators of aldosterone biosynthesis upon separate zones of the adrenal cortex. *European Journal of Clinical Investigation,* **1,** 180–187

NAKAJIN, S., SHINODA, M. and HALL, P. F. (1983) Purification and properties of 17α-hydroxylase from microsomes of pig adrenal: a second C21 side-chain cleavage system. *Biochemical and Biophysical Research Communications,* **111,** 512–517

NEGRO-VILAR, A., SAAD, W. A. and MCCANN, S. M. (1977) Evidence for a role of prolactin in prostate and seminal vesicle growth in immature male rats. *Endocrinology,* **100,** 729–737

NEVILLE, A. M. and O'HARE, M. J. (1978) Cell culture and histopathology in relation to hypercorticalism. In *The Endocrine Function of the Human Adrenal Cortex,* edited by V. H. T. James, M. Serio, G. Giusti and L. Martini, pp. 229–249. London: Academic Press

NEVILLE, A. M. and O'HARE, M. J. (1979) Aspects of structure, function, and pathology. In *The Adrenal Cortex,* edited by V. H. T. James, pp. 1–65. New York: Raven Press

NEVILLE, A. M. and O'HARE, M. J. (1982) *The Human Adrenal Cortex. Pathology and Biology – An Integrated Approach.* Berlin: Springer-Verlag

NICKERSON, P. A. (1975) Quantitative study on the effect of an ACTH-producing pituitary tumor on the ultrastructure of the mouse adrenal gland. *American Journal of Pathology,* **80,** 295–308

NICKERSON, P. A., BROWNIE, A. C. and MOLTENI, A. (1970) Adrenocortical structure and function in rats bearing an adrenocorticotropic hormone, growth hormone, and prolactin-secreting tumor. *Laboratory Investigation,* **23,** 368–375

NIIJIMA, A. and WINTER, D. L. (1968) Baroreceptors in the adrenal gland. *Science,* **159,** 434–435

OGUNSUA, A. O., De NICOLA, A. F., TRAIKOV, H. and BIRMINGHAM, M. K. (1971) Adrenal steroid biosynthesis by different species of mouselike rodents. *Gen. Comp. Endocrinol.,* **16,** 192–199

O'HARE, M. J. and NEVILLE, A. M. (1973) Effects of adrenocorticotrophin on steroidogenesis and proliferation by adult adrenocortical cells in monolayer culture. *Biochemical Society Transactions.* **1,** 1088–1091

O'HARE, M. J., ELLISON, M. L. and NEVILLE, A. M. (1978) Tissue culture in endocrine research: perspectives, pitfalls and potentials. *Current Topics in Experimental Endocrinology,* **3,** 1–50

O'HARE, M. J., NICE, E. C. and NEVILLE, A. M. (1980) Regulation of androgen secretion and sulfoconjugation in the adult human adrenal cortex: studies with primary monolayer cell cultures. In *Adrenal Androgens,* edited by A. R. Genazzani, J. H. H. Thijssen and P. K. Siiteri, pp. 7–25. New York: Raven Press

OHTANI, O., KIKUTA, A., OHTSUKA, A., TAGUCHI, T. and MURAKAMI, T. (1983) Microvasculature as studied by the microvascular corrosion casting/scanning electron microscope method. I. Endocrine and digestive system. *Archivum Histologicum Japonicum,* **46,** 1–42

OTTENWELLER, J. E. and MEIER, A. H. (1982) Adrenal innervation may be an extrapituitary mechanism able to regulate adrenocortical rhythmicity in rats. *Endocrinology,* **111,** 1334–1338

PAPPRITZ, G., TRIEB, G. and DOHM, G. (1972) Autoradiographische Untersuchungen zum Wachstum der Nebennierenrinde der Ratte. *Zeitschrift für Zellforschung,* **126,** 421–430

PARKER, L. N. and ODELL, W. D. (1980) Control of adrenal androgen secretion. *Endocrine Reviews,* **1,** 392–410

PARKER, C. R. JR, CARR, B. R., WINKEL, C. A., CASEY, M. L., SIMPSON, E. R. and MACDONALD, P. C. (1983) Hypercholesterolemia due to elevated low density lipoprotein-cholesterol in newborns with anencephaly and adrenal atrophy. *Journal of Clinical Endocrinology and Metabolism,* **57,** 37–43

PAYET, N. and LEHOUX, J.-G. (1982) Effect of ACTH or zinc treatment on plasma aldosterone and corticosterone levels and on the *in vitro* steroid output from adrenocortical cells. *Canadian Journal of Biochemistry,* **60,** 1058–1064

PAYET, N., LEHOUX, J.-G. and ISLER, H. (1980) Effect of ACTH on the proliferative and secretory activities of the adrenal glomerulosa. *Acta Endocrinologica,* **93,** 365–374

PIRAS, M. M., BINDSTEIN, E. and PIRAS, R. (1973) Regulation of glycogen metabolism in adrenal gland. IV. The effect of insulin on glycogen synthetase, phosphorylase, and related metabolites. *Archives of Biochemistry and Biophysics,* **154,** 263–269

PLZAK, L. (1960) Effect of ACTH on regeneration of adrenal cortex following autografting in hypophysectomized rats. *Proceedings of the Society for Experimental Biology and Medicine,* **103,** 366–368

RACE, G. J. and GREEN, R. F. (1955) Studies on zonation and regeneration of the adrenal cortex of the rat. *Archives of Pathology,* **59,** 578–585

RAINEY, W. E., HORNSBY, P. J. and SHAY, J. W. (1983) Morphological correlates of adrenocorticotropin-stimulated steroidogenesis in cultured adrenocortical cells: differences between bovine and human cells. *Endocrinology,* **113,** 48–54

RAMACHANDRAN, J. and SUYAMA, A. T. (1975) Inhibition of replication of normal adrenocortical cells in culture by adrenocorticotropin. *Proceedings of the National Academy of Sciences of the United States of America,* **72,** 113–117

REYNOLDS, J. W. (1981) Development and function of the human fetal adrenal cortex. In *Fetal Endocrinology,* edited by M. J. Novy and J. A. Resko, pp. 35–52. New York: Academic Press

RICH, B. H., ROSENFIELD, R. L., LUCKY, A. W., HELKE, J. C. and OTTO, P. (1981) Adrenarche: changing adrenal response to adrenocorticotropin. *Journal of Clinical Endocrinology and Metabolism,* **52,** 1129–1136

ROBINSON, P. M., PERRY, K. A., HARDY, K. J., COGHLAN, J. P. and SCOGGINS, B. A. (1977) The innervation of the adrenal cortex in the sheep, *Ovis ovis. Journal of Anatomy,* **124,** 117–129

SAHINEN, F. M. and SODERWALL, A. L. (1965) Accelerated senescent changes in adrenal cortex of the hamster. *Radiation Research,* **24,** 412–422

SAKHATSKAYA, T. S. and ALTUKHOVA, V. I. (1973) Formation of dehydroepiandrosterone sulfate and hydrocortisone in the definitive and fetal adrenal cortex of human fetuses. *Soviet Journal of Developmental Biology,* **4,** 46–50

SCHIEBINGER, R. J., ALBERTSON, B. D., BARNES, K. M., CUTLER, G. B. JR and LORIAUX, D. L. (1981a) Developmental changes in rabbit and dog adrenal function: a possible homologue of adrenarche in the dog. *American Journal of Physiology,* **240,** E694–E699

SCHIEBINGER, R. J., ALBERTSON, B. D., CASSORLA, F. G., BOWYER, D. W. and GEELHOED, G. W. (1981b) The developmental changes in plasma adrenal androgens during infancy and adrenarche are associated with changing activities of adrenal microsomal 17-hydroxylase and 17,20-desmolase. *Journal of Clinical Investigation,* **67,** 1177–1182

SCHMIDTKE, J., WIENKER, T., FLUGEL, M. and ENGEL, W. (1976) *In vitro* inhibition of cyclic AMP phosphodiesterase by cortisol. *Nature,* **262,** 593–594

SCHWEDES, U., WEHNER, H., LEUSCHNER, U., SCHOFFLING, K. and USADEL, K. H. (1974) Development and function of isologous transplants of cell suspensions of fetal adrenal glands in the rat. *Acta Endocrinologica,* Suppl. **184,** 59

SEGAL, B. M. and CHRISTY, N. P. (1968) Potentiation of the biologic activity of ACTH by human plasma. A preliminary study. *Journal of Clinical Endocrinology,* **28,** 1465–1470

SEGAL, B. M., DRUCKER, W. D., BENOVITZ, H., VERDE, A. L. and CHRISTY, N. P. (1970) Further studies of adrenal weight-maintaining activity in the plasma of patients with Cushing's disease. *American Journal of Medicine,* **49,** 34–40

SHIMA, S., KAWASHIMA, Y. and HIRAI, M. (1979) Studies on cyclic nucleotides in the adrenal gland. Effects of ACTH and calcium on cyclic AMP production and steroid output by the zona glomerulosa of the adrenal cortex. *Endocrinologica Japonica,* **26,** 219–225

SHIRE, J. G. M. (1979) Corticosteroids and adrenocortical function in animals. In *Genetic Variation in Hormone Systems,* edited by J. G. M. Shire, **1,** 43–70. Boca Raton: CRC Press

SIMONIAN, M. H. and CAPP, M. W. (1984) Characterization of steroidogenesis in cell cultures of the human fetal adrenal cortex: comparison of definitive and fetal zone cells. *Journal of Clinical Endocrinology and Metabolism,* **59,** 643–651

SIMONIAN, M. H. and GILL, G. N. (1979) Regulation of deoxyribonucleic acid synthesis in bovine adrenocortical cells in culture. *Endocrinology,* **104,** 588–595

SIMONIAN, M. H. and GILL, G. N. (1981) Regulation of the fetal human adrenal cortex: effects of adrenocorticotropin on growth and function of monolayer cultures of fetal and definitive zone cells. *Endocrinology,* **108,** 1769–1779

SIMONIAN, M. H., HORNSBY, P. J., ILL, C. R., O'HARE, M. J. and GILL, G. N. (1979) Characterization of cultured bovine adrenocortical cells and derived clonal lines: regulation of steroidogenesis and culture life span. *Endocrinology,* **105,** 99–108

SIMONIAN, M. H., WHITE, M. L. and GILL, G. N. (1982) Growth and function of cultured bovine adrenocortical cells in a serum-free defined medium. *Endocrinology,* **111,** 919–927

SLAGA, T. J. and KRUM, A. A. (1973) Modification of rabbit adrenal steroid biosynthesis by prolonged ACTH administration, *Endocrinology,* **93,** 517–525

SLANINA, S. M. and FEVOLD, H. R. (1982) The enzyme specificity of ACTH stimulation of rabbit adrenal microsomal 17α-hydroxylase activity. *Journal of Steroid Biochemistry,* **16,** 93–99

SROUGI, M., GITTES, R. F. and UNDERWOOD, R. H. (1980) Influence of exogenous glucocorticoids and ACTH on experimental adrenal autografts. *Investigative Urology,* **17,** 265–268

STACHENKO, J. and GIROUD, C. J. P. (1964) Further observations on the functional zonation of the adrenal cortex. *Canadian Journal of Biochemistry,* **42,** 1777–1785

STOCKER, E, and SCHMID, G. H. (1973) Altersabhängige autoradiographische Untersuchungen über die Nebennierenrinden-Proliferation männlicher Ratten nach [^{3}H]Thymidin-Dauerinfusion. *Virchows Archiv,* **B13,** 247–257

STUDZINSKI, G. P., HAY, D. C. F. and SYMINGTON, T. (1963) Observations on the weight of the human adrenal gland and the effect of preparations of corticotropin of different purity on the weight and morphology of the human adrenal gland. *Journal of Clinical Endocrinology and Metabolism,* **23,** 248–260

SYMINGTON, T. (1982) The adrenal cortex. In *Endocrine Pathology, General and Surgical,* edited by J. M. B. Bloodworth, Jr, pp. 419–440. Baltimore: Williams and Wilkins

SZABO, S., HUTTNER, I., KOVACS, K., HORVATH, E., SZABO, D. and HORNER, H. C. (1980) Pathogenesis of experimental adrenal hemorrhagic necrosis (apoplexy). Ultrastructural, biochemical, neuropharmacologic, and blood coagulation studies with acrylonitrile in the rat. *Laboratory Investigation,* **42,** 533–546

SZABO, S., MCCOMB, D. J., KOVACS, K. and HUTTNER, I. (1981) Adrenocortical hemorrhagic necrosis. The role of catecholamines and retrograde medullary-cell embolism. *Archives of Pathology and Laboratory Medicine,* **105,** 536–539

TAIT, J. F. and TAIT, S. A. (1979) Recent perspectives on the history of the adrenal cortex. The Sir Henry Dale lecture for 1979. *Journal of Endocrinology,* **83,** 3P–24P

ULICK, S. (1976) Diagnosis and nomenclature of the disorders of the terminal portion of the aldosterone biosynthetic pathway. *Journal of Clinical Endocrinology and Metabolism,* **43,** 92–98

UNSICKER, K. and GROSCHEL-STEWART, U. (1977) Distribution of contractile proteins and adrenergic nerves in the adrenal gland of guinea-pig, rat and ox as revealed by immunofluorescence and the glyoxylic acid technique. *Experientia,* **34,** 102–105

USADEL, K. H., ROCKERT, H., OBERT, I. and SHOFFLING, K. (1970) Entwicklung der histiotypischen Organe durch isologe Transplantation von Zellsuspension embryonaler Organanlagen bei der Ratte. *Klinische Wochenschrift,* **48,** 1417–1418

VARGA, B., STARK, E. and FOLLY, G. (1979) Inhibition of the stimulatory effect of ACTH on adrenal and ovarian blood flow by indomethacin in the dog. *Acta Physiologica Academiae Scientarium Hungaricae,* **54,** 123–128

VARON, H. H., TOUCHSTONE, J. C. and CHRISTIAN, J. J. (1966) Biological conditions modifying quality of 17-hydroxycorticoids in mouse adrenal. *Acta Endocrinologica,* **51,** 488–496

VILLEE, D. (1972) Development of steroidogenesis. *American Journal of Medicine,* **53,** 533–544

VOITKEVICH, A. A. (1970) *Regeneratsiia Nadpochechnoi Zhelezy.* Moscow: Meditsina

WALTON, J. (1982) The role of limited cell replicative capacity in pathological age change. A review. *Mechanisms of Ageing and Development,* **19,** 217–244

WEGLICKI, W. B., REICHEL, W. and NAIR, P. P. (1968) Accumulation of lipofuscin-like pigment in the rat adrenal gland as a function of vitamin E deficiency. *Journal of Gerontology,* **23,** 469–475

WHITEHOUSE, B. J., VINSON, G. P. and THODY, A. J. (1982) Dopaminergic control of aldosterone: modulation of the response of rat adrenal zona glomerulosa cells to α-MSH by pretreatment with bromocriptine or metoclopramide. *Steroids,* **39,** 2887–2895

WILKINSON, C. W., SHINSAKO, J. and DALLMAN, M. F. (1981) Return of pituitary–adrenal function after adrenal enucleation or transplantation: diurnal rhythms and responses to ether. *Endocrinology,* **109,** 162–170

WILKINSON, C. W., SHINSAKO, J. and DALLMAN, M. F. (1982) Rapid decreases in adrenal and plasma corticosterone concentrations after drinking are not mediated by changes in plasma adrenocorticotropin concentration. *Endocrinology,* **110,** 1599–1606

WILSON, T. A.., KAISER, D. L., PEACH, M. J., WRIGHT, E. M. and CAREY, R. M. (1983) Possible mechanism of action of metoclopramide-induced aldosterone secretion: *in vivo* and *in vitro* studies in the sheep. *Endocrinology,* **113,** 887–892

WINTER, J. S. D., FUJIEDA, K., FAIMAN, C., REYES, F. I. and THLIVERIS, J. (1980) Control of steroidogenesis by human fetal adrenal cells in tissue culture. In *Adrenal Androgens,* edited by A. R. Genazzani, J. H. H. Thijssen and P. K. Siiteri, pp. 55–62. New York: Raven Press

WRIGHT, N. A. (1971) Cell proliferation in the prepubertal male rat adrenal cortex: an autoradiographic study. *Journal of Endocrinology,* **49,** 599–609

WRIGHT, W. A. (1981) The tissue kinetics in cell loss. In *Cell Death in Biology and Pathology,* edited by I. D. Bowen and R. A. Lockshin, pp. 171–207. London: Chapman and Hall

WRIGHT, N. A., VONCIMA, D. and MORLEY, A. R. (1973) An attempt to demonstrate cell migration from the zona glomerulosa in the prepubertal male rat adrenal cortex. *Journal of Endocrinology,* **59,** 451–459

WRIGHT, N. A., APPLETON, D. R. and MORLEY, A. R. (1974) Effect of dexamethasone on cell population kinetics in the adrenal cortex of the prepubertal male rat. *Journal of Endocrinology,* **62,** 527–536

WYLLIE, A. H., KERR, J. F. and CURRIE, A. R. (1973) Cell death in the normal neonatal rat adrenal cortex. *Journal of Pathology,* **111,** 255–261

WYLLIE, A. H., KERR, J. F., MACASKILL, I. A. and CURRIE, A. R. (1973) Adrenocortical cell deletion: the role of ACTH. *Journal of Pathology,* **111,** 85–94

2 The adrenal cortex in the fetus and neonate

Jeremy S. D. Winter

INTRODUCTION

During fetal life the human adrenal cortex reaches a size that is, relative to body size, 10–20 times larger than that of the adult. The bulk of this enlargement is contributed by a histologically distinct central fetal zone, which involutes rapidly after birth. These developmental peculiarities, which were recognized by the early years of this century (Elliott and Armour, 1911; Thomas, 1911), have stimulated considerable and sometimes fanciful speculation regarding the physiological role of this gland during fetal life and the factors which regulate its growth and function. In recent years attention has been focused on its massive production of dehydroepiandrosterone (DHA) and other Δ^5-3β-hydroxysteroids, and the subsequent metabolism of these steroids by the fetal liver and the placenta. Diczfalusy (1964) developed from such data the concept of a cooperative fetoplacental steroidogenic unit, in which the role of the fetal adrenal was to secrete DHA as an essential substrate for placental oestrogen biosynthesis.

Implicit in this concept is the notion that DHA production must be regulated by some factor of pituitary or placental origin unique to the fetal environment, or that the fetal adrenal cell itself has special steroidogenic properties which disappear spontaneously after birth. This review will present data which indicate instead that the fetal adrenal has much the same intrinsic properties, and responds to the same pituitary regulatory mechanisms, as in postnatal life. These recent observations lead to the alternative hypothesis that the apparent functional and structural peculiarities of the fetal adrenal cortex represent necessary adaptations to metabolic circumstances which derive from the intrauterine environment itself and which disappear at birth.

MORPHOLOGY OF THE FETAL ADRENAL

The primordium of the adrenal cortex appears at about 25 days' gestation in an area of celomic mesothelium just medial to the urogenital ridge and the developing mesonephros. Initially, the gland is composed of apparently immature cells with poorly developed endoplasmic reticulum, and mitotic activity can be observed throughout. The adrenal enlarges rapidly (*Figure 2.1*), and by 6–8 weeks' gestation one can differentiate an inner fetal zone from the outer definitive zone.

Thereafter mitotic activity appears to be restricted to the definitive zone (Crowder, 1957), which is composed of small basophilic cells retaining many of the ultrastructural characteristics of the earlier immature adrenal cells. These cells contribute centripetally via an indistinct transitional zone (Johannisson, 1968) to an enlarging fetal zone, which eventually occupies over four-fifths of the total gland volume. As the adrenal continues to enlarge, mainly due to expansion of the fetal zone, it assumes an extended, flattened form which permits growth without any further increase in total cortical thickness (Dhom, Ross and Widok, 1958).

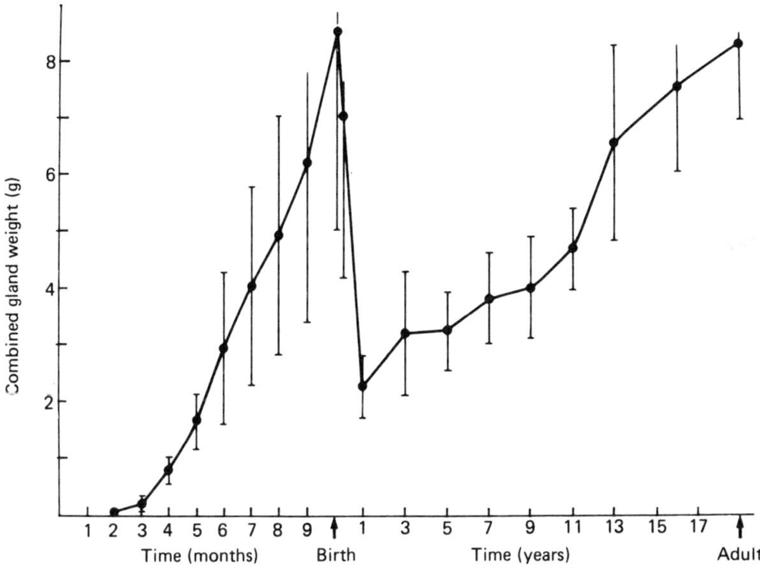

Figure 2.1 Adrenal gland weights during human development. (From Neville and O'Hare, 1982, courtesy of the Publishers, *The Human Adrenal Cortex*)

The large eosinophilic cells of the fetal zone, in contrast to those of the definitive zone, are well differentiated for active steroidogenesis. They show abundant cytoplasm, which is packed with a convoluted network of smooth endoplasmic reticulum, a prominent Golgi apparatus and numerous large mitochondria with a predominantly tubulovesicular internal structure. The development of this fetal zone is clearly ACTH-dependent; it is absent in the anencephalic fetus, but can be restored by administration of ACTH (Johannisson, 1968)

STEROIDOGENESIS BY THE FETAL ADRENAL CORTEX

Substrates

By term, the fetus is producing 100–200 mg of steroid daily, a rate which is several times higher than that of a resting adult (Simmer *et al.*, 1964; Siiteri and MacDonald, 1966). Cholesterol, the obligate precursor for all steroidogenesis, can be synthesized in the fetal adrenal (Bloch and Benirschke, 1959); it is not produced to any degree by the placenta, which uses maternal cholesterol as the substrate for progesterone synthesis. Because *in situ* perfusion studies suggested that neither circulating acetate nor free cholesterol were efficient precursors for fetal steroidogenesis (Solomon *et al.,* 1967), it has been generally accepted that pre-formed steroids of placental origin had to be provided to the fetal adrenal. Thus, placental pregnenolone was considered to be the natural substrate for DHA, while progesterone was considered to be necessary for any cortisol production by the fetus (Diczfalusy, 1969).

Certainly, fetal tissues can utilize both pregnenolone and progesterone, and there is a significant arteriovenous difference in cord serum levels, indicating some metabolism (Harbert *et al.,* 1964; Hagemenas and Kittinger, 1973); but recent evidence indicates that, just as in the adult, the major substrate for the fetal adrenal is circulating low density lipoprotein (LDL), which is assimilated by a process that first involves binding to specific membrane receptors, and then absorptive endocytosis and hydrolysis to release free cholesterol (Simpson *et al.,* 1979; Carr *et al.,* 1980b). Some cholesterol sulfate may also be derived from this process, and can be metabolized directly to sulfated pregnenolone and DHA (Korte, Hemsell and Mason, 1982).

LDL cholesterol is primarily produced in the fetal liver (Carr and Simpson, 1981b). Cord serum LDL concentrations are considerably lower than those of adults (Winkler, Schlag and Goetze, 1977; McConathy and Lane, 1980), which presumably reflects a high rate of clearance and utilization for steroidogenesis, since levels are much higher in anencephaly (Parker *et al.,* 1980). In addition to this circulating cholesterol, up to 30% of fetal adrenal steroidogenesis may be derived from *de novo* cholesterol synthesis within the adrenal itself (Carr and Simpson, 1981a). Pituitary ACTH regulates cholesterol availability both by stimulating *de novo* biosynthesis (Carr, MacDonald and Simpson, 1980) and by increasing the number of LDL receptors (Ohashi, Carr and Simpson, 1981). The fetal zone, as befits its active steroidogenic role, contains more LDL-binding sites per cell and shows a higher rate of *de novo* cholesterol synthesis than the definitive zone (Carr, Ohashi and Simpson, 1982).

Steroidogenic pathways

The fetal adrenal cortex has the same repertoire of steroidogenic enzymes and uses the same pathways for corticoid and androgen biosynthesis as the adult gland (Yoshida *et al.,* 1978). However, there are striking differences in relative enzyme activity, most notably a marked reduction in 3β-hydroxysteroid dehydrogenase (HSD) activity (Solomon *et al.,* 1967), while the activities of the various cytochrome P-450 associated mixed function oxidases that accomplish the subsequent steps in cortisol synthesis are unimpaired (Shibusawa *et al.,* 1978, 1980). Thus, when pre-formed progesterone is available as a precursor, the major end-products are cortisol, corticosterone and 16-hydroxyprogesterone (*Figure 2.2*); some aldosterone can also be produced (Dufau and Villee, 1969). However, the major end-product of pregnenolone or cholesterol metabolism is DHA (*Figure 2.3*). This relative deficiency of 3β-HSD is particularly marked in the inner fetal zone, which also shows a high capacity to sulfurylate the Δ^5-3β-hydroxysteroids that are produced (Seron-Ferre *et al.,* 1978; Korte, Hemsell and Mason, 1982).

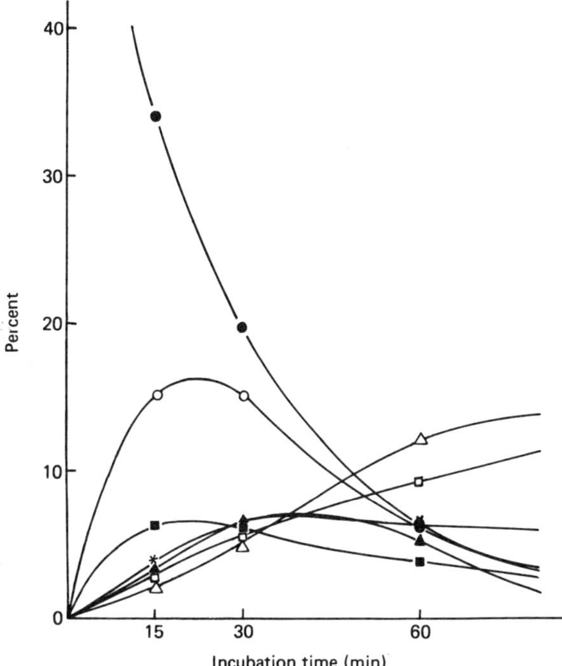

Figure 2.2 Relative amounts of 17OH-progesterone (○——○), 11-desoxycortisol (▲——▲), cortisol (△——△), desoxycorticosterone (■——■), corticosterone (□——□) and 16OH-progesterone (★——★) produced at different time intervals during incubation of human fetal adrenal tissue with ^{14}C-progesterone (●——●). (From Yoshida *et al.*, 1978, courtesy of the Editor and Publishers, *Endocrinologica Japonica*)

Thus in fresh ACTH-stimulated fetal zone cells the principal steroids produced are DHA sulfate, pregnenolone, pregnenolone sulfate and 17-hydroxypregnenolone, while cortisol is secreted in lesser amounts (Mason, Hemsell and Korte, 1983).

Fetoplacental metabolism of adrenal steroids

Corticosteroids

Although quantitatively various Δ^5-3β-hydroxysteroids are the major secretory products of the fetal adrenal, the physiologically significant steroids, as in postnatal life, appear to be Δ^4-3-ketosteroids, such as cortisol and possibly aldosterone. These are the only steroids for which feedback regulatory mechanisms can be demonstrated, and impaired secretion produces fetal or neonatal disease. The effect of gestational age on cord serum total cortisol concentrations is shown in *Figure 2.4*. Note that levels are low in mid-pregnancy, but rise rapidly in late gestation, to a mean of 45±23 (s.d.) ng/ml or approximately 150 nmol/l (Murphy, 1982). These values represent the net result of production from fetal and maternal sources, and clearance through fetal and placental metabolism. It should be noted that fetal serum levels of corticosteroid-binding globulin (CBG) are low

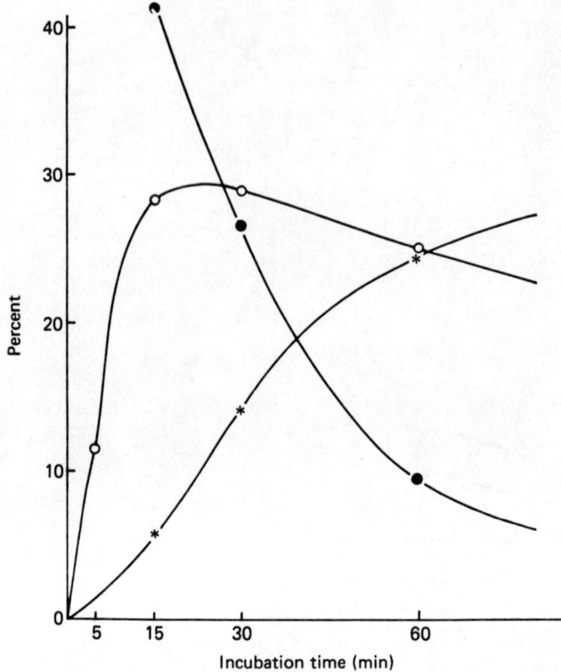

Figure 2.3 Relative amounts of 17OH-pregnenolone (○——○) and dehydroepiandrosterone (★——★) formed at different time intervals during incubation of human fetal adrenal tissue with ^{14}C-pregnenolone (●——●). (From Yoshida *et al.*, 1978, courtesy of the Editor and Publishers, *Endocrinologica Japonica*)

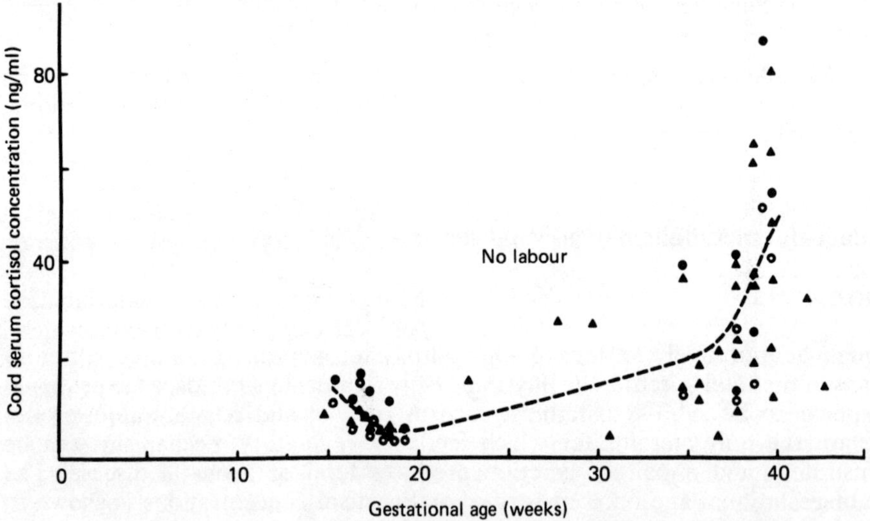

Figure 2.4 Cortisol concentrations in human umbilical arterial (●), venous (○) or mixed arterial and venous (▲) cord serum obtained at delivery by caesarean section in the absence of labour. Results in ng/ml can be converted to nmol/l by multiplying by 2.8. (From Murphy, 1982, courtesy of the Editor and Publishers, *American Journal of Obstetrics and Gynecology*)

(0.25±0.03 μmol/l) relative to maternal values (Hadjian *et al.*, 1975); furthermore, progesterone, the concentrations of which are very high, competes for CBG binding sites and tends to increase levels of free steroid. Mean serum free cortisol concentrations do increase several-fold during the latter half of pregnancy (Campbell and Murphy, 1977), but remain lower than those found in the maternal circulation.

The fetal adrenal is a major source of cortisol production, as evidenced by the positive arteriovenous difference in cord serum cortisol values which exists through to term (Murphy, 1973a; Leong and Murphy, 1976). Studies of chronically catheterized Rhesus monkey fetuses demonstrate that the fetal adrenal actively secretes cortisol at rates which, per unit weight, equal or exceed those of the adult (Walsh, Norman and Novy, 1979; Jaffe *et al.*, 1981). In both monkey and human fetuses, 60–75% of serum cortisol is secreted by the adrenals while the remainder is derived from placental transfer or metabolic conversion of cortisone (Beitins *et al.*, 1973; Jaffe *et al.*, 1981).

Here we see an obvious paradox, in that the fetal adrenal actively secretes cortisol and yet is so deficient in 3β-HSD that its production from cholesterol is remarkably inefficient. While the definitive zone may contain a relatively higher level of 3β-HSD activity per cell (Goldman, Yakovac and Bongiovanni, 1966; Branchaud *et al.*, 1978), the very bulk of the fetal zone and its high rate of total steroidogenesis make it unlikely that, as some have suggested (Seron-Ferre *et al.*, 1978), this zone is completely excluded from cortisol secretion. It seems more likely that instead of such rigid functional specialization there is a declining gradient of cortisol synthesis across the cortex, with lowest rates in the innermost juxtamedullary area.

One resolution of this paradox is the suggestion that the fetal adrenal utilizes only pre-formed placental progesterone for cortisol biosynthesis. *In vitro*, the provision of high concentrations of progesterone does enhance cortisol synthesis (Fujieda *et al.*, 1982). *In vivo* perfusion studies confirm that the fetus can metabolize small amounts of progesterone to cortisol, but this conversion is not enhanced by ACTH (Macnaughton *et al.*, 1977). Pepe and Albrecht (1980) have demonstrated in the fetal baboon that cholesterol rather than placental progesterone is the primary precursor for cortisol synthesis. Although similar data are not available for man, the impressive ACTH-dependent clearance of LDL cholesterol by the fetal adrenal, contrasted with a relatively small arteriovenous difference in cord serum progesterone levels, suggests that the former is the more important precursor.

Maternal cortisol readily crosses the placenta, but the bulk of this material (85%) is converted by placental 11-hydroxysteroid dehydrogenase to cortisone, which is essentially inactive. Jaffe *et al.* (1981) have estimated that up to one-half of the cortisol in the fetal circulation may be of maternal origin, either through direct placental transfer or by metabolic conversion of cortisone. The human fetal liver does not readily reduce cortisone, but it is possible that this conversion may also occur in target cells and thus provide a local source of active cortisol.

The metabolic clearance of cortisol by the fetus and placenta is extremely rapid, with most being converted to cortisone and its further metabolites (Pasqualini *et al.*, 1970). In the fetal monkey approximately 80% is oxidized to cortisone; 60% of the total cortisol produced is ultimately transferred to the mother, either as cortisol or cortisone (Mitchell *et al.*, 1981). The normal human fetus shows a positive arteriovenous gradient in cord blood cortisol levels and appears to be a net exporter of cortisol. However, when fetal cortisol secretion fails, as in the hypopituitary or hypoadrenal fetus, the maternal contribution appears sufficient to sustain reasonably normal fetal health and development.

In addition to cortisol, the fetal adrenal secretes corticosterone, desoxycorticosterone and aldosterone. Little is known about serum concentrations of these steroids during gestation but, by term, levels of desoxycorticosterone and aldosterone are surprisingly high, and respond both to sodium loading and depletion (Beitins *et al.*, 1972).

Δ^5-3β-Hydroxycorticosteroids

Although fetal adrenal steroidogenesis, particularly within the fetal zone, appears to be under maximal stimulatory drive, the major products of this process are not biologically active Δ^4-3-ketosteroids such as cortisol, but rather Δ^5-3β-hydroxycorticosteroids, including pregnenolone, 17-hydroxypregnenolone and DHA, mainly secreted as their sulfate conjugates (Milner and Mills, 1970; Cooke and Taylor, 1971). To some extent this deficiency of 3β-HSD activity relative to that of 17-hydroxylase and 17,20-desmolase is characteristic of the human adrenal at all ages (Whitehouse and Vinson, 1968; Deshpande *et al.*, 1970), but in the fetus the levels of 3β-HSD are so reduced (Shirley and Cooke, 1969) that one must seek some regulatory or inhibitory influence that is unique to the intrauterine situation. Although the specific activity of fetal adrenal 3β-HSD is low, its kinetic properties are similar to those found in adult adrenals, gonads or placenta. The preferred substrates are 17-hydroxypregnenolone and DHA, and the preferred hydrogen acceptor is NAD^+ (Hirato, Yanaihara and Nakayama, 1982). A more than 20-fold enhancement of fetal adrenal 3β-HSD activity can be achieved if the microsomal fraction is separated from cytosol, a phenomenon which points to marked inhibition of this enzyme complex by some soluble cytosolic factor (Serra, Perez-Palacios and Jaffe, 1971), which, as we shall see, is most likely steroidal.

The fetal adrenal and liver both show high levels of 16-hydroxylase and sulfotransferase activity, but fetal tissues contain relatively little sulfatase. Thus, the bulk of the DHA produced is converted to 16α-hydroxy-DHA sulfate, which in turn undergoes placental aromatization and conjugation to oestriol sulfate and glucuronide congeners which are excreted by the mother. Even though the products of this interaction have little or no biological activity, the very magnitude of this pathway has led investigators for years to suggest that it must have some function other than to prevent accumulation of DHA in the fetal circulation, but to date this function remains unclear.

Regulation of fetal adrenal steroidogenesis

A discussion of the control of fetal adrenal function must include an examination of the primary tropic drive which regulates the overall rate of steroidogenesis, and also an analysis of those factors which modulate the adrenal response so as to determine the relative proportions of steroids which are produced. The first is fairly simple, since there is no convincing evidence that any factor other than fetal pituitary ACTH serves to regulate total steroid production. However, in recent years, increasing evidence has been provided to suggest that the activities of the various steroidogenic enzymes, and thus the relative amounts of different steroids secreted, may be greatly influenced by changes in intra-adrenal concentrations of both endogenous and placental steroids.

Fetal adrenal cells *in vitro* are exquisitely sensitive to stimulation by ACTH in concentrations as low as 1 pg/ml (Fujieda *et al.*, 1981); for comparison, the mean human cord plasma ACTH concentration in mid-pregnancy is about 250 pg/ml (Winters *et al.*, 1974), at which level adrenal stimulation is almost maximal. Co-culture with fetal pituitary cells elicits identical responses, an indication that ACTH is probably the only significant adrenal stimulating factor in the fetal pituitary (Goodyer *et al.*, 1977; Fujieda *et al.*, 1981a).

In the human fetus ACTH and other pro-opiomelanocortin-related peptides appear in the anterior and intermediate lobes of the pituitary by 5 weeks' gestation. Since ACTH does not cross the placenta, fetal plasma concentrations depend entirely on the integrity of the fetal hypothalamic–pituitary unit. Fetal concentrations of ACTH, β-lipotropic hormone (LPH) and β-endorphin are higher than those of the adult (Winters *et al.*, 1974; Csontos *et al.*, 1979; Wardlaw *et al.*, 1979), but little is known about changes with gestation, pulsatility or possible diurnal variation in response to changing maternal cortisol levels. At least by mid-gestation negative feedback regulation of fetal ACTH secretion can be demonstrated by the fall in cord cortisol and DHA sulfate levels or maternal oestriol excretion which follows administration of glucocorticoids to the mother (Arai, Kubawara and Okinaga, 1972). Administration of ACTH to the fetus *in vivo* usually increases fetal cortisol production and maternal oestrogen excretion (Jaffe *et al.*, 1977; Strecker *et al.*, 1977; Dell'Acqua *et al.*, 1978; Walsh, Norman and Novy, 1979); the occasional lack of an acute response may indicate that the gland was already maximally stimulated. In the anencephalic fetus, plasma ACTH values are markedly reduced (Allen *et al.*, 1973), and as a result adrenal steroidogenesis, as reflected in fetal serum DHA sulfate levels or maternal oestriol excretion, is negligible. The often-quoted assertion that the anencephalic fetus shows normal adrenal development until mid-pregnancy is not substantiated by careful morphometry, which shows that the adrenals, particularly in the steroidogenic fetal zone, are reduced in size in the youngest fetuses (15 weeks' gestation) that have been studied (Gray and Abramovitch, 1980).

Numerous peptides have been suggested as tropic factors for the fetal adrenal in an attempt to explain its apparently unique pattern of steroidogenesis, but for no factor other than ACTH is the evidence compelling or reproducible. The fetal pituitary contains, and presumably secretes, several other peptides during the processing of pro-opiomelanocortin, including β-lipotropin, β-endorphin, α-melanocyte stimulating hormone ($ACTH_{1-13}$) and a corticotropin-like intermediate lobe peptide (CLIP $ACTH_{18-39}$). However, none of these, nor any other pituitary hormone such as growth hormone, prolactin or thyrotropin, has any stimulatory effect on fetal adrenal cells at physiological concentrations (Fujieda *et al.*, 1981a). An alternative hypothesis is that some placental hormone such as human chorionic gonadotropin (hCG) or somatomammotropin might have a unique and specific effect on the fetal adrenal, either to inhibit 3β-HSD or to stimulate DHA secretion directly (Lehmann and Lauritzen, 1975; Isherwood and Oakey, 1976; Seron-Ferre, Lawrence and Jaffe, 1978; Brown *et al.*, 1981). A major attraction of this view is the ready explanation it provides for the transition from Δ^5-3β-hydroxysteroid to Δ^4-3-ketosteroid production which follows parturition. However, careful studies using physiological amounts of purified preparations free of contaminating

ACTH-like peptides have failed to confirm any effect of these placental hormones on fetal adrenal cells *in vitro* or *in vivo* (Giroud *et al.*, 1979; Voutilainen, Kahri and Salmenpera, 1979; Walsh, Norman and Novy, 1979; Fujieda *et al.*, 1981c).

On balance, therefore, the available evidence indicates that, just as in postnatal life, the dominant and probably exclusive stimulator of fetal adrenal steroid production is ACTH, not only in the definitive zone but also in the steroidogenically more active fetal zone (Seron-Ferre *et al.*, 1978; Branchaud *et al.*,

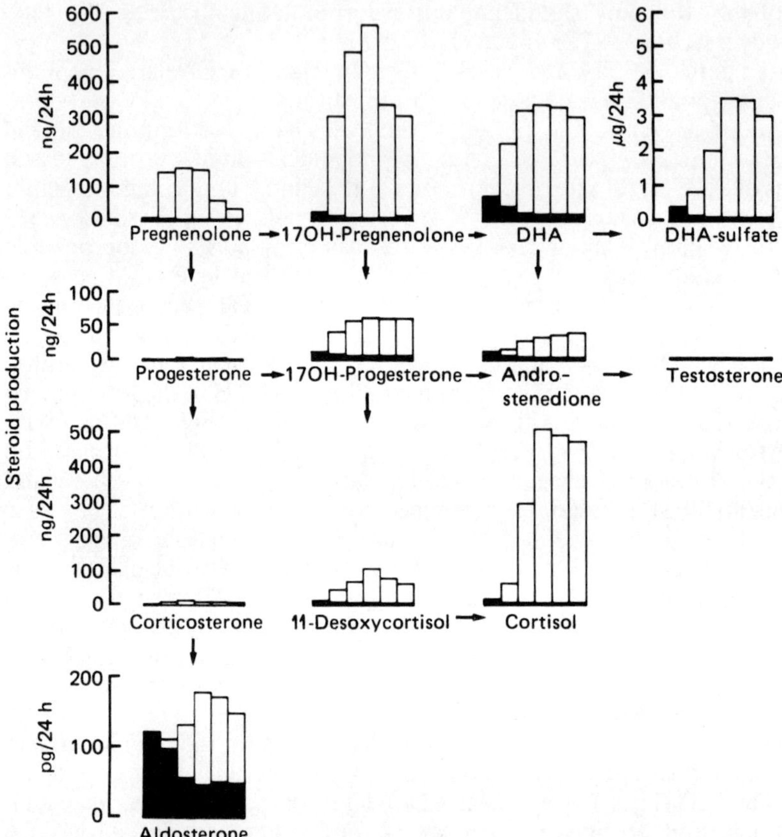

Figure 2.5 The pattern of daily steroid production by human fetal adrenal cells during 6 days of culture in the presence and absence of ACTH. Note the different units used for dehydroepiandrosterone (DHA) sulfate and aldosterone. (□) ACTH 10^3 pg/ml; (■) no ACTH. (From Fujieda *et al.*, 1981a, courtesy of the Editor and Publishers, *Journal of Clinical Endocrinology and Metabolism*)

1978; Carr *et al.*, 1980a; Fujieda *et al.*, 1981b). At the same time, the unique pattern of fetal adrenal secretion makes it obvious that this general stimulatory effect of ACTH must be modulated by some intrauterine inhibitor of 3β-HSD activity. Thus, when fetal adrenal cells are maintained in tissue culture (*Figure 2.5*), they rapidly acquire increased 3β-HSD activity and, in the presence of ACTH, adopt a pattern of Δ^4-3-ketosteroid secretion very similar to that of adult adrenal cells (Kahri, Huhtaniemi and Salmenpera, 1976; Fujieda *et al.*, 1981a; Simonian and Gill, 1981). Regardless of whether the cells are of definitive zone or fetal zone origin, in culture they assume an identical ultrastructural appearance and similar patterns of steroidogenesis (Fujieda *et al.*, 1981b). Clearly, the fetal circulation contains some potent inhibitor(s) of 3β-HSD; removed from this influence, fetal adrenal cells quickly demonstrate this innate ability to produce cortisol and other Δ^4-3-ketosteroids.

To date no pituitary or placental peptide hormone has been shown to have this capacity to inhibit 3β-HSD (Fujieda *et al.*, 1981c). However, it has been known for some time that in all human steroidogenic tissues this enzyme complex is particularly sensitive to inhibition by a wide range of steroids (Kowal, Forchielli and Dorfman, 1964; Villee, 1966; Wiener and Allen, 1967; Ferre *et al.*, 1976).

Table 2.1 A comparison of the inhibitory effects of various ambient steroids on adrenal microsomal 3β-hydroxysteroid dehydrogenase (HSD) and 17,20-desmolase activities*

	Enzyme activity (expressed as % of control)	
Steroid (at 1.0 μmol/l)	*3β-HSD*	*17,20-desmolase*
Cortisol	91	95
Corticosterone	71	100
11-Desoxycortisol	73	100
11-Desoxycorticosterone	69	75
17OH-Progesterone	75	100
Progesterone	73	91
Androstenedione	63	91
11βOH-Androstenedione	91	82
Testosterone	71	98
Dihydrotestosterone	79	59
DHA sulfate	75	100
Oestradiol	6	73
Oestrone	2	90
Oestriol	63	92

* Enzyme activity is expressed as % of the control activity in the absence of added steroid. These results were obtained using 17OH-pregnenolone as substrate for both reactions; similar inhibition of 3β-HSD is observed using dehydroepiandrosterone (DHA) or pregnenolone as substrate. At ambient steroid concentrations of 0.01 μmol/l there is still significant inhibition of 3β-HSD activity, but no effect on 17,20-desmolase

Table 2.1 demonstrates how micromolar concentrations of various steroids can influence the relative activities of 3β-HSD and 17,20-desmolase, two enzymes which utilize 17-OH pregnenolone as their common substrate and which therefore determine the relative amounts of cortisol and DHA produced by each adrenal cell. Note that placental steroids such as oestradiol and oestrone are potent inhibitors of 3β-HSD but have relatively little impact on 17,20-desmolase. This effect of circulating placental Δ^4-3-ketosteroids is in addition to the competitive inhibitory effect of Δ^5-3β-hydroxysteroids such as DHA, pregnenolone, 16OH-pregnenolone and 16OH-DHA, which accumulate within fetal adrenal cells at micromolar concentrations. Thus, the final specific activity of the 3β-HSD complex within any individual fetal adrenal cell is the net result of a variety of kinetic variables, which include the relative intracellular concentrations of the natural substrate (17OH-pregnenolone) and alternative competing Δ^5-3β-hydroxysteroids; both competitive and non-competitive inhibition by Δ^4-3-ketosteroids of placental and adrenal origin; and compensatory ACTH-induced transcriptional increases in synthesis of new enzyme.

Voutilainen and Kahri (1980) demonstrated by co-culture experiments that inhibition of fetal adrenal 3β-HSD activity could be maintained *in vitro* by some soluble factor of placental origin. Fujieda *et al.* (1982) showed that similar results (*Figures 2.6* and *2.7*) could be achieved by culturing cells in the presence of oestradiol and suggested that oestrogens were one soluble placental inhibitory factor. More recently, Smail and Winter (unpublished observations) have shown that this inhibitory effect of oestrogen does not require fetal adrenal cells, but can be replicated with adult adrenal cells.

The concentrations of steroids required to inhibit 3β-HSD activity *in vitro* are in the range of 10^{-8}–10^{-6} mol/l, which is considerably higher than levels found in the

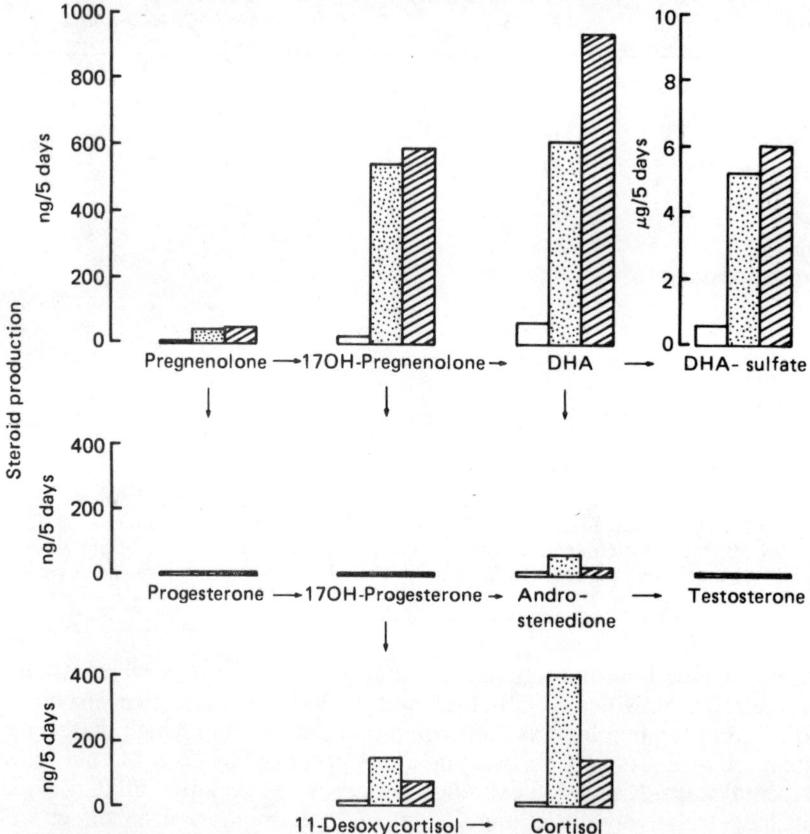

Figure 2.6 The influence of added oestradiol (1 µg/ml) on the pattern of steroids produced by human fetal adrenal cells in the presence of ACTH. Each bar represents the mean total steroid production during 5 days of tissue culture. (□) Control; (▩) ACTH 100 pg/ml; (▨) ACTH 100 pg/ml plus E_2 10^3 ng/ml. (From Fujieda *et al.*, 1982, courtesy of the Editor and Publishers, *Journal of Clinical Endocrinology and Metabolism*)

postnatal circulation. However, in the fetus, plasma concentrations of placental steroids (*Table 2.2*) are within this range. In addition, the micro-environment of each adrenal cell is influenced not only by these circulating steroids but also by high intra-adrenal concentrations of endogenous steroids. Just as in the postnatal adrenal cortex (Dickerman and Winter, unpublished observations), intra-adrenal steroid concentrations are probably higher in the inner areas of the cortex than in the subscapsular area. Such a gradient, imposed by the centripetal blood flow through the adrenal cortex, may explain the apparent differences in 3β-HSD activity between the definitive and fetal zones. Although steroid concentrations within these separate zones of the fetal adrenal have not yet been reported,

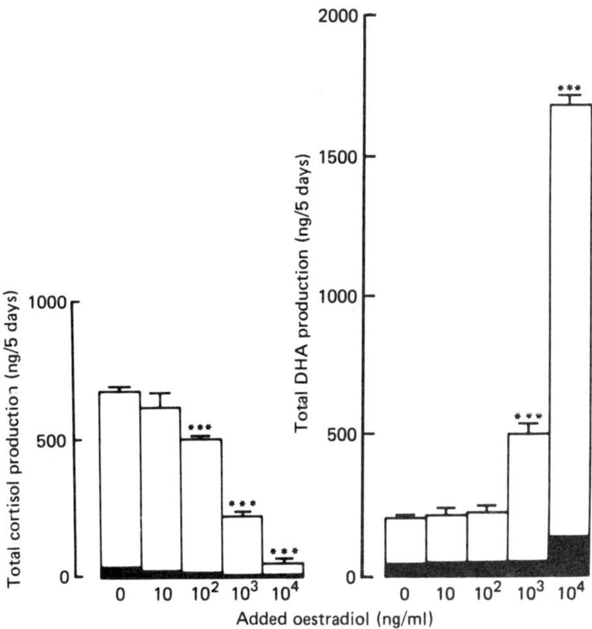

Figure 2.7 The influence of varying concentrations of oestradiol ($0–10^4$ ng/ml) on total cortisol and dehydroepiandrosterone (DHA) production by human fetal adrenal cells during 5 days of culture in the presence of ACTH. (▭) ACTH 10^3 pg/ml; (▬) no ACTH. The asterisks indicate a significant difference from the control cultures ($P<0.001$). (From Fujieda *et al.*, 1982, courtesy of the Editor and Publishers, *Journal of Clinical Endocrinology and Metabolism*)

relatively high mean cortical concentrations of pregnenolone (5×10^{-6} mol/l), cortisol (5×10^{-7} mol/l) and cortisone (3×10^{-7} mol/l) are present (Murphy and Diez d'Aux, 1972; Huhtaniemi, 1973), together with very high concentrations of pregnenolone sulfate (1.6×10^{-5} mol/l), DHA sulfate (3.5×10^{-5} mol/l) and 16αOH-DHA sulfate (1.9×10^{-6} mol/l) (Huhtaniemi, Luukainen and Vihko, 1970). It is possible that one role for the sulfotransferase of the fetal adrenal cortex is to prevent excessive intraglandular accumulation of free steroids, which are much more potent inhibitors of 3β-HSD than their respective sulfate conjugates.

Now that ACTH has been identified as the prime stimulator of fetal adrenal steroid production, and steroids themselves have been recognized as significant

Table 2.2 Mean concentrations of various steroids in fetal plasma at mid-pregnancy (13–20 weeks) and at term

	Mid-gestation (nmol/l)	Term Umbilical artery (nmol/l)	Term Umbilical vein (nmol/l)
Free Δ^4-3-ketosteroids			
progesterone	240	350	530
16αOH-progesterone	47	105	120
17OH-progesterone	50	40	65
11-desoxycorticosterone	NA	17	20
corticosterone	NA	10	13
aldosterone	NA	6	6
androstenedione	3	4	4
cortisol	20	240	215
cortisone	90	140	180
oestrone	23	15	38
oestradiol	11	20	30
oestriol	145	80	350
Free Δ^5-3β-hydroxysteroids			
pregnenolone	NA	36	26
16αOH-pregnenolone	34	89	44
dehydroepiandrosterone (DHA)	NA	28	20
16αOH-DHA	18	20	32
Steroid sulfates			
pregnenolone sulfate	NA	1700	2000
dehydroepiandrosterone sulfate	3200	3000	3400
16αOH-DHA sulfate	3600	8320	6630
oestrone sulfate	22	25	28
oestradiol sulfate	6	8	13
oestriol sulfate	420	3420	2910

Data from Murphy and Diez d'Aux (1972); Tulchinsky and Simmer (1972); Reyes *et al.* (1974); Shutt, Smith and Shearman (1974); Den *et al.* (1979); Sippell *et al.* (1979); Laatikainen *et al.* (1980); Winter, Faiman and Reyes (1981); Parker *et al.* (1980).

NA indicates a lack of sufficient information. Results can be converted to ng/ml (approximately) by multiplying by 0.3

modulators of 3β-HSD activity, it is possible to propose a new and more reasonable model to explain the unique fetal pattern of steroidogenesis (*Figure 2.8*). Presumably the roles of ACTH include not only regulation of the provision of cholesterol substrate but also induction of the enzymes necessary to maintain a high rate of steroidogenesis. The small adrenal characteristic of fetal anencephaly or exposure to exogenous glucocorticoids indicate that ACTH is also essential for the striking hypertrophy of the fetal adrenals. Although ACTH itself, through its effect on cyclic adenosine 3′5′-monophosphate (cAMP) levels, appears to inhibit adrenal replication, its growth-promoting effect on the adrenal *in vivo* may be mediated by changes in blood flow or by mitogenic peptides such as fibroblast growth factor (Crickard, Ill and Jaffe, 1981).

FUNCTIONS OF THE FETAL ADRENAL

The close metabolic interrelationships that exist between the fetal adrenal and liver, the placenta and the maternal circulation (*Figure 2.8*) have in the past frequently been interpreted as evidence for a fetoplacental steroidogenic unit. According to this concept the fetal adrenal is intrinsically deficient in 3β-HSD and therefore entirely dependent on placental progesterone for the synthesis of cortisol. At the same time, since the placenta is deficient in the ACTH-dependent enzymes 17-hydroxylase and 17,20-desmolase, the major purpose of the fetal adrenal was seen to be the provision of DHA sulfate and 16α-OH-DHA sulfate as essential precursors for oestrogen biosynthesis.

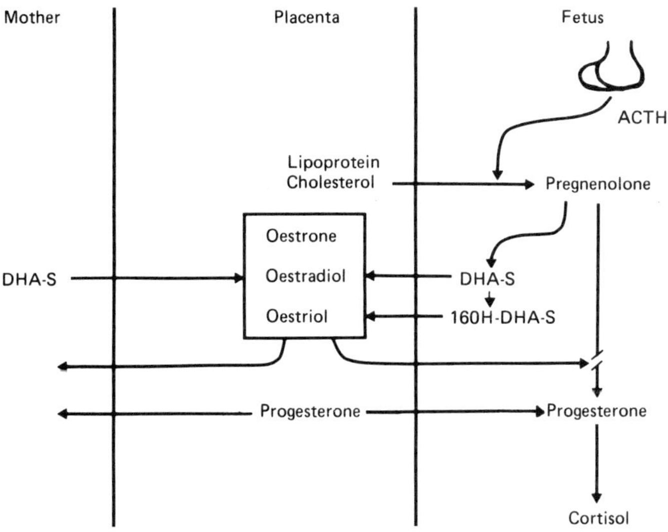

Figure 2.8 A proposed model to describe the interaction of ACTH and placental estrogens in the regulation of fetal adrenal steroidogenesis. Inhibition of adrenal 3β-hydroxysteroid dehydrogenase by estrogens and other steroids, coupled with rapid placental clearance of secreted cortisol, must be counteracted by increased release of ACTH, which in turn leads to adrenal hypertrophy and massive production of dehydroepiandrosterone sulfate (DHA-S). This material plays a permissive but not essential role in placental biosynthesis of oestrone and oestradiol. (From Fujieda *et al.*, 1982, courtesy of the Editor and Publishers, *Journal of Clinical Endocrinology and Metabolism*)

In the light of more recent information, the picture now emerges of a fetal pituitary–adrenal axis which is linked to the placenta more by circumstance than by design. Clearly, cortisol production is carefully regulated by the fetus, and serum concentrations depend on four variables: (1) pituitary ACTH secretion; (2) inhibition of adrenal 3β-HSD by placental and adrenal steroids; (3) rapid placental clearance of cortisol by conversion to cortisone; and (4) placental transport of maternal cortisol. Although fetal LDL cholesterol appears to be the principal substrate, placental progesterone may be utilized to some degree to circumvent the relative 3β-HSD deficiency; unfortunately, no experiment of nature exists which could demonstrate whether progesterone is an essential or merely a fortuitous precursor.

The active 16α-hydroxylase and sulfotransferase enzymes of the fetus have all the earmarks of a waste disposal system, since the fetus lacks sulfatase and cannot metabolize further the steroid conjugates that are formed. The principal product of this fetoplacental interaction is oestriol, which is almost biologically inert and is rapidly excreted by the mother. Suppression of fetal DHA production by exogenous glucocorticoids has little input on the products of conception. It is true that fetal DHA sulfate is also an important precursor for placental synthesis of potent free oestrogens, including oestradiol. However, in placental sulfatase deficiency, the absence of this source and the resulting placental dependence on maternal DHA as substrate presents little or no clinical difficulty except some retardation of cervical ripening. Thus, although obvious metabolic connections exist between the fetal adrenal and the placenta, none of these has yet been demonstrated to be essential for the operations of either component. Indeed, at parturition the fetus demonstrates that it can synthesize essential steroids much more effectively without the influence of its placental connections. Haning *et al.* (1983) have suggested that the fetal adrenal, and particularly DHA sulfate, might play a role in the regulation of placental hCG production; however, it is difficult to visualize fetal DHA sulfate being of any significance in mammalian reproduction when its production is so notably diminished in most non-human species.

While fetal adrenal secretion of DHA sulfate and other Δ^5-3β-hydroxysteroids appears to be largely irrelevant to the successful completion of pregnancy, there is considerable evidence that cortisol plays an important role in fetal development and eventual neonatal survival. In experimental animals fetal glucocorticoids have been shown to stimulate enzymes responsible for gluconeogenesis and hepatic glycogen accumulation; induce enzyme activity in intestinal epithelium and maturation of pancreatic β-cells; promote development of type II alveolar cells and synthesis of pulmonary surfactant; increase phenylethanolamine-*N*-methyltransferase activity in adrenal medulla; and enhance triiodothyronine synthesis (Liggins, 1976). However, this does not mean that fetal adrenal secretion is essential for these maturational processes in man, since significant amounts of maternal cortisol do cross the placenta; indeed, in congenital adrenal hypoplasia or anencephaly the fetus develops relatively normally. Similarly, although the fetal adrenal can secrete other biologically active steroids, such as aldosterone in response to sodium depletion (Beitins *et al.*, 1972), there is no compelling evidence that this secretion is essential for mineral balance *in utero*. This doubt concerning the importance of adrenal secretions for fetal well-being is in sharp contrast to the obvious contribution that adrenal glucocorticoids and mineralocorticoids make to survival immediately after parturition.

Fetal serum and amniotic fluid cortisol concentrations appear to rise during late gestation before the onset of labour, with a further increase during parturition (Murphy, 1983). Before delivery there may not be a parallel increase in plasma ACTH levels (Winters *et al.*, 1974), so the cortisol rise may reflect an increase in ACTH receptors and enhanced steroidogenic efficiency. In several species such a rise in cortisol production provides an essential signal for parturition. In man the relationship is less obvious, although anencephaly is frequently associated with postmaturity, and intra-amniotic cortisol administration may initiate labor in women with prolonged pregnancies (Nwosu, Wallach and Bolognese, 1976). It therefore seems likely that the fetal adrenal does play some role in the initiation of human parturition. However, more information will be needed before one can define exactly the interrelationships which exist between fetal pituitary and adrenal secretion, placental oestrogen and progesterone production, and enhanced prostaglandin generation within the gravid uterus (Siiteri and Seron-Ferre, 1981).

ADAPTATION TO EXTRAUTERINE LIFE

The pituitary–adrenal axis

By term, adrenal steroidogenesis is proceeding at a higher rate than at any other stage of development, and the gland is capable of responding to input signals mediated either by pituitary ACTH or the renin–angiotensin system. Comparison of mean cord plasma ACTH concentrations after vaginal delivery (1538±947 pg/ml) with those observed during caesarean section (31.6±10 pg/ml) demonstrates that the pituitary at term can respond to stress (Pohjavuori and Fyhrquist, 1983). Presumably this rise is mediated by increased neurosecretion of hypothalamic corticotropin-releasing factor, but there is conflicting evidence regarding the extent to which it can be blocked by prenatal administration of betamethasone or dexamethasone (Miyakawa, Ideka and Maeyama, 1974; Huhtaniemi *et al.*, 1982). A parallel rise in cord levels of arginine vasopressin during labour suggests that this hormone may also stimulate ACTH release.

This ACTH surge elicits an immediate response in fetal adrenal secretion of cortisol as well as DHA and pregnenolone (Murphy, 1973b; Arai and Yanaihara, 1977). Fetal plasma concentrations of unbound cortisol rise from 26 nmol/l (0.96±0.11 μg/dl) before labour, a level slightly lower than that of the resting adult, to 43 nmol/l (1.54±0.14 μg/dl) during labour (Predine *et al.*, 1979). Cortisol levels are not reduced in premature infants, but there is evidence that the fetal distress often associated with postmaturity may increase cortisol production (Barnhart, Carlson and Reynolds, 1980). Prenatal administration of glucocorticoid suppresses mean cortisol and aldosterone levels throughout the first week of life (Sippell, Bidlingmaier and Knorr, 1980) but does not block the cortisol response to exogenous ACTH (Ohrlander *et al.*, 1977).

The changes in mean plasma concentrations of various steroids during the neonatal period are summarized in *Table 2.3*. Unconjugated placental steroids such as progesterone and oestradiol are rapidly cleared from the neonatal circulation, but the clearance of sulfate conjugates such as oestriol sulfate is noticeably prolonged. As the concentration of inhibitory steroids falls, there is a striking improvement in adrenal 3β-HSD activity and a corresponding increase in the ability of the neonate to secrete active steroids such as cortisol and aldosterone without excessive production of Δ^5-3β-hydroxysteroids. Basal serum concentrations of free cortisol remain quite stable, but there is a gradual decline in plasma levels of ACTH and DHA. The later increase in total serum cortisol concentrations during infancy reflects a rise in levels of binding protein. Analysis of urinary steroids during infancy confirms the disappearance of progesterone metabolites by the end of the first week, and a more gradual decrease in urinary Δ^5-3β-hydroxysteroids during the subsequent weeks (Shackleton, Gustafsson and Mitchell, 1973; Reynolds, 1980).

Plasma renin activity and aldosterone levels are remarkably high during the first few days of life, perhaps because of competition by progesterone and other steroids for the renal aldosterone receptor. Plasma aldosterone levels throughout infancy remain about five times those of older children; some degree of renal tubular

Table 2.3 Pituitary–adrenal function during infancy

	Mean plasma concentrations						Adult male
	Birth	*1 day*	*1 week*	*1–4 months*	*4–12 months*	*1–2 years*	*range*
Δ^4-3-Ketosteroids (nmol/l)							
progesterone	1920	40.0	1.6	1.0	0.8	1.0	<2
16OH-progesterone	102	3.0	2.1	ND	ND	ND	ND
17OH-progesterone	100	2.7	2.7	2.4	0.9	0.8	<6
11-desoxycorticosterone	19	3.4	0.3	0.1	0.1	0.2	0.1–0.4
corticosterone	29	2.2	1.4	0.8	1.1	2.7	1–50
aldosterone	6	9.0	4.0	1.4	0.8	0.8	0.1–0.8
androstenedione	15	6.3	1.1	1.1	0.4	0.2	1.7–6.9
cortisol	530	100.0	96.0	91.0	174.0	207.0	50–700
cortisone	380	113.0	60.0	55.0	ND	ND	ND
oestradiol	32	ND	0.04	0.05	0.04	0.04	<0.18
Δ-3β-Hydroxysteroids (nmol/l)							
pregnenolone	4290	ND	ND	5.0	1.7	0.8	1.3–7.3
16OH-pregnenolone	73	61.0	32.0	ND	ND	ND	ND
dehydroepiandrosterone (DHA)	32	12.0	12.0	5.0	2.5	1.0	6.2–24.2
16OH-DHA	15	4.0	3.0	ND	ND	ND	ND
Steroid sulfates (nmol/l)							
DHA sulfate	5200	3600	640	145	83	52	2600–10400
oestriol sulfate	4100	2570	1475	ND	ND	ND	ND
pregnenolone sulfate	2750	2150	1350	650	92	42	133
ACTH (pmol/l)	32	14	8	9	9	4	0–16
Renin activity ($pmol \cdot l^{-1} \cdot s^{-1}$)	2.4	5.1	3.2	2.6	2.1	1.9	0.1–0.6

Data from Cacciari *et al.* (1975); de Perètti and Forest (1976, 1978); Dillon *et al.* (1976); Reynolds, Bently and Turnipseed (1977); Sippell *et al.* (1978, 1980); Godard *et al.* (1979); Hammond *et al.* (1979); Sulyok *et al.* (1979); Vincent *et al.* (1980) and de Peretti and Mappus (1983).

ND indicates satisfactory data are not available. Steroid concentrations can be changed approximately to ng/ml by multiplying by 0.3; plasma ACTH can be converted to pg/ml by multiplying by 4.4; plasma renin activity can be converted to $ng \cdot ml^{-1} \cdot h^{-1}$ by multiplying by 4.7

resistance to mineralocorticoids is indicated by the tendency of infants with interstitial pyelonephritis to develop a profound salt-losing condition in spite of extremely high aldosterone levels.

In the first few days of life, administration of exogenous ACTH elicits only a three- to fourfold rise in plasma cortisol levels, and a similar increase in plasma DHA sulfate levels (Ohrlander *et al.*, 1977; Barnhart, Carlson and Reynolds, 1980; Pintor *et al.*, 1980). In sharp contrast, ACTH stimulation between 1 and 4 months of age produces remarkable Δ^4-3-ketosteroid responses, with 30-fold increases in plasma cortisol and 10-fold increases in plasma androstenedione (Forest, 1978), but relatively smaller increments in plasma DHA sulfate (Noguchi and Reynolds, 1978). This incredible sensitivity of the adrenal cortex to ACTH stimulation diminishes after 4 months of age; by 1 year of age, cortisol responses are similar to those of adults, and ACTH elicits only minimal increases in DHA, DHA sulfate and androstenedione (Forest, de Peretti and Bertrand, 1978). Irregular fluctuations in serum cortisol concentrations can be observed in all infants; it would appear that the usual circadian rhythm is established by at least 6 months of age (Onishi *et al.*, 1983).

Postnatal involution of the adrenal cortex

At term, each adrenal gland weighs 4–5 g, more than 80% of which consists of the inner, hyperaemic fetal zone. By 1 month of age half of this weight has been lost, and by 1 year of age the average gland weight is only about 1 g. The histological correlates of this process were first described in routine autopsy material, with insufficient allowance for the effects of disease and *ante mortem* stress. Various authors described extensive degeneration and necrosis within the fetal zone, with heterophagocytosis, interstitial oedema and haemorrhage (Benner, 1940; Mäusle, 1972). These changes began immediately after birth, and by 6 months of age little if any of the original fetal zone remained. Meanwhile, the outer definitive zone appeared to enlarge and extend inwards to replace the involuting fetal zone; this zone was considered to provide the adult glomerulosa and fasciculata (Dhom, 1965).

Re-examination of these necessarily cross-sectional morphological studies suggests that any necrosis was an artefact caused by *ante mortem* stress (De Sa, 1978). Similarly, one cannot observe the increased mitotic activity necessary to permit the apparent extension of the definitive zone. Rather, what appears to happen during the first weeks and months of life is a gradual remodelling of fetal zone cells into a zona fasciculata, without any significant necrosis or replacement. McNulty (1981) studied this process in healthy Rhesus monkeys during the first six months of life, and observed a centripetal wave of transformation of fetal zone cells to zona fasciculata, with only rare mitotic figures and no evidence of collapse, haemorrhage or fibrosis. Pre-term hypophysectomy or dexamethasone administration induced precocious involution and transformation, while prolonged infusion of ACTH caused cellular hypertrophy and an apparent expansion of the fetal zone to the gland capsule. Ultrastructural study of this process suggests that the transformation of fetal zone cells occurs within a band of small, dense cells which are deficient in smooth endoplasmic reticulum (McClellan and Brenner, 1981). Although these studies were carried out in macaques, there is every reason to believe that the process of postnatal adrenal involution is similar in the healthy human neonate.

CONCLUSIONS

Miriam Benner wrote in 1940 that 'the function of the fetal cortex is still doubtful and none of the theories so far presented in the literature explain all of the facts observed'. Subsequent demonstrations of the metabolic interconversions possible *in utero* led to the concept of the fetoplacental steroidogenic unit, in which the fetal adrenal, regulated by some unknown androgen-stimulating hormone, served as an essential intermediate in placental oestrogen biosynthesis. It is now appropriate to re-examine this concept in the light of new biochemical and morphological information.

It is quite clear that a similar fetoplacental steroidogenic unit cannot be found in most other species. This observation raises important questions about its possible significance, since vital reproductive processes tend to be conserved through evolution. Furthermore, removal of the fetal contribution to this fetoplacental unit has little impact on the course of human pregnancy, although maternal oestriol excretion becomes negligible. Finally, no specific intrauterine DHA-stimulating hormone has been identified to regulate this process, while *in vitro* studies suggest that it is placental steroid rather than some unique property of the fetal adrenal itself which underlies its massive secretion of Δ^5-3β-hydroxysteroids.

It seems more reasonable to conclude that the essential role of the fetal adrenal is to be able to secrete sufficient amounts of cortisol and aldosterone to permit survival immediately after parturition. The major factors which militate against the maintenance of appropriate free cortisol levels *in utero* include inhibition of 3β-HSD activity by placental and intra-adrenal steroids, and rapid placental and peripheral metabolism of cortisol to cortisone. The characteristic zonal morphology of the fetal cortex appears to reflect merely a response to the necessarily high levels of plasma ACTH, since a reduction in ACTH release is consistently followed by rapid adrenal involution. In the normal course of events this occurs after parturition, as the fetal zone becomes transformed to what we recognize as zona fasciculata. However, until this involution is well advanced the neonatal adrenal remains able to mount a more striking steroidogenic response to ACTH stimulation than at any other time in postnatal life.

In conclusion, then, it would appear that neither the fetal pituitary nor the adrenal cortex possesses unique properties and functions, but rather each is forced to perform its usual tasks in a somewhat difficult environment. This concept may disappoint those who would prefer a more marvellous or mysterious explanation, but at least it provides a rational explanation for all the verified observations and thus a basis for future research. As Diczfalusy (1969), in his seminal review of the fetoplacental unit, quoted Claude Bernard:

'Il n'y a pas des théories fausses et des théories vraies, il y a des théories fécondes et des théories stériles.'

Acknowledgements

The author would like to acknowledge the research contributions made to this study by Drs Kenji Fujieda, Peter Smail, Zvi Dickerman and Geoffrey Byrne; the technical assistance of Mrs Yvette Perry; and the secretarial assistance of Miss Emma Gulbis.

References

ALLEN, J. P., COOK, D. M., KENDALL, T. W. and MCGILVRA, R. (1973) Maternal–fetal ACTH relationship in man. *Journal of Clinical Endocrinology and Metabolism,* **37,** 230–234

ARAI, K., KUBAWARA, Y. and OKINAGA, S. (1972) The effect of adrenocorticotropic hormone and dexamethasone, administered to the fetus *in utero*, upon maternal and fetal estrogens. *American Journal of Obstetrics and Gynecology,* **113,** 316–322

ARAI, K. and YANAIHARA, T. (1977) Steroid hormone changes in fetal blood during labor. *American Journal of Obstetrics and Gynecology,* **127,** 879–883

BARNHART, B. J., CARLSON, C. V. and REYNOLDS, J. W. (1980) Adrenal cortical function in the postmature fetus and newborn infant. *Pediatric Research,* **14,** 1367–1369

BEITINS, I. Z., BAYARD, F., ANCES, I. G., KOWARSKI, A. and MIGEON, C. J. (1973) The metabolic clearance rate, blood production, interconversion and transplacental passage of cortisol and cortisone in pregnancy near term. *Pediatric Research,* **7,** 509–519

BEITINS, I. Z., BAYARD, F., LEVISTKY, B. L., ANCES, I. G., KOWARSKI, A. and MIGEON, C. J. (1972) Plasma aldosterone concentration at delivery and during the newborn period. *Journal of Clinical Investigation,* **51,** 386–394

BENNER, M. C. (1940) Studies on the involution of the fetal cortex of the adrenal glands. *American Journal of Pathology,* **16,** 787–798

BLOCH, E. and BENIRSCHKE, K. (1959) Synthesis *in vitro* of steroids by human fetal adrenal gland slices. *Journal of Biological Chemistry,* **234,** 1085–1089

BRANCHAUD, C. T., GOODYER, C. G., HALL, C. ST. G., ARATO, J. S., SILMAN, R. E. and GIROUD, C. J. P. (1978) Steroidogenic activity of hACTH and related peptides on the human neocortex and fetal adrenal cortex in organ culture. *Steroids,* **31,** 557–571

BROWN, T. J., GINZ, B., MILNE, C. M. and OAKEY, R. E. (1981) Stimulation by polypeptides of dehydroepiandrosterone sulphate synthesis in human fetal adrenal slices. *Journal of Endocrinology,* **91,** 111–122

CACCIARI, E., CICOGNANI, A., PIRAZZOLI, P., DALLACASA, P., MOZZARACCHIO, M. A., TASSONI, P., BERNARDI, F., SALARDI, S. and ZAPPULLA, F. (1975) Plasma ACTH values during the first seven days of life in infants of diabetic mothers. *Journal of Pediatrics,* **87,** 943–945

CAMPBELL, A. L. and MURPHY, B. E. P. (1977) The maternal–fetal cortisol gradient during pregnancy and at delivery. *Journal of Clinical Endocrinology and Metabolism,* **45,** 435–440

CARR, B. R. and SIMPSON, E. R. (1981a) *De novo* synthesis of cholesterol by the human fetal adrenal gland. *Endocrinology,* **108,** 2154–2162

CARR, B. R. and SIMPSON, E. R. (1981b) Synthesis of cholesterol in the human fetus: 3-hydroxy-3-methylglutaryl coenzyme A reductase activity of liver microsomes. *Journal of Clinical Endocrinology and Metabolism,* **53,** 810–812

CARR, B. R., MACDONALD, P. C. and SIMPSON, E. R. (1980) The regulation of *de novo* synthesis of cholesterol in the human fetal adrenal gland by low density lipoprotein and adrenocorticotropin. *Endocrinology,* **107,** 1000–1006

CARR, B. R., OHASHI, M. and SIMPSON, E. R. (1982) Low density lipoprotein binding and *de novo* synthesis of cholesterol in the neocortex and fetal zones of the human fetal adrenal gland. *Endocrinology,* **110,** 1994–1998

CARR, B. R., PARKER, C. R., MILEWICH, L., PORTER, J. C., MACDONALD, P. C. and SIMPSON, E. R. (1980a) Steroid secretion by ACTH-stimulated human fetal adrenal tissue during the first week in organ culture. *Steroids,* **36,** 563–574

CARR, B. R., PORTER, J. C., MACDONALD, P. C. and SIMPSON, E. R. (1980b) Metabolism of low density lipoprotein by human fetal adrenal tissue. *Endocrinology,* **107,** 1034–1040

COOKE, B. A. and TAYLOR, P. D. (1971) Site of dehydroepiandrosterone sulphate biosynthesis in the adrenal gland of the previable fetus. *Journal of Endocrinology,* **51,** 547–556

CRICKARD, K., ILL, C. R. and JAFFE, R. B. (1981) Control of proliferation of human fetal adrenal cells *in vitro*. *Journal of Clinical Endocrinology and Metabolism,* **53,** 790–796

CROWDER, R. E. (1957) The development of the adrenal gland in man, with special reference to origin and ultimate location of cell types and evidence in favor of the cell migration theory. *Carnegie Contributions to Embryology,* **36,** 193–217

CSONTOS, K., RUST, M., HÖLLT, V., MAHR, W., KROMER, W. and TESCHEMACHER, H. J. (1979) Elevated plasma beta-endorphin levels in pregnant women and their neonates. *Life Sciences,* **25,** 835–844

DELL'ACQUA, S., LUCISANO, A., TORTOROLO, G., ZUPPA, A. and ARNO, E. (1978) Adrenal function in the foetus. In *The Endocrine Function of the Human Adrenal Cortex,* edited by V. H. T. James, M. Serio, G. Giusti and L. Martini, pp. 529–545. London: Academic Press

DEN, K., HARUYAMA, T., HAGIWARA,, H., FUJU, K. T., KAMBEGAWA, A., TAKAMI, Y., YOSHIDA, T. and TAKAGI, S. (1979) Plasma levels of 16α-hydroxypregnenolone, 16α-hydroxyprogesterone and 16α-hydroxydehydroepiandrosterone in the fetus and neonates. *Endocrinologica Japonica,* **26,** 439–447

DE PERETTI, E. and FOREST, M. G. (1976) Unconjugated dehydroepiandrosterone plasma levels in normal subjects from birth to adolescence in humans: the use of a sensitive radioimmunoassay. *Journal of Clinical Endocrinology and Metabolism,* **43,** 982–991

DE PERETTI, E. and FOREST, M. G. (1978) Pattern of plasma dehydroepiandrosterone sulfate levels in humans from birth to adulthood: evidence for testicular production, *Journal of Clinical Endocrinology and Metabolism,* **47,** 572–577

DE PERETTI, E. and MAPPUS, E. (1983) Pattern of plasma pregnenolone sulfate levels in humans from birth to adulthood. *Journal of Clinical Endocrinology and Metabolism,* **57,** 550–556

DE SA, D. J. (1978) Stress response and its relationship to cystic pseudofollicular change in the definitive cortex of the adrenal gland in stillborn infants. *Archives of Disease in Childhood,* **53,** 769–776

DESHPANDE, N., JENSEN, V., CARSON, P., BULBROOK, R. D. and DOOUSS, T. W. (1970) Adrenal function in breast cancer: biogenesis of androgens and cortisol by the human adrenal gland *in vivo. Journal of Endocrinology,* **47,** 231–242

DHOM, G. (1965) *Die Nebennierenrinde im Kindesalter.* Berlin: Springer

DHOM, G., ROSS, W. and WIDOK, K. (1958) Die Nebennieren des Feten und des Neugeborernen. Eine quantitative und qualitative Analyse. *Beiträge zur Pathologischen Anatomie,* **119,** 177–195

DICZFALUSY, E. (1964) Endocrine function of the human fetoplacental unit. *Federation Proceedings,* **23,** 791–795

DICZFALUSY, E. (1969) Steroid metabolism in the foeto-placental unit. *Excerpta Medica International Congress Series,* **183,** 65–109

DILLON, M. T., GILLON, M. E. A., RYNESS, J. M. and DE SWIET, M. (1976) Plasma renin activity and aldosterone concentration in the human newborn. *Archives of Disease in Childhood,* **51,** 537–540

DUFAU, M. L. and VILLEE, D. B. (1969) Aldosterone biosynthesis by human fetal adrenal *in vivo. Biochimica et Biophysica Acta,* **176,** 637–641

ELLIOTT, J. R. and ARMOUR, R. G. (1911) The development of the cortex in the human suprarenal gland and its condition in hemicephaly. *Journal of Pathology and Bacteriology,* **15,** 481–496

FERRE, F., BREUILLER, M., CEDARD, L., DUCHESNE, M. J., SAINTOT, M., DESCOMPS, B. and CRASTES DE PAULET, A. (1976) Human placental Δ^5-3β-hydroxysteroid dehydrogenase activity (Δ^5-3β-HSDH): intra-cellular distribution, kinetic properties, retroinhibition and influence of membrane delipidation. *Steroids,* **26,** 551–570

FOREST, M. G. (1978) Age-related response to plasma testosterone, Δ^4-androstenedione, and cortisol to adrenocorticotropin in infants, children, and adults. *Journal of Clinical Endocrinology and Metabolism,* **47,** 931–937

FOREST, M. G., DE PERETTI, E. and BERTRAND, J. (1978) Developmental patterns of the plasma levels of testosterone, Δ^4-androstenedione, 17α-hydroxyprogesterone, dehydroepiandrosterone and its sulfate in normal infants and prepubertal children. In *The Endocrine Function of the Human Adrenal Cortex,* edited by V. H. T. James, M. Serio, G. Giusti and L. Martini, pp. 561–582. London: Academic Press

FUJIEDA, K., FAIMAN, C., REYES, F. I. and WINTER, J. S. D. (1981a) The control of steroidogenesis in human fetal adrenal cells in tissue culture. I. Responses to adrenocorticotropin. *Journal of Clinical Endocrinology and Metabolism,* **53,** 34–38

FUJIEDA, K., FAIMAN, C., REYES, F. I., THLIVERIS, J. and WINTER, J. S. D. (1981b) The control of steroidogenesis by human fetal adrenal cells in tissue culture. II. Comparison of morphology and steroid production in cells of the fetal and definitive zones. *Journal of Clinical Endocrinology and Metabolism,* **53,** 401–405

FUJIEDA, K., FAIMAN, Ç., REYES, F. I. and WINTER, J. S. D. (1981c) The control of steroidogenesis by human fetal adrenal cells in tissue culture. III. The effects of various hormonal peptides. *Journal of Clinical Endocrinology and Metabolism,* **53,** 690–693

FUJIEDA, K., FAIMAN, C., REYES, F. I.. and WINTER, J. S. D. (1982) The control of steroidogenesis by human fetal adrenal cells in tissue culture. IV. The effect of exposure to placental steroids. *Journal of Clinical Endocrinology and Metabolism,* **54,** 89–94

GIROUD, C. J. P., GOODYER, C., HALL, G. and BRANCHAUD, C. (1979) Studies of the human fetal pituitary–adrenal axis in tissue culture. In *Interaction Within the Brain–Pituitary–Adrenocortical System,* edited by M. T. Jones, M. F. Dullman, B. Gilham and S. Chattopadhyay, pp. 235–246. New York: Academic Press

GODARD, C., GEERING, J. M., GEERING, K. and VALLOTTON, M. B. (1979) Plasma renin activity related to sodium balance, renal function and urinary vasopressin in the newborn infant. *Pediatric Research,* **13,** 742–745

GOLDMAN, A. S., YAKOVAC, W. C. and BONGIOVANNI, A. M. (1966) Development of activity of 3β-hydroxysteroid dehydrogenase in human fetal tissues and in two anancephalic newborns. *Journal of Clinical Endocrinology and Metabolism,* **26,** 14–22

GOODYER, C. G., HALL, C. ST. G., BRANCHAUD, C. and GIROUD, C. J. P. (1977) Exploration of the human fetal pituitary adrenal axis: stimulation of cortisol and dehydroepiandrosterone sulfate biosynthesis by homologous pituitary in organ culture. *Steroids,* **29,** 407–416

GRAY, E. S. and ABRAMOVITCH, D. R. (1980) Morphologic features of the anencephalic adrenal gland in early pregnancy. *American Journal of Obstetrics and Gynecology,* **137,** 491–495

HADJIAN, A. J., CHEDIN, M., COCHET, C. and CHAMBAZ, E. M. (1975) Cortisol binding to proteins in plasma in the human neonate and infant. *Pediatric Research,* **9,** 40–45

HAGEMENAS, F. C. and KITTINGER, G. W. (1973) The influence of fetal sex on the levels of plasma progesterone in the human fetus. *Journal of Clinical Endocrinology and Metabolism,* **36,** 389–391

HAMMOND, G. L., KOIVISTO, M., KOUVALAINEN, K. and VIHKO, R. (1979) Serum steroids and pituitary hormones in infants with particular reference to testicular activity. *Journal of Clinical Endocrinology and Metabolism,* **49,** 40–45

HANING, R. V., JR., CHOI, L., CURET, L. B., HENDERSON, P. A., KIGGENS, A. J., LEIFHEIT, T. L., OLSON, R. W., SCHWABE, M. G., SCHWARTZ, D. B. and TRAVER, M. (1983) Interrelationships among human chorionic gonadotropin (hCG), 17β-estradiol, progesterone, and estriol in maternal serum: evidence for an inhibitory effect of the fetal adrenal on secretion of hCG. *Journal of Clinical Endocrinology and Metabolism,* **56,** 1188–1194

HARBERT, G. M. JR, MCGAUGHEY, H. S. JR, SCOGGIN, W. A. and THORTON, W. M. (1964) Concentration of progesterone in newborn and maternal circulation at delivery. *Obstetrics and Gynecology,* **23,** 413–426

HIRATO, K., YANAIHARA, T. and NAKAYAMA, T. (1982) A study of Δ^5-3β-hydroxysteroid dehydrogenase in human foetal adrenal glands. *Acta Endocrinologica (Kbh),* **99,** 122–128

HUHTANIEMI, I. (1973) Identification and quantification of unconjugated neutral steroids in adrenal and liver tissue of early and mid-term human fetuses. *Steroids,* **21,** 511–519

HUHTANIEMI, I., KOIVISTO, M., PAKARINEN, A., TUIMALA, R. and KAUPPILA (1982) Pituitary–adrenal and testicular function in preterm infants after prenatal dexamethasone treatment. *Acta Paediatrica Scandinavica,* **71,** 425–429

HUHTANIEMI, I., LUUKAINEN, T. and VIHKO, R. (1970) Identification and determination of neutral steroid sulphates in human foetal adrenal and liver tissue. *Acta Endocrinologica (Kbh),* **64,** 273–286

ISHERWOOD, D. M. and OAKEY, R. E. (1976) Control of oestrogen production in human pregnancy: effect of trophic hormones on steroid biosynthesis by the foetal adrenal gland *in vitro. Journal of Endocrinology,* **68,** 321–329

JAFFE, R. B., SERON-FERRE, M., CRICKARD, K., KORITNIK, D., MITCHELL, B. F. and HUHTANIEMI, I. (1981) Regulation and function of the primate fetal adrenal gland and gonad. *Recent Progress in Hormone Research,* **37,** 41–96

JAFFE, R. B., SERON-FERRE, M., HUHTANIEMI, I. and KORENBROT, C. (1977) Regulation of the primate fetal adrenal gland and testis *in vitro* and *in vivo. Journal of Steroid Biochemistry,* **8,** 479–490

JOHANNISSON, E. (1968) The foetal adrenal cortex in the human. Its ultrastructure at different stages of development and in different functional states. *Acta Endocrinologica,* **58** (Suppl.), 130

KAHRI, A. I., HUHTANIEMI, I. and SALMENPERA, M. (1976) Steroid formation and differentiation of cortical cells in tissue culture of human fetal adrenals in the presence and absence of ACTH. *Endocrinology,* **98,** 33–41

KORTE, K., HEMSELL, P. G. and MASON, J. I. (1982) Sterol sulfate metabolism in the adrenals of the human fetus, anencephalic newborn, and adult, *Journal of Clinical Endocrinology and Metabolism,* **55,** 671–675

KOWAL, J., FORCHIELLI, E.. and DORFMAN, R. I. (1964) The Δ^5-3β-hydroxysteroid dehydrogenases of corpus luteum and adrenal. II. Interaction of C_{19} and C_{21} substrates and products. *Steroids,* **4,** 77–100

LAATIKAINEN, T., PELKONEN, J., APTER, D. and RANTA, T. (1980) Fetal and maternal serum levels of steroid sulfates, unconjugated steroids, and prolactin at term pregnancy and in early spontaneous labor. *Journal of Clinical Endocrinology and Metabolism,* **50,** 489–494

LEHMANN, W. D. and LAURITZEN, C. (1975) HCG + ACTH stimulation of *in vitro* dehydroepiandrosterone production in human fetal adrenals from precursor cholesterol and Δ^5-pregnenolone. *Journal of Perinatal Medicine,* **3,** 231–236

LEONG, M. K. H. and MURPHY, B. E. P. (1976) Cortisol levels in maternal venous and umbilical cord arterial and venous serum at vaginal delivery. *American Journal of Obstetrics and Gynecology,* **126,** 471–473

LIGGINS, G. C. (1976) Adrenocortical-related maturational events in the fetus. *American Journal of Obstetrics and Gynecology,* **126,** 931–941

MACNAUGHTON, M. C., TAYLOR, T., MCNALLY, E. M. and COUTTS, J. R. T. (1977) The effect of synthetic ACTH on the metabolism of [4-^{14}C]progesterone by the previable human fetus. *Journal of Steroid Biochemistry,* **8,** 499–504

MASON, J. I., HEMSELL, P. G. and KORTE, K. (1983) Steroidogenesis in dispersed cells of the human fetal adrenal. *Journal of Clinical Endocrinology and Metabolism,* **56,** 1057–1062

MÄUSLE, E. (1972) Ultrastruktur des Involutionsprozesses der Cortex fetalis beim menschlichen Neugeborenen. *Verhandlungen der Deutschen Gesellschaft für Pathologie,* **55,** 147–158

MCCLELLAN, M. C. and BRENNER, R. M. (1981) Development of the fetal adrenals in nonhuman primates: electron microscopy. In *Fetal Endocrinology,* edited by M. J. Novy and J. A. Resko, pp. 383–403. New York: Academic Press

MCCONATHY, W. J. and LANE, D. M. (1980) Studies on the apolipoproteins and lipoproteins of cord serum. *Pediatric Research,* **14,** 757–761

MCNULTY, W. P. (1981) Postnatum evolution of the adrenal glands of rhesus macaques. In *Fetal Endocrinology,* edited by M. J. Novy and J. A. Resko, pp. 53–64. New York: Academic Press

MILNER, A. J. and MILLS, I. H. (1970) Patterns of steroid biosynthesis by human adrenals incubated *in vitro* with [7α-^{3}H] pregnenolone: changes with (a) gestational age, (b) incubation period and (c) weight of incubated tissue. *Journal of Endocrinology,* **47,** 369–378

MITCHELL, B. F., SERON-FERRE, M., HESS, D. L. and JAFFE, R. B. (1981) Cortisol production and metabolism in the late gestation Rhesus monkey fetus. *Endocrinology,* **108,** 916–924

MIYAKAWA, I., IKEDA, I. and MAEYAMA, M. (1974) Transport of ACTH across human placenta. *Journal of Clinical Endocrinology and Metabolism,* **39,** 440–442

MURPHY, B. E. P. (1973a) Steroid arteriovenous differences in umbilical cord plasma: evidence of cortisol production by the human fetus in mid-gestation. *Journal of Clinical Endocrinology and Metabolism,* **35,** 678–683

MURPHY, B. E. P. (1973b) Does the fetal adrenal play a role in parturition? *American Journal of Obstetrics and Gynecology,* **115,** 521–525

MURPHY, B. E. P. (1982) Human fetal serum cortisol levels related to gestational age: evidence of a midgestational fall and a steep late gestational rise, independent of sex or mode of delivery. *American Journal of Obstetrics and Gynecology,* **144,** 276–282

MURPHY, B. E. P. (1983) Human fetal serum cortisol levels at delivery: a review. *Endocrine Reviews,* **4,** 150–154

MURPHY, B. E. P. and DIEZ D'AUX, R. C. (1972) Steroid levels in the human fetus: cortisol and cortisone. *Journal of Clinical Endocrinology and Metabolism,* **35,** 678–683

NEVILLE, A. M. and O'HARE, M. J. (1982) *The Human Adrenal Cortex,* p. 12. Berlin: Springer-Verlag

NOGUCHI, A. and REYNOLDS, J. W. (1978) Serum cortisol and dehydroepiandrosterone sulfate responses to adrenocorticotropin stimulation in premature infants. *Pediatric Research,* **12,** 1057–1061

NWOSU, U. A., WALLACH, E. E. and BOLOGNESE, R. J. (1976) Initiation of labor by intraamniotic cortisol instillation in prolonged human pregnancy. *Obstetrics and Gynecology,* **47,** 137–142

OHASHI, M., CARR, B. R. and SIMPSON, E. R. (1981) Effects of adrenocorticotropic hormone on low density lipoprotein receptors of human fetal adrenal tissue. *Endocrinology,* **108,** 1237–1242

OHRLANDER, S., GENSSER, G., NILSSON, K. O. and ENEROTH, P. (1977) ACTH test to neonates after administration of corticosteroids during gestation. *Obstetrics and Gynecology,* **49,** 691–694

ONISHI, S., MIYAZAWA, G., NISHIMURA, Y., SUGIYAMA, S., YAMAKAWA, T., INAGAKI, H., KATOH, T., ITOH, S. and ISOBE, K. (1983) Postnatal development of circadian rhythm in serum cortisol levels in children. *Pediatrics,* **72,** 399–404

PARKER, C. R., JR, SIMPSON, E. R., BILHEIMER, D. W., LEVENO, K., CARR, B. R. and MACDONALD, P. C. (1980) Inverse relationship between LDL-cholesterol and dehydroisoandrosterone sulfate in human fetal plasma. *Science,* **208,** 512–513

PASQUALINI, J. R., NGUYEN, B. L., UHRICH, F., WIQVIST, N. and DICZFALUSY, E. (1970) Cortisol and cortisone metabolism in the human foeto-placental unit at midgestation. *Journal of Steroid Biochemistry,* **1,** 209–219

PEPE, G. J. and ALBRECHT, E. D. (1980) The utilization of placental substrates for cortisol synthesis by the baboon fetus near term. *Steroids,* **35,** 591–597

PINTOR, C., GENAZZANI, A. R., BAGNOLI, F., CANTARINI, A., DETTORI, M. and CODA, R. (1980) The first 48 hours of life: cortisol and dehydroepiandrosterone sulphate under basal conditions and after ACTH stimulation. In *Problems in Pediatric Endocrinology,* edited by C. La Cauza and A. W. Root, pp. 305–310. London: Academic Press

POHJAVUORI, M. and FYRHQUIST, F. (1983) Vasopressin, ACTH and neonatal haemodynamics. *Acta Paediatrica Scandinavica,* Suppl. 305, 79–83

PREDINE, J., MERCERON, L., BARRIER, G., SUREAU, C. and MILGROM, E. (1979) Unbound cortisol in umbilical cord plasma and maternal plasma: a reinvestigation. *American Journal of Obstetrics and Gynecology,* **135,** 1104–1108

REYES, F. I., BORODITSKY, R. S., WINTER, J. S. D. and FAIMAN, C. (1974) Studies on human sexual development. II. Fetal and maternal serum gonadotropin and sex steroids concentrations. *Journal of Clinical Endocrinology and Metabolism,* **38,** 612–617

REYNOLDS, J. W. (1980) Fetal and neonatal steroid metabolism. In *Problems in Pediatric Endocrinology,* edited by C. La Cauza and A. W. Root, pp. 239–251. London: Academic Press

REYNOLDS, J. W., BENTLEY, K. and TURNIPSEED, M. R. (1977) Serum total estriol in abnormal newborn infants. *Journal of Steroid Biochemistry,* **8,** 853–858

SERON-FERRE, M., LAWRENCE, C. C. and JAFFE, R. B. (1978) Role of hCG in regulation of the fetal zone of the human fetal adrenal gland. *Journal of Clinical Endocrinology and Metabolism,* **46,** 834–837

SERON-FERRE, M., LAWRENCE, C. C., SIITERI, P. K. and JAFFE, R. B. (1978) Steroid production by definitive and fetal zones of the human fetal adrenal gland. *Journal of Clinical Endocrinology and Metabolism,* **47,** 603–609

SERRA, G. B., PEREZ-PALACIOS, G. and JAFFE, R. B. (1971) Enhancement of 3β-hydroxysteroid dehydrogenase-isomerase in the human fetal adrenal by removal of the soluble cell fraction. *Biochimica et Biophysica Acta,* **244,** 186–190

SHACKLETON, C. H. L., GUSTAFSSON, J-A. and MITCHELL, F. L. (1973) Steroids in newborns and infants. The changing pattern of urinary steroid excretion during infancy. *Acta Endocrinologica (Kbh),* **74,** 157–167

SHIBUSAWA, H., SANO, Y., OKINAGA, S. and ARAI, K. (1980) Studies on 11β-hydroxylase of the human fetal adrenal gland. *Journal of Steroid Biochemistry,* **13,** 881–887

SHIBUSAWA, H., SANO, Y., YOSHIDA, N., OKINAGA, S. and ARAI, K. (1978) Studies of the human fetal adrenal gland – properties of 17α-hydroxylase and C_{17}-C_{20} lyase in the biosynthesis of dehydroepiandrosterone from pregnenelone. *Journal of Steroid Biochemistry,* **9,** 1125–1132

SHIRLEY, I. M. and COOKE, B. A. (1969) Metabolism of dehydroepiandrosterone by the separated zones of the human foetal and newborn adrenal cortex. *Journal of Endocrinology,* **44,** 411–419

SHUTT, D. A., SMITH, I. D. and SHEARMAN, R. P. (1974) Fetal plasma steroids in relation to parturition. III. The effect of parity and method of delivery upon umbilical plasma oestrone and oestradiol levels. *Journal of Obstetrics and Gynaecology of the British Commonwealth,* **81,** 968–970

SIITERI, P. K. and MACDONALD, P. C. (1966) Placental estrogen biosynthesis during human pregnancy. *Journal of Clinical Endocrinology and Metabolism,* **26,** 751–761

SIITERI, P. K. and SERON-FERRE, M. (1981) Some new thoughts on the fetoplacental unit and parturition in primates. In *Fetal Endocrinology,* edited by M. J. Novy and J. A. Resko, pp. 1–34. New York: Academic Press

SIMMER, H. H., EASTERLING, W. E., PION, R. J. and DIGNAM, W. J. (1964) Neutral C_{19}-steroids and steroid sulphates in human pregnancy. *Steroids,* **4,** 125–137

SIMONIAN, M. H. and GILL, G. N. (1981) Regulation of the fetal human adrenal cortex: effects of adrenocorticotropin on growth and function of monolayer cultures of fetal and definitive zone cells. *Endocrinology,* **108,** 1769–1779

SIMPSON, E. R., CARR, B. R., PARKER, C. R. JR, MILEWICH, L., PORTER, J. C. and MACDONALD, P. C. (1979) The role of serum lipoproteins in steroidogenesis by the human fetal adrenal cortex. *Journal of Clinical Endocrinology and Metabolism,* **49,** 146–148

SIPPELL, W. G., BECKER, H., VERSMOLD, H. T., BIDLINGMAIER, F. and KNORR, D. (1978) Longitudinal studies of plasma aldosterone, corticosterone, deoxycorticosterone, progesterone, 17-hydroxyprogesterone, cortisol and cortisone determined simultaneously in mother and child at birth and during the early neonatal period. I. Spontaneous delivery. *Journal of Clinical Endocrinology and Metabolism,* **46,** 971–985

SIPPELL, W. G., BIDLINGMAIER, F. and KNORR, D. (1980) Development of endogenous glucocorticoids, mineralocorticoids and progestins in the human fetal and perinatal period. *European Journal of Clinical Pharmacology,* **18,** 95–104

SIPPELL, W. G., DORR, H. G., BECKER, H., BIDLINGMAIER, F., MICKAN, H. and HOLZMANN, K. (1979) Simultaneous determination of seven unconjugated steroids in maternal venous and umbilical arterial and venous serum in elective and emergency cesarean section at term. *American Journal of Obstetrics and Gynecology,* **135,** 530–542

SIPPELL, W. G., DORR, H. G., BIDLINGMAIER, F. and KNORR, D. (1980) Plasma levels of aldosterone, corticosterone, 11-deoxycorticosterone, progesterone, 17-hydroxyprogesterone, cortisol and cortisone during infancy and childhood. *Pediatric Research,* **14,** 39–46

SOLOMON, S., BIRD, C. E., LING, W., IWAMIYA, M. and YOUNG, P. C. M. (1967) Formation and metabolism of steroids in the fetus and placenta. *Recent Progress in Hormone Research,* **23,** 297–335

STRECKER, J. R., LAURITZEN, C., DAHLEN, H., JONATHA, W., GOSSLER, W. TETTENBORN, U. (1977) Injections of ACTH and HCG into the fetus during mid-pregnancy legal abortion performed by intra-amniotic instillation of prostaglandin. Influence on maternal oestrogens and testosterone. *Hormone and Metabolic Research,* **9,** 409–414

SULYOK, E., NEMETH, M., TENYI, I., CSABA, I., GYORY, E., ERTL, T. and VARGA, F. (1979) Postnatal development of renin–angiotensin–aldosterone system, RAAS, in relation to electrolyte balance in premature infants. *Pediatrics Research,* **13,** 817–820

THOMAS, E. (1911) Ueber die Nebenniere des Kindes und ihre Veränderung bei Infektionskrankheiten. *Beiträge zur Pathologischen Anatomie und zur Allgemeinen Pathologie,* **1,** 283–316

TULCHINSKY, D. and SIMMER, H. H. (1972) Sources of plasma 17α-hydroxyprogesterone in human pregnancy. *Journal of Clinical Endocrinology and Metabolism,* **35,** 799–808

VILLEE, D. B. (1966) Effects of progesterone on enzyme activity of adrenals in organ culture. *Advances of Enzyme Regulation,* **4,** 269–280

VINCENT, M., DUSSART, Y., ANNAT, G., SASSARD, J., FRANCOIS, R. and CIER, J. F. (1980) Plasma renin activity, aldosterone and dopamine β-hydroxylase activity as a function of age in normal children. *Pediatric Research,* **14,** 894–895

VOUTILAINEN, R. and KAHRI, A. I. (1980) Placental origin of the suppression of 3β-hydroxysteroid dehydrogenase in the fetal zone cells of human fetal adrenals. *Journal of Steroid Biochemistry,* **13,** 39–43

VOUTILAINEN, R., KAHRI, A. I. and SALMENPERA, M. (1979) The effects of progesterone, pregnenolone, estriol, ACTH and hCG on steroid secretion of cultured human fetal adrenals. *Journal of Steroid Biochemistry,* **10,** 695–700

WALSH, S. W., NORMAN, R. L. and NOVY, M. J. (1979) *In utero* regulation of Rhesus monkey fetal adrenals: effects of dexamethasone, adrenocorticotropin, thyrotropin-releasing hormone, prolactin, human chorionic gonadotropin and α-melanocyte stimulating hormone on fetal and maternal plasma steroids. *Endocrinology,* **104,** 1805–1813

WARDLAW, A. L., STARK, R. I., BOXI, L. and FRANTZ, A. G. (1979) Plasma β-endorphin and β-lipotropin in the human fetus at delivery: correlation with arterial pH and Po_2. *Journal of Clinical Endocrinology and Metabolism,* **49,** 888–891

WHITEHOUSE, B. J. and VINSON, G. P. (1968) Corticosteroid biosynthesis from pregnenolone and progesterone by human adrenal tissue *in vitro.* A kinetic study. *Steroids,* **11,** 245–264

WIENER, M. and ALLEN, S. H. G. (1967) Inhibition of placental steroid synthesis by steroid metabolites: possible feedback control. *Steroids,* **9,** 567–582

WINKLER, L., SCHLAG, B. and GOETZE, E. (1977) Concentration and composition of the lipoprotein classes in human umbilical cord serum. *Clinica Chimica Acta,* **76,** 187–191

WINTER, J. S. D., FAIMAN, C. and REYES, F. I. (1981) Sexual endocrinology of fetal and perinatal life. In *Mechanisms of Sex Differentiation in Animals and Man,* edited by R. G. Edwards and C. R. Austin, pp. 206–253. London: Academic Press

WINTERS, A. J., OLIVER, C., COLSTON, C., MACDONALD, P. C. and PORTER, J. C. (1974) Plasma ACTH levels in the human fetus and neonate as related to age and parturition. *Journal of Clinical Endocrinology and Metabolism,* **39,** 269–273

YOSHIDA, N., SEKIBA, K., SHIBUSAWA, H., SANO, Y., YANAIHARA, T., OKINAGA, S. and ARAI, K. (1978) Biosynthetic pathways for corticoids and androgen formation in human fetal adrenal tissue *in vivo..* *Endocrinologica Japonica,* **25,** 191–195

3
Cellular mechanisms involved in the acute and chronic actions of ACTH

Michael R. Waterman and Evan R. Simpson

INTRODUCTION

The adrenal cortex is the target tissue for the peptide hormone ACTH, produced by the anterior pituitary. The actions of ACTH on the adrenal cortex can be divided into two types on a temporal basis, namely: short-term or acute actions, which occur in a matter of minutes or even seconds; and long-term or chronic actions which require a period of hours or even days to be expressed. In the present discussion, current concepts as to the mechanisms of both the acute and chronic actions of ACTH will be reviewed. In the adrenal cortex, and other steroidogenic tissues as well, the rate-limiting step in steroid hormone biosynthesis and the primary site of trophic hormone action is the first step in cholesterol utilization unique to the steroidogenic pathway, namely the removal of the six-carbon fragment from the side chain of cholesterol to form pregnenolone (Stone and Hechter, 1954). The acute stimulation of this reaction in the adrenal cortex by ACTH occurs within minutes, and is inhibited by inhibitors of messenger RNA translation such as cycloheximide. Upon addition of cycloheximide to ACTH-stimulated adrenal cells, steroidogenesis decays with a half-life of only a few minutes, which has led to the postulate that a 'labile protein factor' mediates the acute response to ACTH. In addition to this effect, ACTH is also known to have a long-term effect on the maintenance of the levels of enzymes involved in cholesterol side-chain cleavage in particular and steroidogenesis in general. For example, the levels of both mitochondrial and microsomal steroid hydroxylases in the rat adrenal cortex fall dramatically after hypophysectomy, but are restored toward control levels following several days of ACTH administration to hypophysectomized animals (Purvis *et al.*, 1973).

It seems to us that these two temporally distinct actions of ACTH are related, or coupled by some as yet undescribed mechanism. Recent studies in several laboratories have aimed at elucidating the mechanisms of one or the other of the chronic or acute responses. Examination of these results reveals an underlying thread, namely cholesterol, which runs through most of the results of these studies. It is our aim to review initially the details of the mechanism of the acute adrenocortical response to ACTH, and then the details of the mechanism of the chronic response. Finally, we will attempt to relate the two mechanisms to one another, and to illustrate that perhaps they are distinct only in a temporal sense.

PROPERTIES OF THE ADRENOCORTICAL STEROID HYDROXYLASES AND RELATED ENZYMES

As shown in the simplified diagram in *Figure 3.1*, the pathway of steroidogenesis from cholesterol to cortisol in the adrenal cortex involves four distinct forms of cytochrome P-450. These steroid hydroxylases catalyze reactions of the mixed-function oxidase type, whereby one atom of molecular oxygen is added to the substrate, yielding an hydroxylated product, while the other atom of oxygen is used

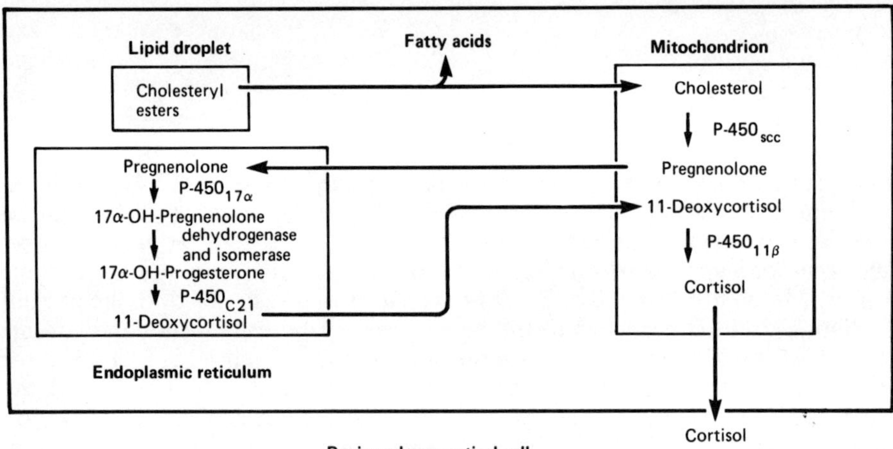

Figure 3.1 Pathway of biosynthesis of cortisol in the adrenal cortex. In this diagram, emphasis is placed on the various forms of cytochrome P-450 involved in the biosynthesis of cortisol: P-450_{SCC} (side-chain cleavage cytochrome P-450); P-$450_{17\alpha}$ (17α-hydroxylase cytochrome P-450); P-450_{C21} (21-hydroxylase cytochrome P-450); P-$450_{11\beta}$ (11β-hydroxylase cytochrome P-450). (From Simpson and Waterman, 1983, courtesy of the Editor and Publishers, *Canadian Journal of Biochemistry*)

to form water. The cytochrome P-450 family of enzymes is large and diverse, being found in most, if not all, mammalian tissues as well as in invertebrates, yeast, bacteria and plants. Many of the members of this family of enzymes hydroxylate xenobiotic or exogenous compounds. The steroid hydroxylases of the adrenal cortex represent the best-studied forms of this family which metabolize endogenous substrates.

The conversion of cholesterol to pregnenolone occurs in the mitochondria of steroid-producing tissues and, in common with other mixed-function oxidases, requires molecular oxygen and NADPH. The reaction sequence has been presented most frequently as follows:

cholesterol → (20S)-20-hydroxycholesterol → (20R,22R)-20,22-dihydroxy-cholesterol → pregnenolone.

This became the accepted version of the reaction, but was seriously challenged by the work of Burstein and colleagues (Burstein *et al.*, 1972; Burstein, Middleditch and Gut, 1974; Burstein, Gut and Hahn, 1975), who conducted a detailed kinetic analysis of the metabolism of cholesterol in acetone powders of bovine adrenal cortex mitochondria. These authors concluded that the major mechanism of pregnenolone formation was via an initial hydroxylation at the 22-position of cholesterol to yield (22R)-22-hydroxycholesterol, which was subsequently converted to the dihydroxy intermediate and then to pregnenolone. Further support for this reaction pathway came from studies of single turnover cycles of the purified cholesterol side-chain cleavage enzyme. Using this technique, Hume and Boyd (1978) found that after one turnover cycle cholesterol was converted primarily to a product identified as (22R)-22-dihydroxycholesterol. On repeating the cycle several times, it appeared that this product was converted to (20R,22R)-20,22-dihydroxycholesterol, which was then converted to pregnenolone.

Measurement of the stoichiometry of the cholesterol side-chain cleavage reaction indicates that 3 moles of oxygen and 3 moles of NADPH are utilized for every mole of pregnenolone formed (Shikita and Hall, 1974), implying that after the second hydroxylation the cleavage of the carbon–carbon bond between positions 20 and 22 of the cholesterol side-chain is an oxidative step. However, cleavage of the carbon–carbon bond between positions 20 and 22 of (20R,22R)-20,22-dihydroxycholesterol in an atmosphere containing oxygen-18 does not result in introduction of this isotope into either pregnenolone or isocapraldehyde (Takemoto *et al.*, 1968). It remains to be determined, therefore, how oxygen is utilized to cleave the vicinal diol.

A role for cytochrome P-450 as the terminal oxidase for the cholesterol side-chain cleavage reaction has been clearly established. The evidence is based on the following criteria:

(1) Inhibition of pregnenolone formation in the presence of a gas phase consisting of mixtures of oxygen and carbon monoxide (Simpson and Boyd, 1967a; Wilson and Harding, 1970).
(2) Reversal of the carbon monoxide inhibition of pregnenolone formation by light of 450 nm wavelength (Simpson and Boyd, 1967a; Wilson and Harding, 1970). This latter criterion is based on the photochemical action spectrum methodology pioneered by Warburg (1949) and subsequently adopted by Estabrook and colleagues in their studies of various cytochrome P-450-catalyzed reactions (Estabrook, Cooper and Rosenthal, 1963; Rosenthal and Cooper, 1967).
(3) Purification of the enzyme system catalyzing pregnenolone formation (Simpson and Boyd, 1967b; Bryson and Sweat, 1968). These studies have indicated that the cholesterol side-chain cleavage system consists of three protein components – a cytochrome P-450-type hemoprotein, a flavin adenine dinucleotide (FAD)-containing flavoprotein called NADPH-adrenodoxin reductase, and an Fe_2–S_2 type iron-sulphur protein which has been variously dubbed 'adrenodoxin' or 'testodoxin' depending on the tissue of origin.

Application of similar methodology in studies of the steroid 11β-hydroxylase and 18-hydroxylase (Omura *et al.*, 1966; Greengard *et al.*, 1967; Kimura and Suzuki, 1967; Nakamura, Otsuka and Tamaoki, 1969; Chu and Kimura, 1973), both of which occur within the mitochondria of the adrenal cortex, has yielded similar results. A number of laboratories have reported on the separation of two forms of cytochrome P-450 from bovine adrenal cortex mitochondria (Jefcoate, Hume and Boyd, 1970; Takemori *et al.*, 1975a,b; Wang and Kimura, 1976; Tilley, Watanuki and Hall, 1977). One of these catalyzes cholesterol side-chain cleavage (P-450_{SCC}), whereas the other catalyzes 11β-hydroxylation, and apparently also 18-hydroxylation (P-$450_{11β}$). These two forms of cytochrome P-450 can be distinguished not only by their different behavior with regard to purification techniques but also by virtue of their different optical and electron paramagnetic resonance (EPR) spectral properties, and different amino acid compositions (Mitani, 1979). The absolute optical spectrum of cytochrome P-$450_{11β}$ isolated from bovine adrenal cortex mitochondria is characterized by absorption maxima at 418, 539 and 570 nm, indicating a low-spin ferric hemoprotein. On addition of deoxycorticosterone (DOC) the spectrum shifts to that of a high-spin ferric hemoprotein, with absorption maxima at 394, 510 and 645 nm. Such low-spin to high-spin transitions of ferric cytochrome P-450 characteristically occur when the hemoprotein binds to substrate. On the other hand, isolated cytochrome P-450_{SCC} displays optical spectral characteristics of either a high-spin ferric hemoprotein or a mixed spin state form. This preparation always contains variable amounts of endogenous cholesterol, and conversion of this cholesterol to pregnenolone occurs when NADPH and cytochrome P-450 reductase proteins are added. Such conversion results in a high-spin to low-spin transition of the ferric hemoprotein, indicating the loss of bound substrate.

High- and low-spin ferric forms of cytochrome P-450 (i.e. substrate-bound and substrate-free forms) can also be distinguished by means of EPR spectrometry. Low-spin ferric cytochrome P-450 of rat adrenal mitochondria has an EPR spectrum characterized by g-values of 2.42, 2.25 and 1.92 (Brownie *et al.*, 1972; Jefcoate *et al.*, 1973; Simpson and Williams-Smith, 1976). On the other hand, the EPR spectrum of high-spin ferric cytochrome P-450 has g-values around 8, 4 and 2. More specifically, the high-spin ferric forms of cytochromes P-450_{SCC} and P-$450_{11β}$ of rat adrenal mitochondria can be distinguished by EPR spectrometry, since high-spin ferric cytochrome P-450_{SCC} is characterized by an EPR spectrum with g-values of 8.1 and 3.5 whereas cytochrome P-$450_{11β}$ has an EPR spectrum with g-values of 7.8 and 4.0 (Simpson and Williams-Smith, 1976; Alfano *et al.*, 1973). The EPR spectra of intact rat adrenal mitochondria indicate a mixture of low- and high-spin forms of ferric cytochrome P-450, suggesting the presence within the mitochondria of both substrate-free and substrate-bound forms. In addition, examination of the $g = 8$ region of the EPR spectrum reveals the presence of the high-spin forms of both types of cytochrome P-450 (Brownie *et al.*, 1973). These observations indicate the presence of cytochrome P-450_{SCC} bound to cholesterol and cytochrome P-$450_{11β}$ bound to endogenous substrate within rat adrenal mitochondria.

In contrast to these mitochondrial steroid hydroxylases, the microsomal enzymes display different properties. Although the microsomal hydroxylases also comprise distinct species of cytochrome P-450 and, in addition, utilize an NADPH-cytochrome P-450 flavoprotein, in the case of the microsomal systems electrons are passed directly from the flavoprotein to the cytochrome P-450. No adrenodoxin-like iron-sulfur protein is present in the endoplasmic reticulum. It is assumed that there is only one form of this reductase in the endoplasmic reticulum of different tissues and that the same enzyme services drug-metabolizing forms of cytochrome P-450 in the liver and steroid-hydroxylating forms of cytochrome P-450 in the adrenal cortex. We have far less information about the properties of adrenocortical microsomal forms of cytochrome P-450 than about the mitochondrial forms. In fact, it is only very recently that the microsomal steroid 21-hydroxylase cytochrome P-450 (P-450_{C21}) and 17α-hydroxylase cytochrome P-450 (P-$450_{17\alpha}$) have been purified from adrenal cortex (Kominami *et al.*, 1980; Bumpus and Dus, 1982; Kominami, Shinzawa and Takemori, 1982; Nakajin, Shinoda and Hall, 1983). We can expect to learn a great deal more about the nature and action of these microsomal steroid hydroxylases in the next few years. It is noted in passing that cytochrome P-450_{C21} plays a pivotal role in the history of the study of cytochrome P-450, for it is this enzyme that Estabrook, Cooper and their colleagues used first to demonstrate a cytochrome P-450 photochemical action spectrum with a maximum at 450 nm (Estabrook, Cooper and Rosenthal, 1963). Both cytochromes P-450_{SCC} and P-$450_{17\alpha}$ have been isolated from sources other than adrenal cortex. Cytochrome P-450_{SCC} was purified from bovine corpus luteum by Salhanick and his colleagues (Kashiwagi, Dafeldecker and Salhanick, 1980). To date, the properties of this enzyme have been found to be indistinguishable from those of the enzyme from bovine adrenal cortex (Nakajin, Shinoda and Hall, 1983). It seems likely that cytochrome P-450_{SCC} is the same protein in all steroidogenic tissues. Hall and his colleagues have purified a species of cytochrome P-450 from neonatal pig testes which has both 17α-hydroxylase and $C_{17,20}$-lyase activity (Nakajin and Hall, 1981; Nakajin *et al.*, 1981). These studies have provided convincing evidence that one form of cytochrome P-450 catalyzes both reactions. Similar results have been obtained with this enzyme purified from guinea pig adrenal microsomes (Kominami, Shinzawa and Takemori, 1982) and pig adrenal microsomes (Nakajin, Shinoda and Hall, 1983). If it is true that one enzyme contains both activities in the adrenal cortex, then it can be anticipated that an as yet undiscovered regulatory mechanism controls the flow of pregnenolone and/or progesterone toward glucocorticoids on the one hand and androgens on the other. Finally, it should be noted that not all species have adrenals which contain the four cytochrome P-450 steroid hydroxylases shown in *Figure 3.1*. Species in which corticosterone is the only glucocorticoid produced (i.e. the rat) presumably contain no adrenocortical 17α-hydroxylase activity.

Studies of reconstituted cholesterol side-chain cleavage enzyme systems have been conducted largely in the laboratories of Lambeth and colleagues (Seybert *et al.*, 1979; Lambeth, Seybert and Kamin, 1979; Lambeth, Kamin and Seybert, 1980; Lambeth, 1981) and of Jefcoate and colleagues (Hanukoglu and Jefcoate, 1979; Hanukoglu *et al.*, 1980). Kinetic studies on reconstituted cholesterol side-chain cleavage systems from both these laboratories indicate that adrenodoxin serves as a mobile electron shuttle between NADPH-adrenodoxin reductase and cytochrome P-450_{SCC}. The binding of the adrenodoxin or cholesterol to cytochrome P-450_{SCC} causes an increase in the affinity for the other substrate of the cytochrome (Lambeth, Seybert and Kamin, 1979; Hanukoglu *et al.*, 1980).

When the cholesterol-free cytochrome P-450_{SCC} was incorporated into a phospholipid liposome, the preparation had a low-spin optical spectrum (Hanukoglu and Jefcoate, 1979; Lambeth, Seybert and Kamin, 1979; Lambeth, Kamin and Seybert, 1980). When cholesterol was added to the vesicles, the fraction of hemoprotein in the high-spin form was proportional to the cholesterol:phospholipid ratio. Adrenodoxin caused a further conversion to the high-spin form. Virtually all of the cholesterol within the vesicle was accessible to metabolism by the enzyme. On the other hand, cholesterol contained within a separate vesicle was not accessible to the enzyme (Seybert *et al.*, 1979).

The rate of cholesterol metabolism in such vesicles also depended on the fatty acid composition of the phospholipids; in particular, a requirement for unsaturation in the fatty acid chain was suggested. Cardiolipin was found to be a more effective activator than phosphatidylcholine (Seybert *et al.*, 1979; Lambeth, 1981). Both diphosphoinositide and cardiolipin were observed to increase the V_{max} of the side-chain cleavage reaction in reconstituted systems while cardiolipin also decreased the K_m for cholesterol (Lambeth, 1981). These results were interpreted in terms of a model in which membrane phospholipids increase the affinity of cytochrome P-450_{SCC} for cholesterol by binding to one or more effector site(s) on cytochrome P-450_{SCC} rather than by causing alterations in membrane fluidity. Far fewer investigations have been carried out on the reconstitution of the 11β-hydroxylase system. Initial results indicated that maximal cytochrome P-$450_{11\beta}$ activity occurred in vesicles composed of phosphatidylcholine containing saturated fatty acyl groups (Seybert, 1983). Cardiolipin was found to inhibit cytochrome P-$450_{11\beta}$ activity, leading to the interesting possibility that the action of lipid to modulate cytochrome P-$450_{11\beta}$ activity is opposite to that which modulates cytochrome P-450_{SCC} activity.

INITIATION OF ACTH ACTION

It is generally believed that the response of an adrenocortical cell to ACTH is initiated by the hormone binding to specific receptors on the cell surface. Recent work by Buckley and Ramachandran (1981) provides for the first time conclusive evidence for the existence of a single class of high-affinity specific receptors for ACTH in membrane fractions of adrenal cells. Current evidence indicates that the binding of ACTH to its receptors results in an activation of adenylate cyclase and subsequent increase in intracellular levels of adenosine 3′,5′-monophosphate (cyclic AMP). Early results suggested a dissociation between cyclic AMP formation and steroidogenesis in that stimulation of steroidogenesis could be observed at low concentrations of ACTH which did not result in any apparent change in cyclic AMP levels (*for a more detailed review, see* Schulster, 1976; Gill, 1976). However, it is now apparent that this is due simply to the fact that the dose–response relationships for steroidogenesis on the one hand, and cyclic AMP formation on the other, are different, possibly because of the presence of spare receptors (Buckley and Ramachandran, 1981). Furthermore, the amount of cyclic AMP bound to the regulatory subunit of cyclic AMP-dependent protein kinase is also increased at all concentrations of ACTH which result in stimulation of steroidogenesis (Sala *et al.*, 1979). It seems reasonable to assume, therefore, that at both low and high concentrations the actions of ACTH are mediated by activation of cyclic AMP-dependent protein kinases. This discussion will focus on events which presumably occur after phosphorylation of proteins by cyclic AMP-dependent protein kinases.

THE ACUTE ACTION OF ACTH

As the cholesterol side-chain cleavage reaction is the rate-limiting step in steroidogenesis, increased rates of flux through subsequent steps in the steroidogenic pathway largely result from an increased entry of cholesterol into the pathway. Thus, study of the acute action of ACTH must of necessity concentrate on the cholesterol side-chain cleavage reaction.

Spectroscopic studies

There is ample evidence that isolated adrenal mitochondria from rats pretreated with ACTH have an increased capacity to synthesize pregnenolone compared to mitochondria prepared from control rats (Koritz and Kumar, 1970; Simpson *et al.*, 1972; Alfano *et al.*, 1973; Brownie *et al.*, 1973; Johnson, Ruhmann-Weinhold and Nelson, 1973; Jefcoate, Simpson and Boyd, 1974). This is true for ether-stressed rats compared to quiescent and cycloheximide-injected controls (Simpson *et al.*, 1972; Jefcoate, Simpson and Boyd, 1974); for intact rats injected with ACTH compared to quiescent controls (Koritz and Kumar, 1970); and for short-term hypophysectomized rats injected with ACTH compared to vehicle-injected controls (Alfano *et al.*, 1973; Simpson and Williams-Smith, 1976). Similar findings were observed in rat testis mitochondria prepared from animals injected with and without LH (Van der Wusse *et al.*, 1975), and in ovarian mitochondria from Parlow rats injected with and without LH (Arthur and Boyd, 1976). These observations imply that trophic hormone action results in an increase in the capacity of mitochondria to convert cholesterol to pregnenolone. This change is stable enough to survive tissue homogenization and isolation of mitochondria.

EPR spectra of adrenal mitochondria from ACTH-treated or ether-stressed rats are characterized by an increase in the size of the g = 8.1 signal (indicative of the cholesterol complex of cytochrome P-450_{SCC}) relative to the spectra of the control group (Alfano *et al.*, 1973; Brownie *et al.*, 1973; Simpson, McCarthy and Peterson, 1978). Parallel studies have been made of optical difference spectra of rat adrenal mitochondria. Addition of pregnenolone to these mitochondria results in an inverted type I difference spectrum, indicating displacement of endogenous substrate, in this case cholesterol (Simpson *et al.*, 1972; Alfano *et al.*, 1973; Brownie *et al.*, 1973; Koritz and Moustafa, 1976). Consequently, addition of pregnenolone can be used to titrate the amount of endogenous cholesterol bound to the enzyme. Addition of pregnenolone to adrenal mitochondria from ACTH-treated or ether-stressed rats results in formation of an inverted type I difference spectrum which is greater than the corresponding spectrum in adrenal mitochondria from control animals. These effects may be due to a difference in the amount of cholesterol bound to cytochrome P-450_{SCC} in the two groups of mitochondria (Jefcoate, Simpson and Boyd, 1974; Paul *et al.*, 1976). It may be concluded, therefore, that ACTH causes an increase in the amount of mitochondrial cholesterol which is specifically bound to cytochrome P-450_{SCC}. The rate of this association appears to be the rate-limiting step in cholesterol side-chain cleavage, since utilization of enzyme-bound cholesterol leads to a drastic reduction in the rate of pregnenolone formation in spite of the fact that the total mitochondrial cholesterol pool is decreased to only a relatively minor extent. Consequently, the action of ACTH to increase the association of cholesterol with cytochrome P-450_{SCC} results in the conversion of this cholesterol to pregnenolone, and hence in a stimulation of steroidogenesis.

In order to observe these ACTH-induced changes in isolated rat adrenal mitochondria it is necessary that the adrenal glands be in an anaerobic state prior to homogenization (Jefcoate and Orme-Johnson, 1975). This observation has been documented in a study of EPR spectra of single whole adrenal glands from hypophysectomized rats (Williams-Smith *et al.*, 1976). In excised glands frozen in liquid N_2 within 5 seconds of resection of the adrenal blood supply, there were no differences in the EPR spectra regardless of whether the glands were from control

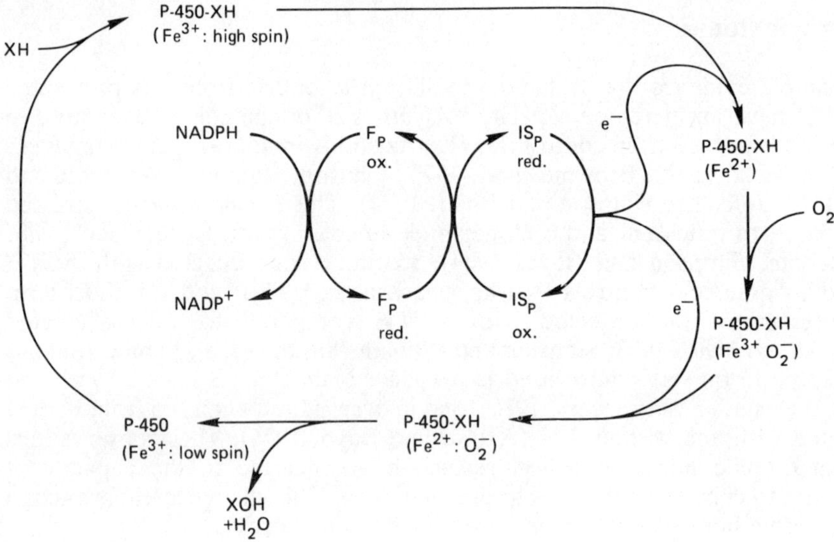

Figure 3.2 Reaction mechanism of hydroxylations catalyzed by cytochrome P-450 of adrenal cortex mitochondria. XH = substrate; XOH = product; F_P = flavoprotein; IS_P = iron-sulfur protein. (From Simpson and Waterman, 1983, courtesy of the Editor and Publishers, *Canadian Journal of Biochemistry*)

or ACTH-injected rats. However, when excised glands were allowed to become oxygen-depleted before freezing, a 40% decrease in the size of the EPR spectrum of low-spin ferric cytochrome P-450 relative to controls was observed in the case of glands removed from ACTH-injected rats. Since the total cytochrome P-450 was unchanged, depletion of oxygen in the ACTH-stimulated glands resulted in the conversion of low-spin ferric cytochrome P-450 to an EPR-undetectable form, i.e. to the ferrous form.

The explanation for this phenomenon is apparent from the reaction mechanism of cytochrome P-450 (*Figure 3.2*). It can be seen that reduction of cytochrome P-450 can only occur after binding of substrate (presumably because of the difference in redox potential between the substrate-bound and substrate-free forms). Clearly, upon depletion of oxygen, the ferrous form of cytochrome P-450 will accumulate only if substrate is present. Since this occurred after ACTH administration, it is concluded once again that ACTH causes an increase in the availability of substrate to the cytochrome P-450. The fact that in anaerobic glands no accumulation of ferrous cytochrome P-450 was seen after ACTH administration implies that in the normal operation of the cycle, when oxygen is present, the bound substrate is metabolized as quickly as the substrate complex is formed. These observations provide additional evidence that the rate-limiting step in steroidogenesis is the binding of substrate cholesterol to cytochrome P-450_s 450_{scc}.

Locus of the acute action of ACTH

Activation of cholesterol ester hydrolase

Increased intracellular cyclic AMP levels cause an activation of protein kinase activity which in turn is believed to result in phosphorylation of the enzyme cholesterol ester hydrolase, converting it from an inactive to an active form

(Trzeciak and Boyd, 1973; Beckett and Boyd, 1977; Naghshimah *et al.*, 1978). The active cholesterol ester hydrolase catalyzes the hydrolysis of stored cholesterol esters in the lipid droplets, and thereby increases the supply of free cholesterol to the mitochondria, which in turn could result in an increase in the amount of cholesterol bound to cytochrome P-450_{SCC}. It is worthy of note at this point that γ-melanocyte stimulating hormone (MSH) has been shown to stimulate cholesterol ester hydrolase activity and to act synergistically with ACTH to stimulate corticosterone production by the rat adrenal (Pederson and Brownie, 1980; Pederson, Brownie and Ping, 1980). The mechanism of this response does not appear to involve cyclic AMP.

Transport of cholesterol to the mitochondria

There is evidence that steroidogenesis in adrenal cells of the rat is impaired by inhibitors of microtubule and microfilament formation (Kraicer and Milligan, 1971; Mrotek and Hall, 1977; Crivello and Jefcoate, 1978, 1980). These results have been interpreted in terms of inhibition of cholesterol transport to the mitochondria. The locus of inhibition by these agents is clearly different from that of inhibitors of protein synthesis, which act to inhibit steroidogenesis at an intramitochondrial site. On the other hand, cytochalasin B (an inhibitor of microtubule formation) stimulates steroidogenesis in Y-1 adrenal tumor cells (Cortese and Wolff, 1978) and in bovine adrenal cortex cells maintained in the presence of lipoproteins (Crivello and Jefcoate, 1980). These findings have been interpreted to mean that the network of microfilaments and microtubules is required to transfer cholesterol from lipid droplets to the mitochondria, but actually inhibits transfer from lipoproteins to mitochondria.

Another mechanism which has been invoked to explain the transfer of cholesterol from lipid droplets involves sterol carrier proteins (SCP) similar to SCP_2 of rat liver (Chanderbhan *et al.*, 1982; Vahouny *et al.*, 1983). SCP_2 stimulates the utilization of cholesterol contained in lipid droplets for pregnenolone formation by adrenal mitochondria in a concentration-dependent fashion. Since sterol carrier proteins have been described in steroidogenic tissues, namely adrenal, ovary and testes (Kan and Ungar, 1973; Lefevre, Morera and Saez, 1978), a role for such a protein in transport of cholesterol to mitochondria remains a possibility. A role for ACTH to stimulate cholesterol transport utilizing either cytoskeletal elements or SCP remains to be determined. In addition, it should be noted that chronic stimulation of adrenal cells with ACTH results in a decrease in the cellular content of actin (Cheitlin and Ramachandran, 1981).

Intramitochondrial location of ACTH action

The acute action of ACTH is envisaged to involve two loci (*Figure 3.3*). One such locus is the activation of cholesterol ester hydrolase as detailed above. Cholesterol ester hydrolase activity also may be stimulated by γ-MSH, via mechanisms independent of cyclic AMP (Pederson and Brownie, 1980; Pederson, Brownie and Ping, 1980). This is sufficient to ensure an increased supply of cholesterol to the mitochondria, but inadequate to allow increased metabolism of this cholesterol. In addition, ACTH initiates a sequence of events within the mitochondria which results in a redistribution of mitochondrial cholesterol so that more cholesterol binds to the cytochrome P-450_{SCC}. This second step is cycloheximide-sensitive and therefore involves the 'labile protein factor', whereas the pathways required to increase the supply of extramitochondrial cholesterol to the mitochondria are not inhibited by cycloheximide. In *Figure 3.3* the 'labile protein factor' is envisaged as being synthesized or activated in response to ACTH, perhaps via cAMP-dependent protein kinase.

Evidence for this scheme has been provided by experiments in which hypophysectomized rats were injected with ACTH or cycloheximide plus ACTH (Mahafee, Reitz and Ney, 1974; Arthur, Mason and Boyd, 1976; Simpson, McCarthy and Peterson, 1978). In the absence of cycloheximide, ACTH caused an

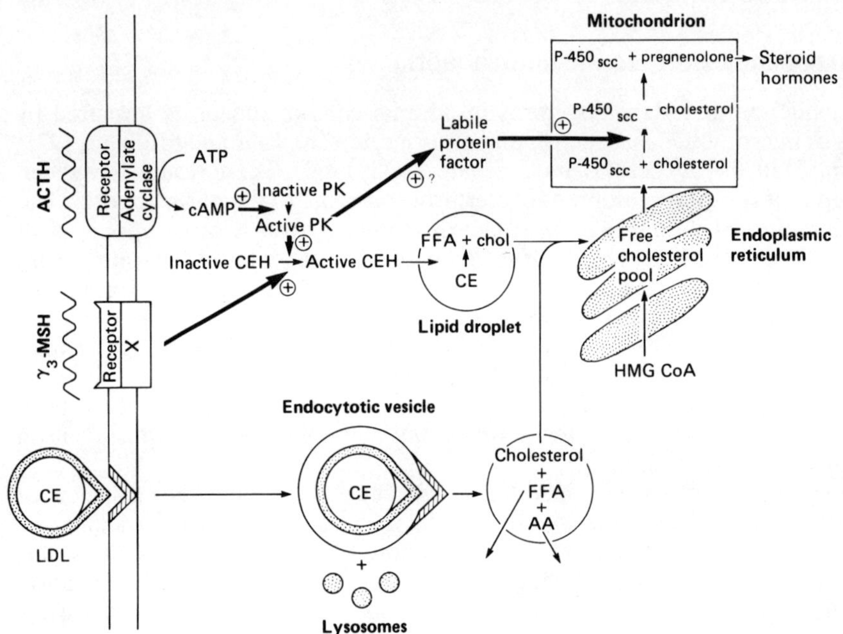

Figure 3.3 Acute effects of ACTH on cholesterol metabolism in the zona fasciculata of the adrenal cortex. The thin arrows denote pathways of metabolism. The heavy arrows denote regulatory pathways; ⊕ = activation; AA = amino acids; CE = cholesteryl esters; CEH = cholesterol ester hydrolase; chol. = cholesterol; FFA = free fatty acids; LDL = low density lipoprotein; HMG CoA = 3-hydroxy-3-methylglutaryl coenzyme A; PK = protein kinase. (From Simpson and Waterman, 1983, courtesy of the Editor and Publishers, *Canadian Journal of Biochemistry*)

enhancement of mitochondrial pregnenolone formation and an increase in size of the EPR signal in the g = 8.1 region. However, there was no change in mitochondrial cholesterol content relative to the cholesterol content of control mitochondria. On the other hand, when ACTH and cycloheximide were administered simultaneously, the resulting adrenal mitochondria were very similar to control mitochondria in terms of their ability to synthesize pregnenolone and in size of the g = 8.1 EPR signal. However, adrenal mitochondria from rats injected with cycloheximide plus ACTH contained twice as much cholesterol as control mitochondria. Consequently, in the presence of cycloheximide, ACTH caused an increased translocation of cholesterol to the mitochondria, but this cholesterol could not bind to cytochrome P-450$_{SCC}$ and therefore was not metabolized to pregnenolone. Addition of cholesterol to adrenal mitochondria *in vitro*, either dissolved in acetone (Simpson, McCarthy and Peterson, 1978) or contained in lipid droplets prepared from adrenal tissue (Farese, 1978), did not mimic the *in vivo* administration of ACTH, since it was found that adrenal mitochondria from ACTH-treated rats could metabolize added cholesterol more rapidly than control adrenal mitochondria.

These experiments provide evidence that the primary event in ACTH action is intramitochondrial, and depends on the integrity of the mitochondrial membranes. The 'labile protein factor' is likely to be involved in the binding of this mitochondrial cholesterol to cytochrome P-450$_{SCC}$. Disruption of the mitochondrial membranes, either by vigorous homogenization, hypo-osmotic shock or addition of high concentrations of calcium, obviates the need for the 'labile protein factor' required during ACTH action and, in the case of adrenal mitochondria prepared from rats injected with both cycloheximide and ACTH, allows utilization of the increased mitochondrial free cholesterol present as a result of the pretreatment. In addition to the mitochondrial locus of action, ACTH also stimulates cholesterol ester hydrolysis in the cytoplasmic lipid droplets, thus ensuring a continuous supply of cholesterol to the mitochondria in order to meet the demands of activated pregnenolone biosynthesis.

It should be emphasized that other sources of cholesterol are available to the mitochondrial cholesterol side-chain cleavage system, namely *de novo* synthesis from 2-carbon units or uptake and degradation of plasma lipoproteins (Gwynne *et al.*, 1976; Anderson and Dietschy, 1977; Balasubramanian *et al.*, 1977; Faust, Goldstein and Brown, 1977; Carr and Simpson, 1981a, b). Cholesterol ester synthesis in the adrenal is stimulated during uptake of lipoproteins, whereas endogenous cholesterol biosynthesis is low. In the absence of lipoproteins, the cholesterol ester stores become depleted, but the rate of cholesterol biosynthesis increases greatly. Regardless of the source of cholesterol, i.e. whether from endogenous biosynthesis or from lipoprotein uptake, ACTH addition causes a stimulation of steroidogenesis, but the maximum rate which can be attained is greater when cholesterol is supplied from lipoprotein uptake or cholesterol ester hydrolysis, rather than from endogenous biosynthesis (Faust, Goldstein and Brown, 1977; Brownie, Kramer and Gallant, 1979; Carr and Simpson, 1981b). These results imply that when pregnenolone formation is fully activated by ACTH, a continuing supply of cholesterol from extramitochondrial sources to the mitochondria becomes the rate-limiting step in steroidogenesis, contrary to the situation in the absence of ACTH. Furthermore, cholesterol can be supplied to the mitochondria at a faster rate from lipoprotein uptake or cholesterol ester hydrolysis than from *de novo* biosynthesis (*see Figure 3.3*). In the case of bovine adrenocortical cells in monolayer culture, it appears that *de novo* cholesterol synthesis is extremely low whether or not ACTH is present, and the cells have virtually a complete dependence on low density lipoprotein (LDL) as source of cholesterol (Ohashi *et al.*, 1982).

Activation of cholesterol side-chain cleavage by ACTH

From the foregoing discussion it is obvious that an understanding of tropic hormone action requires a basic knowledge of the topological relationship of the cholesterol side-chain cleavage enzymes to the structure of the mitochondrion. It now appears that the flavoprotein component of the mitochondrial steroid hydroxylation system is located primarily in the mitochondrial matrix, whereas the iron-sulphur protein is probably located on the matrix side of the inner mitochondrial membrane (Estabrook *et al.*, 1973). On the other hand, cytochrome P-450 is clearly located within the inner mitochondrial membrane (Sartre, Vignais and Idelman, 1969; Yago *et al.*, 1970). The question whether cytochrome P-450 is exposed to the outer or inner aspect of this membrane, whether it spans the membrane, or is entirely buried in the interior, is important for elucidation of the mitochondrial events involved in tropic stimulation of cholesterol side-chain cleavage. The fact that addition of cholesterol to adrenal mitochondria from control hypophysectomized rats (Simpson, McCarthy and Peterson, 1978) causes practically no stimulation of pregnenolone formation might be taken as evidence that cytochrome P-450_{SCC} also is not exposed to the outer aspect of the inner mitochondrial membrane. Evidence for the localization of cytochrome P-450_{SCC} has been provided by Mitani *et al.* (1982), who used immunochemical staining methods together with electron microscopy to locate this enzyme at the matrix side of the inner mitochondrial membrane.

It is clear that the action of ACTH at the mitochondrial level involves changes which affect the interaction of cholesterol with cytochrome P-450_{SCC}. Since both of these molecules are membrane constituents, it follows that the structure of the mitochondrial membranes may play an important role in mediating this response. This is clearly indicated by studies showing that the integrity of the mitochondrial membranes is essential for the stimulation of pregnenolone formation observed after pretreatment with the tropic hormone. Disruption of mitochondrial structural integrity by osmotic or ultrasonic shock (Jefcoate, Simpson and Boyd, 1974) results in loss of this tropic hormone effect, primarily as a result of an increase in the pregnenolone-synthesizing capacity of the control mitochondria. Associated with this increase in cholesterol side-chain cleavage activity following shock treatment, there is a corresponding increase in the magnitude of the pregnenolone-induced inverted type I difference spectrum. It appears, therefore, that the structure of the intact mitochondrial inner membrane imposes a constraint upon the interaction of cholesterol with cytochrome P-450_{SCC} which is partially overcome by the intracellular processes mediating the trophic hormone signal (Jefcoate, Simpson and Boyd, 1974). Furthermore, since removal of this constraint by the tropic stimulus is inhibited by pretreatment with cycloheximide, it follows that the action of the 'labile protein factor' is related to changes occurring in the mitochondrial membranes.

Intermembrane cholesterol transport

There is evidence that most of the cholesterol in rat-liver mitochondria is located in the outer mitochondrial membrane (Colbeau, Nachbaur and Vignais, 1971), whereas in hog adrenal cortex mitochondria it is more uniformly distributed between inner and outer membranes (Yago *et al.*, 1970). Presumably, cholesterol coming into the mitochondria from extramitochondrial sources is first of all associated with the outer membrane. As discussed earlier, cholesterol in a vesicle not containing cytochrome P-450_{SCC} is not readily metabolized by cytochrome P-450_{SCC} present in another vesicle. If extrapolation can be made to the mitochondrial membranes, this suggests that cholesterol may not readily traverse from the outer to the inner membranes. In recent studies, Privalle, Crivello and Jefcoate (1983) determined the distribution of cholesterol in outer and inner mitochondrial membranes of rat adrenal, separated by discontinuous Ficoll gradient centrifugation. They observed that in the presence of aminoglutethimide, ether anesthesia stress (ACTH) caused an accumulation of cholesterol in the inner mitochondrial membrane. This cholesterol was available to cytochrome P-450_{SCC}, as demonstrated by its conversion to pregnenolone *in vitro* on removal of the aminoglutethimide. This increase in inner membrane cholesterol was blocked by simultaneous administration of cycloheximide. It was concluded that cycloheximide blocks transfer of cholesterol from outer to inner mitochondrial membranes of the rat adrenal, but does not affect side-chain cleavage activity of the inner membrane *per se*. These results suggest that the 'labile protein factor' is involved in transferring cholesterol from the outer membrane to the inner membrane of the adrenal mitochondrion.

Cholesterol binding to cytochrome P-450_{SCC}

Although cholesterol can move rapidly in the plane of a bilayer, controversy exists whether movement of cholesterol from one side of the bilayer to the other (i.e. flip-flop motion) is rapid or slow (Kirby and Green, 1977; Poznansky and Lange, 1978). Cytochrome P-450_{SCC} inserted into a phospholipid/cholesterol vesicle apparently has access to all of the cholesterol in the vesicle (Seybert *et al.*, 1979); whether the cytochrome is inserted into the vesicle membrane in a vectorial manner so that its active center has access to only one side is unclear at this time. Since the cytochrome P-450_{SCC} is added to preformed vesicles (Seybert *et al.*, 1979), it may be that incorporation occurs only into the outer face. On the other hand, the active center of cytochrome P-450_{SCC} in the inner mitochondrial membrane may well have access to only the inner aspect of this membrane. The data of Privalle, Crivello and Jefcoate (1983) suggest that not all of the cholesterol in the inner membrane is readily available for side-chain cleavage. Consequently, it is not yet clear whether the pool of cholesterol which is rapidly converted to pregnenolone consists of cholesterol bound to cytochrome P-450_{SCC} plus that present on the matrix side of the inner mitochondrial membrane, or whether it consists of cholesterol bound to cytochrome P-450_{SCC} plus the entire pool of inner membrane cholesterol.

Nonetheless, the data of Lambeth (1981) indicate that certain phospholipids, including phosphatidylcholine containing unsaturated fatty acids, but especially cardiolipin, cause an increase in the binding of cholesterol to cytochrome P-450_{SCC} in lipid vesicles. Consequently, if the inner mitochondrial membrane were enriched in such phospholipids, the binding of cholesterol to cytochrome P-450_{SCC}, and hence cholesterol side-chain cleavage activity, would be enhanced.

Phospholipid turnover and ACTH action

The work of Farese and colleagues (Farese and Sabir, 1979, 1980; Farese, Sabir and Larson, 1980, 1981a, b, c; Farese *et al.*, 1980) has provided evidence that ACTH does cause a rapid increase in synthesis of phosphatidylinositol. This leads, in turn, to an increase in the synthesis of diphospho- and triphosphoinositide, but not of cardiolipin. It was further observed that both diphospho- and triphosphoinositides, as well as cardiolipin, stimulated pregnenolone formation in mitochondria of adrenals from control rats, but not in mitochondria of adrenals from ACTH-treated rats (Farese, Sabir and Larson, 1981b). The actions of these phospholipids on pregnenolone formation were accompanied by an increase in the proportion of cytochrome P-450_{SCC} in the high-spin state, similar to the actions of ACTH *in vivo* (Farese and Sabir, 1979), and certain phospholipids in reconstituted vesicles (Lambeth, Kamin and Seybert, 1980). Farese *et al.* (1980) examined the site of ACTH action on phosphatidylinositol metabolism and concluded that ACTH and cyclic AMP stimulate *de novo* phosphatidate synthesis. Furthermore, such synthesis is inhibited by cycloheximide, and the time course of such inhibition is consistent with a role for the 'labile protein factor' in phosphatidate synthesis.

Calcium and the action of ACTH

Because of its important role in stimulation–secretion coupling mechanisms in general, many investigators have considered the possibility that calcium might serve as a mediator of the tropic hormone response of steroidogenic tissues. While calcium does seem to be involved at the level of the interaction of ACTH with its plasma membrane receptor and the adenylate cyclase system, a further action at the level of the mitochondria remains a possibility but has not been proven.

The fact that high concentrations of calcium can stimulate NADPH-supported steroid hydroxylation reactions in adrenal mitochondria has been known for some time (Peron and McCarthy, 1968), but this effect is clearly the result of large-amplitude swelling and disruption of the mitochondrial membranes, resulting in increased accessibility of added pyridine nucleotides to the steroid hydroxylating enzymes. However, at much lower concentrations of calcium (100 μmol/l–1 mmol/l), there is specific activation of cholesterol side-chain cleavage but no effect on 11β-hydroxylation (Simpson, Waters and Williams-Smith, 1975; Simpson and Williams-Smith, 1975). Associated with this stimulation there is an increase in the magnitude of the EPR spectrum in the g = 8.1 region (Simpson and Williams-Smith, 1976), suggesting that the calcium-induced stimulation of pregnenolone formation is caused by a redistribution of cholesterol within the mitochondrial membranes and a consequent increase in the proportion of cholesterol which is bound to cytochrome P-450_{SCC}. Comparison of adrenal mitochondria from ACTH-injected and control rats revealed that 1 mmol/ℓ calcium stimulated pregnenolone formation in both groups of mitochondria, but the difference between the two groups was maintained (Arthur, Mason and Boyd, 1976; Farese and Prudente, 1977; Simpson, McCarthy and Peterson, 1978).

In the above experiments the mitochondria were prepared from adrenal glands homogenized in 0.25M-sucrose. When mitochondria were prepared from adrenal glands homogenized in the presence of EDTA and albumin, the ACTH effect on pregnenolone formation was diminished or absent (Farese and Prudente, 1977). Addition of calcium restored the ACTH effect so that in the presence of calcium the difference between the two groups of mitochondria in terms of pregnenolone-synthesizing capacity was maximal. Similar effects of calcium have been observed in testicular mitochondria prepared from rats injected with LH (Van der Wusse *et al.*, 1976). In more recent studies, Hall, Osawa and Thomasson (1981) examined the role of calmodulin in the regulation of steroid synthesis by mouse Y-1 adrenal tumor cells. It was found that addition of trifluoperazine (an inhibitor of calmodulin) to the cells inhibited ACTH- or dibutyryl cyclic AMP-stimulated steroidogenesis. It also inhibited the accumulation of cholesterol by the inner mitochondrial membrane induced by aminoglutethimide and the increase in mitochondrial pregnenolone formation caused by ACTH or dibutyryl cyclic AMP. On the other hand, when calmodulin and Ca^{2+} were presented to the cells in liposomes, steroid synthesis was stimulated, as was accumulation of cholesterol in the inner mitochondrial membrane in the presence of aminoglutethimide, and pregnenolone formation in mitochondria isolated from these cells. These results strongly suggest that calmodulin and calcium play a role in ACTH-induced cholesterol transport into the inner mitochondrial membrane.

Characterization of the 'labile protein factor'

There have been several attempts to characterize this hypothetical mediator of ACTH action (Farese, 1967; Kan and Ungar, 1973; Ray and Strott, 1981; Chanderbhan *et al.*, 1982; Neher *et al.*, 1982). In addition, it has been demonstrated that polylysine has many of the properties predicted for the endogenous activator (Kido and Kimura, 1981). Very recently, however, new evidence has been presented for a peptide whose properties are those of the 'labile protein factor'. Pedersen and Brownie (1982) subjected an organic acid extract of adrenal cytosol (from ACTH-treated rats) to gel filtration followed by high performance liquid chromatography. The purified material stimulated a four- to fivefold increase in pregnenolone formation by adrenal mitochondria from rats treated with ACTH plus cycloheximide, and also caused an increase in the inverted type 1 spectrum induced by pregnenolone. Little or no activator peptide activity was present in cytosol fractions prepared from adrenals of rats treated with cycloheximide. Activity was also less in adrenal cytosol fractions from hypophysectomized rats, and absent in liver cytosol fractions. The material had an apparent molecular weight of 2200 (which is smaller than that of SCPs) and resisted attempts at Edman degradation, suggesting that the N-terminal end was blocked. In addition, a much larger protein, 28 000 daltons, whose synthesis is rapidly induced by ACTH and which is sensitive to cycloheximide, has been isolated from rat adrenals (Krueger and Orme-Johnson, 1983). This protein meets several of the other requirements of a 'labile protein factor'.

There is thus strong evidence for the existence of this long-sought substance. Although the mechanism whereby ACTH regulates the activity of this factor is not known, it has been speculated that an inactive precursor is converted to the active form by limited proteolysis, and that ACTH regulates this, possibly via cyclic AMP-mediated phosphorylation (Pederson and Brownie, 1982). Perhaps this explanation will resolve the apparent discrepancy with respect to size of the two proteins described above.

Conclusions

A considerable body of knowledge has emerged in recent years regarding both the properties of the cholesterol side-chain cleavage enzyme system and the nature of the acute response of the adrenal cortex cell to ACTH. However, knowledge of the molecular mechanisms of the acute action of ACTH still remains elusive. The concept that the primary site of the acute action of ACTH is intramitochondrial has gained strength, as has the further concept that this intramitochondrial event involves increased binding of cholesterol to cytochrome P-450_{SCC}.

There are two principal models to explain the increased binding of cholesterol to cytochrome P-450_{SCC}. In the first, binding of inner mitochondrial membrane cholesterol to the enzyme is increased by a mechanism involving the simultaneous association of specific phospholipids with the cytochrome. The synthesis of such phospholipids is stimulated by ACTH. In this model, calcium is involved insofar as it stimulates ACTH-induced phosphatidylinositol synthesis. However, calcium could also be involved at the level of the mitochondrion. In this context, it is interesting to note that the di- and triphospho-phosphatidylinositols readily bind Ca^{2+}.

In the second model, the transfer of cholesterol to the inner mitochondrial membrane is stimulated by ACTH and requires the 'labile protein factor'. This could come about in one of two ways: cholesterol could be transported to the outer mitochondrial membrane by a process involving cytoskeletal elements, calmodulin/Ca^{2+} and perhaps also a sterol binding protein. Transfer of cholesterol to the inner mitochondrial membrane would then require the 'labile protein factor'. Alternatively, cholesterol could be transported directly from the cytoplasm to the inner mitochondrial membrane, without interacting with the outer mitochondrial membrane, by a process involving the 'labile protein factor'. The fact that cycloheximide does not block cholesterol accumulation by mitochondria isolated from adrenals of rats treated with ACTH, but does prevent the subsequent ability of such mitochondria to convert this cholesterol to pregnenolone *in vitro*, would tend to argue in favor of the former mechanism.

If this is correct, then the 'labile protein factor' presumably must play some role in the transport of cholesterol, shipping it from the outer membrane to sites within the inner mitochondrial membrane. Clearly, further advances in understanding the mechanism of the acute action of ACTH must await characterization of the mechanism of action of the 'labile protein factor'.

THE CHRONIC ACTION OF ACTH

While the acute action of ACTH on the adrenal cortex is confined primarily to the cholesterol side-chain cleavage reaction, the chronic effects of this peptide hormone are found to be directed at almost every step in the steroidogenic pathway. As noted earlier, the contents of both mitochondrial and microsomal cytochromes P-450 in rat adrenal cortex fall dramatically after hypophysectomy (Doering and Clayton, 1969; Kimura, 1969; Pfeiffer *et al.*, 1972; Purvis *et al.*, 1973). The same is also true for adrenodoxin content, 11β-hydroxylase activity, 21-hydroxylase activity and NADPH-cytochrome P-450 reductase activity (Purvis *et al.*, 1973). In all these cases, administration of ACTH to hypophysectomized animals results in partial restoration of cytochrome P-450 content and these enzyme activities. Furthermore, Mitani *et al.* (1982), using immunofluorescence techniques, noted a decrease in the levels of cytochromes P-450_{SCC} and P-$450_{11\beta}$ in rat adrenal mitochondria after hypophysectomy, and an increase after ACTH administration. It was thus considered possible that chronic ACTH administration has secondary effects on several, if not all, components of the steroidogenic pathway. However, the nature of this regulatory action has not been examined in detail until recently. It is well known that the amounts of several hepatic forms of cytochrome P-450 can be increased by treatment with xenobiotic agents (usually substrates for cytochrome P-450), the levels of specific forms of hepatic cytochrome P-450 being increased by treatment with specific agents. This induction process could involve synthesis of new cytochrome P-450, activation of already existing cytochrome P-450, alteration in the turnover rate of existing cytochrome P-450, or a combination of one or more of these processes (Omura, 1979). Furthermore, if synthesis of new cytochrome P-450 were to occur, it could result from increased synthesis of cytochrome P-450 mRNA sequences, stabilization of such sequences, or more efficient use of cytochrome P-450 mRNA in protein synthesis, i.e. induction could be due to transcriptional or translational events or both. Recently, it has been demonstrated that induction of cytochrome P-450 in liver by xenobiotics involves, at least in part, synthesis of mRNA specific for cytochrome P-450 (Adesnik, 1981; Bresnick *et al.*, 1981; Nebert and Negishi, 1981). One can view the results obtained with hypophysectomized rats described above as being indicative of induction of synthesis of adrenocortical forms of cytochrome P-450 and related enzymes. Thus, an explanation of the chronic effect of ACTH on steroidogenesis will be based on a description of the mechanism by which induction takes place. As will be seen below, a picture is now beginning to emerge as to how ACTH exerts its chronic action to regulate steroidogenic capacity in the adrenal cortex.

Induction of synthesis of cytochrome P-450$_{SCC}$

Using bovine adrenocortical cells in monolayer culture (a system devised by Gospodarowicz *et al.* 1977), DuBois *et al.* (1981a) showed that the rate of synthesis of cytochrome P-450$_{SCC}$ increased severalfold after administration of ACTH. Following pulse radiolabeling of cultured cells with [^{35}S]methionine, and using an antibody specific for bovine adrenocortical cytochrome P-450$_{SCC}$, these investigators showed that ACTH treatment led to an increased rate of synthesis of cytochrome P-450$_{SCC}$. In addition, it was found that RNA isolated from cells maintained in the presence of ACTH directed cytochrome P-450$_{SCC}$ synthesis in a rabbit reticulocyte lysate cell-free translation system to a greater extent than did RNA isolated from control (untreated) cells. In each case, it was found that up to 12 hours after initiation of ACTH treatment of cultured cells there was no increase in the rate of synthesis or translatability of RNA. However, beyond 12 hours an increase was observed in each case, with a maximum occurring at 36 hours.

Interestingly, ACTH treatment continued beyond 36 hours led to diminished rates of cytochrome P-450$_{SCC}$ synthesis. In the same study, it was found that ACTH caused a biphasic increase in pregnenolone production from endogenous cholesterol in cultured cells. The initial increase occurred rapidly and was attributed to the acute action of ACTH. A second increase in pregnenolone formation occurred, which reached a maximum value at 36 hours after ACTH treatment. Following longer periods of exposure to ACTH, pregnenolone production decreased in a similar fashion to the synthesis of cytochrome P-450$_{SCC}$. The results of this study have been interpreted to indicate that ACTH treatment leads to an increase in transcriptional synthesis of mRNA specific for cytochrome P-450$_{SCC}$, which in turn leads to an increase in the rate of synthesis of apocytochrome P-450$_{SCC}$. As indicated by the increase in pregnenolone production, the increased synthesis causes an increase in enzymatic activity. Thus it is concluded that ACTH has a long-term action to regulate the synthesis and thus the level of cytochrome P-450_{SCC}.

Induction of synthesis of the other mitochondrial proteins involved in steroidogenesis

The observed increase in cytochrome P-450_{SCC} synthesis in response to ACTH treatment raises the possibility that ACTH treatment also leads to increased synthesis of the other components of the cholesterol side-chain cleavage system, adrenodoxin and adrenodoxin reductase. Kowal, Simpson and Estabrook (1970) noted that ACTH caused an increase in adrenodoxin content of mouse Y-1 adrenocortical tumor cells as measured by EPR spectroscopy. Asano and Harding (1976) showed that ACTH treatment acted to increase the rate of adrenodoxin biosynthesis in Y-1 cells in culture. In this study, an increase in the rate of [^{3}H]leucine incorporation into adrenodoxin was observed within two hours following initiation of ACTH treatment. Examination of adrenodoxin biosynthesis in bovine adrenocortical cells in monolayer culture revealed that the temporal profile of adrenodoxin synthesis, determined by pulse radiolabeling of cell protein or by translation of cellular RNA, was the same as that described above for cytochrome P-450_{SCC} (Kramer *et al.*, 1982a). Additionally, continued ACTH treatment beyond 36 hours led to decreased rates of adrenodoxin synthesis. A two-fold increase in adrenodoxin content of cultured cells was also observed by EPR spectroscopy 36 hours after administration of ACTH. It was thus concluded that the synthesis of adrenodoxin is induced by ACTH coordinately with that of cytochrome P-450_{SCC}.

Little information is available about the effect of ACTH on synthesis of adrenodoxin reductase. In this laboratory, preliminary evidence has been obtained that ACTH induces the synthesis of adrenodoxin reductase in a fashion similar to that of cytochrome P-450_{SCC} and adrenodoxin (Kramer *et al.*, 1982b). The coordinated regulation of cytochrome P-450_{SCC}, adrenodoxin and adrenodoxin reductase synthesis by ACTH may provide the adrenocortical cell with a means of maintaining appropriate molar ratios of these proteins, and thus optimizing the efficiency of cholesterol side-chain cleavage.

ACTH has also been found to exert a long-term effect on 11β-hydroxylase activity in mouse Y-1 adrenal tumor cells in culture (Kowal, 1969; Kowal, Simpson and Estabrook, 1970). The synthesis of cytochrome P-$450_{11\beta}$ has recently been examined in cultured bovine adrenocortical cells in a fashion similar to that described above for cytochrome P-450_{SCC}, adrenodoxin and adrenodoxin reductase. In response to ACTH, the synthesis of cytochrome P-$450_{11\beta}$ is induced, and in both cell radiolabeling studies and RNA translation experiments the temporal profile of synthesis follows that reported for cytochrome P-450_{SCC} and adrenodoxin (Kramer, Simpson and Waterman, 1983a). Thus, it is found that the ACTH-mediated induction of synthesis of all the mitochondrial components of the adrenocortical steroid hydroxylase pathway occurs in a coordinated fashion.

The synthesis of the mitochondrial components (cytochromes P-450_{SCC} and P-$450_{11\beta}$ and adrenodoxin) of the steroidogenic pathway in bovine adrenocortical cells in monolayer culture has been found to be increased by agents other than

ACTH. Cyclic AMP analogs (dibutyryl cyclic AMP and 8-bromo cyclic AMP) are found to mimic the chronic effect of ACTH both with respect to the increase in synthesis of these enzymes and in terms of the temporal pattern of the increase (Kramer *et al.*, 1984). No additive effect of ACTH plus the cyclic AMP analog was observed on the synthesis of these enzymes. Furthermore, treatment with cholera toxin and prostaglandins E_2 and $F_{2\alpha}$, effectors which lead to increased levels of cyclic AMP in adrenocortical cells, also leads to increased synthesis of these enzymes (Boggaram, Simpson and Waterman, 1984). This increase is observed in both cell radiolabeling studies and RNA translation studies. It may be concluded from these observations that the chronic action of ACTH to regulate the synthesis of the adrenocortical steroid hydroxylases is mediated via cyclic AMP.

Precursor forms of mitochondrial steroidogenic proteins

An interesting aspect of the synthesis of the mitochondrial components of the steroidogenic pathway is that cytochromes P-450_{SCC} and P-$450_{11\beta}$, adrenodoxin and adrenodoxin reductase are all synthesized *in vitro* as precursors of higher molecular weight (Nabi and Omura, 1980; Nabi *et al.*, 1980; DuBois *et al.*, 1981b; Kramer *et al.*, 1982c). These precursors are encoded by nuclear genes and are synthesized in the cytoplasm as higher molecular weight precursors which are presumably processed proteolytically on insertion into the mitochondrion. In the case of adrenodoxin, processing of the precursor protein to the mature form has been reported (Nabi and Omura, 1980; Omura *et al.*, 1983). The two proteins localized in the mitochondrial matrix contain precursor segments of quite distinct sizes, with molecular weights of 2000 in the case of adrenodoxin reductase and 7000 in the case of adrenodoxin. However, both cytochrome P-450_{SCC} and cytochrome P-$450_{11\beta}$, which are localized in the inner mitochondrial membrane, contain precursor segments of molecular weights 5500, larger than the mature proteins. This is interesting in view of the fact that both cytochromes presumably react with adrenodoxin in the same way. Perhaps homology exists between the precursor segments for these two proteins so that they can be inserted into the inner mitochondrial membrane in the same orientation. On the other hand, microsomal components of the steroid hydroxylase pathway are not synthesized as higher molecular weight precursor forms. This has been shown for both cytochrome P-450_{C21} and NADPH-cytochrome P-450 reductase (Nabi *et al.*, 1980; Waterman, 1982; Funkenstein *et al.*, 1983). It has previously been observed that liver microsomal forms of cytochrome P-450 and its reductase are synthesized as the mature form (Colbert *et al.*, 1979; DuBois and Waterman, 1979; Bar-Nun *et al.*, 1980; Gonzalez and Kasper, 1980). Thus it is to be expected that in all steroidogenic tissues mitochondrial steroid hydroxylases will be synthesized as higher molecular weight precursors, while microsomal steroid hydroxylases will be synthesized as the mature form.

Induction of synthesis of microsomal steroidogenic enzymes

During the course of the investigation of steroidogenesis in cultured bovine adrenocortical cells, it was noticed that after administration of ACTH equivalent amounts of corticosterone and cortisol were secreted. However, by 24 hours cortisol predominated and by 36 hours the corticosterone secretion was below the basal level. This shift in the relative production of corticosterone and cortisol could be attributed to an induction and/or activation of 17α-hydroxylase (Kramer *et al.*, 1983). By measuring 17α-hydroxylase activity in post-mitochondrial supernatant fractions prepared from cells cultured for different times in the presence of ACTH, it was found that 17α-hydroxylase activity was induced by ACTH treatment (McCarthy *et al.*, 1983). At 36 hours after initiation of ACTH treatment, 17α-hydroxylase activity was 15 times that in control cells – the largest change yet noted for any enzyme activity in cultured bovine adrenocortical cells after administration of ACTH. In addition, the content of microsomal cytochrome P-450 had increased five-fold 36 hours after initiation of ACTH treatment. These findings suggest that ACTH treatment induces synthesis of cytochrome P-$450_{17\alpha}$, thus leading to an increase in 17α-hydroxylase activity. A similar conclusion was reached by Fevold, Wilson and Slalina (1978) with regard to the action of ACTH on the activity of rabbit adrenal 17α-hydroxylase.

The synthesis of cytochrome P-450_{C21} has also been reported to be induced by ACTH (Funkenstein *et al.*, 1983). In both cell radiolabeling experiments and *in vitro* translation studies, synthesis of cytochrome P-450_{C21} reached a maximum value 24 hours after initiation of ACTH treatment. As observed earlier in the study of mitochondrial steroid hydroxylases, the rate of synthesis of cytochrome P-450_{C21} decreased toward control values after longer periods of ACTH treatment. Although ACTH mediated induction of cytochrome P-450_{C21} synthesis, no increase in 21-hydroxylase activity could be observed in either intact cells or postmitochondrial supernatant fractions derived from cells. In a study of rabbit adrenals, Fevold and Brown (1978) reached a similar conclusion regarding the failure of ACTH to stimulate steroid 21-hydroxylase activity. In our studies using bovine adrenocortical cells, 21-hydroxylase activity was high relative to other activities in the steroidogenic pathway and thus not likely ever to be rate-limiting. However, the dilemma raised by an increase in synthesis of cytochrome P-450_{C21} in the absence of an increase in 21-hydroxylase activity remains unresolved. Perhaps an additional factor is required for 21-hydroxylase activity which is not induced by ACTH over the time frame of this experiment (Greenfield *et al.*, 1980; Ponticorvo *et al.*, 1980). Preliminary studies indicate that NADPH-cytochrome P-450 reductase, the flavoprotein necessary for both 17α-hydroxylase and 21-hydroxylase activities, is induced by ACTH. The temporal pattern of this induction has yet to be resolved; however, it appears that this cannot be the missing factor for 21-hydroxylase activity. It has also been noted that cholera toxin and prostaglandins E_2 and $F_{2\alpha}$ lead to increased synthesis of cytochrome P-450_{C21} in cultured adrenocortical cells, indicating that the effect of ACTH is mediated via cyclic AMP (Boggaram, Simpson and Waterman, 1984).

Thus the synthesis of all of the mitochondrial components and several of the microsomal components of the adrenocortical steroidogenic pathway is controlled by ACTH. Recent preliminary evidence from this laboratory indicates that the increase in 17α-hydroxylase activity is due in part to an increase in synthesis of cytochrome P-$450_{17\alpha}$. If this enzyme is identical with the $C_{17,20}$-lyase, as suggested by Hall and colleagues (Nakajin and Hall, 1981; Nakajin, Hall and Onada, 1981), an increase in this activity in response to ACTH is to be expected. In addition to the hydroxylase enzymes, there is evidence that ACTH induces the synthesis of the 3β-hydroxysteroid dehydrogenase in rat adrenals (Rybak and Ramachandran, 1982). The temporal profile of induction of this microsomal enzyme has yet to be examined, as has the effect of ACTH on this enzyme in bovine adrenocortical cells.

Specificity of the action of ACTH

It seems likely that ACTH will be found to induce the synthesis of all of the enzymes involved in the pathway of steroidogenesis from cholesterol in the adrenocortical cell. The question then arises whether this is a specific action of ACTH, restricted to these particular enzymes, or whether it is part of a general pleiotropic response of the adrenal cell to ACTH. In order to address this issue we have measured the activities of a number of housekeeping enzymes in bovine adrenocortical cells maintained in the presence or absence of ACTH, including enzymes of glycolysis and the TCA cycle, as well as glucose-6-phosphate dehydrogenase and malic enzyme. ACTH had no effect on the activities of any of these enzymes. However, it did cause an increase in the number of LDL receptors in these cells, and has been shown to cause an increase in the activity of HMG CoA reductase, the rate-controlling enzyme in cholesterol synthesis, of other types of adrenal cells. We conclude, therefore, that the action of ACTH to induce synthesis of proteins is specific to enzymes involved in the steroidogenic pathway from cholesterol, as well as to enzymes and proteins involved in optimizing the supply of cholesterol to this pathway.

COMPARISON OF ACUTE AND CHRONIC ACTIONS OF ACTH

The results cited in this chapter clearly indicate that ACTH regulates steroidogenesis in the adrenal cortex at two levels. As a result of the acute action of ACTH, increased amounts of cholesterol are presented to cytochrome P-450_{SCC}, leading to an increase in the synthesis of steroid hormones. This response is of obvious physiological importance since it leads to increased production of cortisol in response to stress, although the factors which regulate the flow of steroid precursors in the direction of either glucocorticoids or androgens remain to be determined. The chronic action of ACTH involves regulation of the synthesis of enzymes involved in the steroidogenic pathway. It is unlikely that the levels of steroidogenic enzymes fluctuate greatly under normal conditions *in vivo*, but more likely that ACTH acts to maintain an optimal level of these enzymes.

Both the acute and the chronic actions of ACTH are mediated by cyclic AMP. Thus, it can be imagined that ACTH binds to a cell surface receptor, leading to activation of adenylate cyclase and elevated levels of intracellular cyclic AMP. Among the responses to elevated levels of cyclic AMP is the activation of cholesterol ester hydrolase, which leads to increased levels of free cholesterol in the adrenocortical cells. The association of this cholesterol with cytochrome P-450_{SCC} is increased in the acute response to ACTH. We consider it likely that an increase in intracellular free cholesterol is also important in modulating the synthesis of cytochrome 450_{SCC} (*Figure 3.4*). We have observed that factors other than ACTH which would be expected to increase concentrations of intracellular free cholesterol also lead to increased synthesis of cytochrome P-450_{SCC}. For example, inhibition of the cholesterol side-chain cleavage reaction with aminoglutethimide leads to increased synthesis of cytochrome P-450_{SCC} in the presence of ACTH, as does the supplementation of culture medium with low density lipoprotein. It is interesting to consider in this context that induction of virtually all members of the cytochrome P-450 family of enzymes occurs in response to substrates of the particular form of cytochrome P-450 induced. It is not yet clear whether cholesterol might be important in regulating the synthesis of other enzymes in the steroidogenic pathway or whether it is specific for the side-chain cleavage enzyme complex. In this case, the synthesis of other steroidogenic enzymes could be regulated by their own substrates. Whatever the case may be, we consider it possible that cholesterol could mediate the action of ACTH to stimulate the synthesis of cytochrome P-450$_{SCC}$, and thus both the acute and chronic actions of ACTH to stimulate adrenocortical steroidogenesis may be coupled through a common mediator, namely cholesterol (*Figure 3.4*).

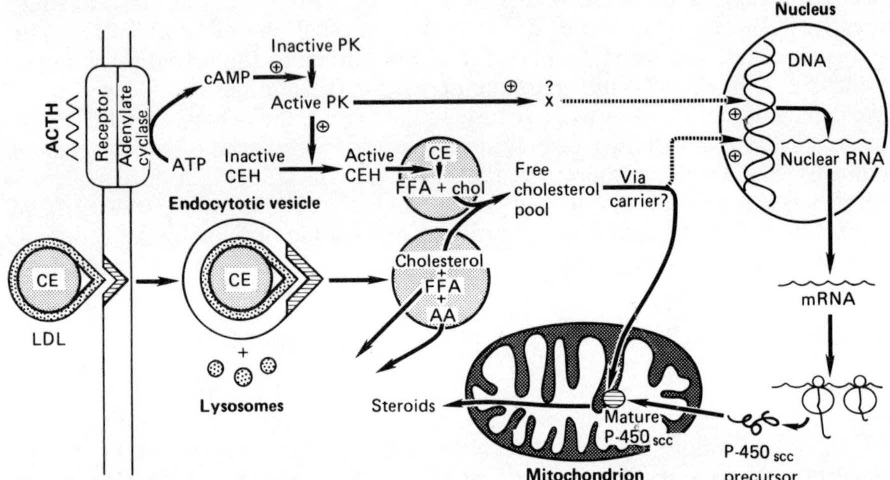

Figure 3.4 Proposed long-term effects of ACTH on the synthesis of cytochrome P-450$_{SCC}$ in the zona fasciculata of the bovine adrenal cortex. *For key see Figure 3.3.* X represents unknown steps involved in mediating the action of cAMP to initiate gene transcription

Several fundamental questions remain to be answered about both the acute and the chronic effects of ACTH. The precise mechanism by which ACTH mobilizes cholesterol to the site of side-chain cleavage is of key importance in elucidating the acute response. The nature of the agent(s) which regulate the synthesis of the steroid hydroxylation enzymes, and the arrangement and structure of the genes coding for these enzymes, are important in understanding the mechanism of the chronic effect of ACTH. It is anticipated that much progress will be made in the next few years in elucidating the mechanisms by which ACTH regulates steroid hormone biosynthesis in the adrenal cortex.

Acknowledgements

The authors gratefully acknowledge the support of USPHS grants AM28350 and HD13234.

References

ADESNIK, M., BAR-NUN, S., MASCHIO, F., ZUNICH, M., LIPPMAN, A. and BARD, E. (1981) Mechanism of induction of cytochrome P-450 by phenobarbital. *Journal of Biological Chemistry,* **256,** 10340–10345

ALFANO, J., BROWNIE, A. C., ORME-JOHNSON, W. H. and BEINERT, H. (1973) Adrenal mitochondrial cytochrome P-450 and cholesterol side-chain cleavage activity. *Journal of Biological Chemistry,* **248,** 7860–7864

ANDERSON, J. M. and DIETSCHY, J. M. (1977) Regulation of sterol synthesis in 15 tissues of rat. *Journal of Biological Chemistry,* **252,** 3652–3659

ARTHUR, J. R. and BOYD, G. S. (1976) The effect of inhibitors of protein synthesis on cholesterol side-chain cleavage in the mitochondria of luteinized rat ovaries. *European Journal of Biochemistry,* **49,** 117–127

ARTHUR, J. R., MASON, J. I. and BOYD, G. S. (1976) The effect of calcium ions on the metabolism of exogenous cholesterol by rat adrenal mitochondria. *FEBS Letters,* **66,** 206–209

ASANO, K. and HARDING, B. W. (1976) Biosynthesis of adrenodoxin in mouse adrenal tumor cells. *Endocrinology,* **99,** 977–987

BALASUBRAMANIAN, S., GOLDSTEIN, J. L., FAUST, J. R., BRUNSCHEDE, G. Y. and BROWN, M. S. (1977) Lipoprotein-mediated regulation of 3-hydroxy-3-methylglutaryl coenzyme A reductase activity and cholesteryl ester metabolism in the adrenal gland of the rat. *Journal of Biological Chemistry,* **252,** 1771–1779

BAR-NUN, S., KREIBACH, G., ADESNIK, M., ALTERMAN, L., NEGISHI, M. and SABATINI, D. D. (1980) Synthesis and insertion of cytochrome P-450 into endoplasmic reticulum membranes. *Proceedings of the National Academy of Sciences (USA),* **77,** 965–969

BECKETT, G. J. and BOYD, G. S. (1977) Purification and control of bovine adrenal cortical cholesterol ester hydrolase and evidence for the activation of the enzyme by a phosphorylation. *European Journal of Biochemistry,* **72,** 223–233

BOGGARAM, V., SIMPSON, E. R. and WATERMAN, M. R. (1984) Induction of synthesis of bovine adrenocortical cytochromes $P\text{-}450_{SCC}$, $P\text{-}450_{11\beta}$, $P\text{-}450_{C21}$ and adrenodoxin by prostaglandins E_2 and $F_{2\alpha}$ and cholera toxin. *Archives of Biochemistry and Biophysics,* **231,** 271–279

BRESNICK, E., BROSSEAU, M., LEVIN, W., REICK, L., RYAN, D. E. and THOMAS, P. E. (1981) Administration of 3-methycholanthrene to rats increases the specific hybridizable mRNA coding for cytochrome P-450c. *Proceedings of the National Academy of Sciences (USA),* **78,** 4083–4087

BROWNIE, A. C., ALFANO, J., JEFCOATE, C. R., ORME-JOHNSON, W. H., BEINERT, H. and SIMPSON, E. R. (1973) Effect of ACTH on adrenal mitochondrial P-450 in the rat. *Annals of the New York Academy of Sciences,* **212,** 344–360

BROWNIE, A. C., SIMPSON, E. R., JEFCOATE, C. R., BOYD, G. S., ORME-JOHNSON, W. H. and BEINERT, H. (1972) Effect of ACTH on cholesterol side-chain cleavage in rat adrenal mitochondria. *Biochemical and Biophysical Research Communications,* **46,** 483–490

BROWNIE, A. C., KRAMER, R. E. and GALLANT, S. (1979) The cholesterol side chain cleavage system of the rat adrenal cortex and its relationship to the circadian rhythm. *Endocrinology,* **104,** 1266–1269

BRYSON, M. J. and SWEAT, M. L. (1968) Cleavage of cholesterol side chain associated with cytochrome P-450, flavoprotein, and nonheme iron-protein derived from the bovine adrenal cortex. *Journal of Biological Chemistry,* **243,** 2799–2804

BUCKLEY, D. I. and RAMACHANDRAN, J. (1981) Characterization of corticotropin receptors on adrenocortical cells. *Proceedings of the National Academy of Sciences (USA),* **78,** 7431–7435

BUMPUS, J. A. and DUS, K. M. (1982) Bovine adrenocortical microsomal hemeproteins $P\text{-}450_{17\alpha}$ and $P\text{-}450_{C\text{-}21}$. *Journal of Biological Chemistry,* **257,** 12696–12704

BURSTEIN, S., DINH, T., CO, N., GUT, M., SCHLEYER, H., COOPER, D. Y. and ROSENTHAL, O. (1972) Kinetic studies on substrate–enzyme interaction in the adrenal cholesterol side-chain cleavage system. *Biochemistry,* **11,** 2883–2891

BURSTEIN, S., GUT, M. and HAN, G. (1975) Cholesterol kinetics and mechanism of the conversion of pregnenolone in bovine adrenal cortex. In *Proceedings of the 57th Annual Meeting of the Endocrine Society,* p. 56

BURSTEIN, S., MIDDLEDITCH, B. S. and GUT, M. (1974) Enzymatic formation of (20R,22R)-20,22-dihydroxy-cholesterol from cholesterol and a mixture of $^{16}O_2$ and $^{18}O_2$: random incorporation of oxygen atoms. *Biochemical and Biophysical Research Communications,* **61,** 642–647

CARR, B. R. and SIMPSON, E. R. (1981a) *De novo* synthesis of cholesterol by the human fetal adrenal gland. *Endocrinology,* **108,** 2154–2162

CARR, B. R. and SIMPSON, E. R. (1981b) Lipoprotein utilization and cholesterol synthesis by the human fetal adrenal gland. *Endocrine Reviews,* **2,** 306–326

CHANDERBHAN, R., NOLAND, B. J., SCALLEN, T. J. and VAHOUNY, G. V. (1982) Sterol carrier protein$_2$. *Journal of Biological Chemistry,* **257,** 8928–8934

CHEITLIN, R. and RAMACHANDRAN, J. (1981) Regulation of actin in rat adrenocortical cells by corticotropin. *Journal of Biological Chemistry,* **256,** 3156–3158

CHU, J. and KIMURA, T. (1973) Studies on adrenal steroid hydroxylases. *Journal of Biological Chemistry,* **248,** 2089–2094

COLBEAU, A., NACHBAUR, J. and VIGNAIS, P. M. (1971) Enzymic characterization and lipid composition of rat liver subcellular membranes. *Biochimica et Biophysica Acta,* **249,** 462–492

COLBERT, R. A., BRESNICK, E., LEVIN, W., RYAN, D. E. and THOMAS, P. E. (1979) Synthesis of liver cytochrome P-450b in a cell free protein synthesizing system. *Biochemical and Biophysical Research Communication,* **91,** 886–891

CORTESE, F. and WOLFF, J. (1978) Cytochalasin-stimulated steroidogenesis from high density lipoproteins. *Journal of Cell Biology,* **77,** 507–516

CRIVELLO, J. F. and JEFCOATE, C. R. (1978) Mechanism of corticotropin action in rat adrenal cells. *Biochimica et Biophysica Acta,* **542,** 315–329

CRIVELLO, J. F. and JEFCOATE, C. R. (1980) Intracellular movement of cholesterol in rat adrenal cells. *Journal of Biological Chemistry,* **55,** 8144–8151

DOERING, C. H. and CLAYTON, R. B. (1969) Cholesterol side-chain cleavage activity in the adrenal gland of the young rat: development and responsiveness to adrenocorticotropic hormone. *Endocrinology,* **85,** 500–511

DUBOIS, R. N., SIMPSON, E. R., KRAMER, R. E. and WATERMAN, M. R. (1981a) Induction of synthesis of cholesterol side chain cleavage cytochrome P-450 by ACTH in cultured bovine adrenocortical cells. *Journal of Biological Chemistry,* **256,** 7000–7005

DUBOIS, R. N., SIMPSON, E. R., TUCKEY, J., LAMBETH, J. D. and WATERMAN, M. R. (1981b) Evidence for a higher molecular weight precursor of cholesterol side chain cleavage cytochrome P-450 and induction of mitochondrial and cytosolic proteins by ACTH in adult bovine adrenal cells. *Proceedings of the National Academy of Sciences (USA),* **78,** 1028, 1032

DUBOIS, R. N.. and WATERMAN, M. R. (1979) Effect of phenobarbital administration to rats on the level of the *in vitro* synthesis of cytochrome P-450 directed by total rat liver RNA. *Biochemical and Biophysical Research Communications,* **90,** 150–157

ESTABROOK, R. W., COOPER, D. Y. and ROSENTHAL, O. (1963) The light-reversible carbon monoxide inhibition of the steroid C-21 hydroxylase system of the adrenal cortex. *Biochemische Zeitschrift,* **338,** 741–755

ESTABROOK, R. W., SUZUKI, K., MASON, J. I., BARON, J., TAYLOR, W. E., SIMPSON, E. R., PURVIS, J. and MCCARTHY, J. (1973) An iron-sulphur protein of adrenal cortex mitochondria. In *Iron-Sulfur Proteins,* edited by W. Lovenberg, pp. 193–223. New York: Academic Press

FARESE, R. V. (1967) Adrenocorticotrophin induced changes in the steroidogenic activity of adrenal cell-free preparations. *Biochemistry,* **6,** 2052–2065

FARESE, R. V. (1978) Further localization of ACTH effects on cholesterol side-chain cleavage (CSCC) in adrenal mitochondria supplemented with cholesterol-rich lipoproteins. *Proceedings of the 60th Annual Meeting of the Endocrine Society,* p. 222

FARESE, R. V. and PRUDENTE, W. J. (1977) A role for calcium in the stimulation of cholesterol side-chain cleavage by ACTH. In *Proceedings of the 59th Annual Meeting of the Endocrine Society,* p. 60

FARESE, R. V. and SABIR, M. A. (1979) Polyphosphorylated glycerolipids mimic adrenocorticotropin-induced stimulation of mitochondrial pregnenolone synthesis. *Biochimica et Biophysica Acta,* **575,** 299–304

FARESE, R. V. and SABIR, A. M. (1980) Polyphosphoinositides: stimulator of mitochondrial cholesterol side chain cleavage and possible identification as an adrenocorticotropin-induced, cycloheximide-sensitive, cytosolic, steroidogenic factor. *Endocrinology,* **106,** 1869–1879

FARESE, R. V., SABIR, A. M. and LARSON, R. E. (1980) On the mechanism whereby ACTH and cyclic AMP increase adrenal polyphosphoinositides. *Journal of Biological Chemistry,* **255,** 7232–7237

FARESE, R. V., SABIR, M. A. and LARSON, R. E. (1981a) Kinetic aspects of cycloheximide-induced reversal of adrenocorticotropin effects on steroidogenesis and phospholipid metabolism in rat adrenal sections *in vitro. Endocrinology,* **109,** 1424–1427

FARESE, R. V., SABIR, M. A. and LARSON, R. E. (1981b) Adrenocorticotropin and adenosine 3′,5′-monophosphate stimulate *de novo* synthesis of adrenal phosphatidic acid by a cycloheximide-sensitive, Ca^{++}-dependent mechanism. *Endocrinology,* **109,** 1895–1901

FARESE, R. V., SABIR, M. A. and LARSON, R. E. (1981c) Effects of adrenocorticotropin and cycloheximide on adrenal diglyceride kinase. *Biochemistry,* **20,** 6047–6051

FARESE, R. V., SABIR, A. M., VANDOR, S. L. and LARSON, R. E. (1980) Are polyphosphoinositides the cycloheximide-sensitive mediator in the steroidogenic actions of adrenocorticotropin and adenosine 3′,5′-monophosphate? *Journal of Biological Chemistry,* **255,** 5728–5734

FAUST, J. R., GOLDSTEIN, J. L. and BROWN, M. S. (1977) Receptor-mediated uptake of low density lipoprotein and utilization of its cholesterol for steroid synthesis in cultured mouse adrenal cells. *Journal of Biological Chemistry,* **252,** 4861–4871

FEVOLD, H. R. and BROWN, R. L. (1978) The apparent lack of stimulation of rabbit adrenal 21-hydroxylase activity by ACTH. *Journal of Steroid Biochemistry,* **9,** 583–584

FEVOLD, H. R., WILSON, P. L. and SLALINA, S. M. (1978) Stimulated rabbit adrenal 17α-hydroxylase kinetic properties and a comparison with those of 3β-hydroxysteroid dehydrogenase. *Journal of Steroid Biochemistry,* **9,** 1033–1041

FUNKENSTEIN, B., MCCARTHY, J. L., DUS, K. M., SIMPSON, E. R. and WATERMAN, M. R. (1983) Effect of adrenocorticotropin on steroid 21-hydroxylase synthesis and activity in cultured bovine adrenocortical cells. *Journal of Biological Chemistry,* **258,** 9398–9405

GILL, G. N. (1976) ACTH regulation of the adrenal cortex. *Pharmacology and Therapeutics,* **B2,** 313–339

GONZALEZ, F. J. and KASPER, C. B. (1980) Phenobarbital induction of NADPH-cytochrome C (P-450) oxidoreductase messenger ribonucleic acid. *Biochemistry,* **19,** 1970–1976

GOSPODAROWICZ, D., ILL, C. R., HORNSBY, P. J. and GILL, G. N. (1977) Control of bovine adrenal cortical cell proliferation by fibroblast growth factor. Lack of effect of epidermal growth factor. *Endocrinology,* **100,** 1080–1089

GREENFIELD, N., PONTICORVO, L., CHASALOW, F. and LIEBERMAN, S. (1980) Activation and inhibition of the adrenal steroid 21-hydroxylation system by cytosolic constituents: influence of glutathione, glutathione reductase, and ascorbate. *Archives of Biochemistry and Biophysics,* **200,** 232–244

GREENGARD, P., PSYCHOYOS, S., TALLAN, H. H., COOPER, D. Y., ROSENTHAL, O. and ESTABROOK, R. W. (1967) Aldosterone synthesis by adrenal mitochondria. III. Participation of cytochrome P-450. *Archives of Biochemistry and Biophysics,* **121,** 298–303

GWYNNE, J. T., MAHAFFEE, D., BREWER, H. G. JR. and NEY, R. L. (1976) Adrenal cholesterol uptake from plasma lipoproteins: regulation by corticotropin. *Proceedings of the National Academy of Sciences (USA),* **73,** 4329–4333

HALL, P. F., OSAWA, S. and THOMASSON, C. L. (1981) A role for calmodulin in the regulation of steroidogenesis. *Journal of Cell Biology,* **90,** 402–407

HANUKOGLU, I. and JEFCOATE, C. R. (1979) Mitochondrial cytochrome P-450_{SCC}. *Journal of Biological Chemistry,* **25,** 3057–3061

HANUKOGLU, I., SPITZBERG, V., BUMPUS, J. A., DUS, K. M. and JEFCOATE, C. R. (1980) Adrenal mitochondrial cytochrome P-450_{SCC}. *Journal of Biological Chemistry,* **256,** 4321–4328

HUME, R. and BOYD, G. S. (1978) Cholesterol metabolism and steroid-hormone production. *Biochemical Society Transactions,* **6,** 893–898

JEFCOATE, C. R., HUME, R. and BOYD, G. S. (1970) Separation of two forms of cytochrome P-450 adrenal cortex mitochondria. *FEBS Letters,* **9,** 41–44

JEFCOATE, C. R. and ORME-JOHNSON, W. H. (1975) Cytochrome P-450 of adrenal cortex. *Journal of Biological Chemistry,* **250,** 4671–4677

JEFCOATE, C. R., SIMPSON, E. R., BOYD, G. S., BROWNIE, C. and ORME-JOHNSON, W. H. (1973) The detection of different states of the P-450 cytochromes in adrenal mitochondria – changes induced by ACTH. *Annals of the New York Academy of Sciences,* **212,** 243–261

JEFCOATE, C. R., SIMPSON, E. R. and BOYD, G. S. (1974) Spectral properties of rat adrenal-mitochondrial cytochrome P-450. *European Journal of Biochemistry,* **42,** 539–551

JOHNSON, L. R., RUHMANN-WEINHOLD, A. and NELSON, D. H. (1973) The *in vivo* effect of ACTH on utilization of reducing energy for pregnenolone synthesis in adrenal mitochondria. *Annals of the New York Academy of Sciences,* **212,** 307–318

KAN, K. W. and UNGAR, F. (1973) Characterization of an adrenal activator of cholesterol side-chain cleavage. *Journal of Biological Chemistry,* **248,** 2868–2875

KASHIWAGI, K., DAFELDECKER. W. P. and SALHANICK, A. A. (1980) Purification and characterization of mitochondrial cytochrome P-450 associated with cholesterol side chain cleavage from bovine corpus luteum. *Journal of Biological Chemistry,* **255,** 2606–2611

KIDO, T. and KIMURA, T. (1981) Stimulation of cholesterol binding to steroid-free cytochrome P-450_{SCC} by poly-L-lysine. *Journal of Biological Chemistry,* **256,** 8561–8568

KIMURA, T. (1969) Effects of hypophysectomy and ACTH administration on the level of adrenal cholesterol side-chain desmolase. *Endocrinology,* **85,** 492–499

KIMURA, T. and SUZUKI, K. (1967) Components of the electron transport system in adrenal steroid hydroxylase. *Journal of Biological Chemistry,* **242,** 485–491

KIRBY, C. J. and GREEN, C. (1977) Transmembrane migration ('flip-flop') of cholesterol in erythrocyte membranes. *Biochemical Journal,* **168,** 575–577

KOMINAMI, S., OCHI, H., KOBAYASHI, Y. and TAKEMORI, S. (1980) Studies on the steroid hydroxylation system in adrenal cortex microsomes. *Journal of Biological Chemistry,* **255,** 3386–3394

KOMINAMI, S., SHINZAWA, K. and TAKEMORI, S. (1982) Purification and some properties of cytochrome P-450 specific for steroid 17α-hydroxylation and C_{17}-C_{20} bond cleavage from guinea pig adrenal microsomes. *Biochemical and Biophysical Research Communications,* **109,** 916–921

KORITZ, S. B. and KUMAR, A. M. (1970) On the mechanism of action of the adrenocorticotrophic hormone. *Journal of Biological Chemistry,* **245,** 152–159

KORITZ, S. B. and MOUSTAFA, A. M. (1976) Some characteristics of adrenal steroidogenesis and their possible relationships to the action of the adrenocorticotropic hormone. *Archives of Biochemistry and Biophysics,* **174,** 20–26

KOWAL, J. (1969) Adrenal cells in tissue culture. III. Effect of adrenocorticotropin and 3′,5′-cyclic adenosine monophosphate on 11β-hydroxylase and other steroidogenic enzymes. *Biochemistry,* **8,** 1821–1831

KOWAL, J., SIMPSON, E. R. and ESTABROOK, R. W. (1970) Adrenal cells in tissue culture: on the specificity of the stimulation of 11β-hydroxylation by adrenocorticotropin. *Journal of Biological Chemistry,* **245,** 2438–2443

KRAICER, J. and MILLIGAN, J. V. (1971) Effect of colchicine on *in vitro* ACTH release induced by high K^+ and by hypothalamus-stalk, median-eminence extract. *Endocrinology,* **89,** 408–412

KRAMER, R. E., ANDERSON, C. M., PETERSON, J. A., SIMPSON, E. R. and WATERMAN, M. R. (1982a) Adrenodoxin biosynthesis by bovine adrenal cells in monolayer culture: induction by ACTH. *Journal of Biological Chemistry,* **257,** 14921–14925

KRAMER, R. E., ANDERSON, C. M., MCCARTHY, J. L., SIMPSON, E. R. and WATERMAN, M. R. (1982b) Coordinate induction of synthesis of adrenal mitochondrial steroid hydroxylases by ACTH. *Federation Proceedings,* **41,** 1298

KRAMER, R. E., DUBOIS, R. N., SIMPSON, E. R., ANDERSON, C. M., KASHIWAGI, K., LAMBETH, J. D., JEFCOATE, C. R. and WATERMAN, M. R. (1982c) Cell-free synthesis of precursor forms of mitochondrial steroid hydroxylase enzymes of the bovine adrenal cortex. *Archives of Biochemistry and Biophysics,* **215,** 478–485

KRAMER, R. E., MCCARTHY, J. L., SIMPSON, E. R. and WATERMAN, M. R. (1983) Effects of ACTH on steroidogenesis in bovine adrenocortical cells in primary culture: increased secretion of 17α-hydroxylated steroids associated with a refractoriness in total steroid output. *Journal of Steroid Biochemistry,* **18,** 715–723

KRAMER, R. E., RAINEY, W. E., FUNKENSTEIN, B., DEE, A., SIMPSON, E. R. and WATERMAN, M. R. (1984) Induction of synthesis of mitochondrial steroidogenic enzymes of bovine adrenocortical cells by analogs of cyclic AMP. *Journal of Biological Chemistry,* **259,** 707–713

KRAMER, R. E., SIMPSON, E. R. and WATERMAN, M. R. (1983) Induction of 11β-hydroxylase by corticotropin in primary cultures of bovine adrenocortical cells. *Journal of Biological Chemistry,* **258,** 3000–3005

KRUEGER, R. J. and ORME-JOHNSON, N. R. (1983) Acute adrenocorticotropic hormone stimulation of adrenal corticosteroidogenesis. *Journal of Biological Chemistry,* **258,** 10159–10167

LAMBETH, J. D. (1981) Cytochrome $P\text{-}450_{SCC}$. *Journal of Biological Chemistry,* **256,** 4757–4762

LAMBETH, J. D., KAMIN, H. and SEYBERT, D. W. (1980) Phosphatidylcholine vesicle reconstituted cytochrome $P\text{-}450_{SCC}$. *Journal of Biological Chemistry,* **255,** 8282–8288

LAMBETH, J. D., SEYBERT, D. W. and KAMIN, H. (1979) Phospholipid vesicle-reconstituted cytochrome $P\text{-}450_{SCC}$. *Journal of Biological Chemistry,* **255,** 138–143

LEFEVRE, A., MORERA, A.-M. and SAEZ, J. M. (1978) Adrenal cholesterol-binding protein: properties and partial purification. *FEBS Letters,* **89,** 287–292

MAHAFEE, D., REITZ, R. C. and NEY, R. L. (1974) The mechanism of action of adrenocorticotropic hormone. *Journal of Biological Chemistry,* **249,** 227–233

MCCARTHY, J. L., KRAMER, R. E., FUNKENSTEIN, B., SIMPSON, E. R. and WATERMAN, M. R. (1983) Induction of 17α-hydroxylase (cytochrome $P\text{-}450_{17\alpha}$) activity by adrenocorticotropin in bovine adrenocortical cells maintained in monolayer culture. *Archives of Biochemistry and Biophysics,* **222,** 590–598

MITANI, F. (1979) Cytochrome P-450 in adrenocortical mitochondria. *Molecular and Cellular Biochemistry,* **24,** 21–42

MITANI, F., SHIMIZU, T., UENO, R., ISHIMURA, Y., IZUMI, S., KOMATSU, N. and WATANABE, K. (1982) Cytochrome $P\text{-}450_{11\beta}$ and $P\text{-}450_{SCC}$ in adrenal cortex. *Journal of Histochemistry and Cytochemistry,* **20,** 1066–1074

MROTEK, J. J. and HALL, P. F. (1977) Response of adrenal tumor cells to adrenocorticotropin: site of inhibition by cytochalasin B. *Biochemistry,* **16,** 3177–3181

NABI, N. and OMURA, T. (1980) *In vitro* synthesis of adrenodoxin and adrenodoxin reductase: existence of a putative large precursor form of adrenodoxin. *Biochemical and Biophysical Research Communications,* **97,** 680–687

NABI, N., KOMINAMI, S., TAKEMORI, S. and OMURA, T. (1980) *In vitro* synthesis of mitochondrial cytochrome P-450 (scc) and P-450 (11β) and microsomal cytochrome P-450 (C-21) by both free and bound polysomes isolated from bovine adrenal cortex. *Biochemical and Biophysical Research Communications,* **97,** 687–693

NAGHSHIMAH, S., TREADWELL, C. R., GALLO, L. L. and VAHOUNY, G. V. (1978) Protein kinase-mediated phosphorylation of a purified sterol ester hydrolase from bovine adrenal cortex. *Journal of Lipid Research,* **19,** 561–569

NAKAJIN, S. and HALL, P. F. (1981) Microsomal cytochrome P-450 from neonatal pig testis. *Journal of Biological Chemistry,* **256,** 3871–3876

NAKAJIN, S., HALL, P. F. and ONADA, M. (1981) Testicular microsomal cytochrome P-450 for C_{21}-steroid side chain cleavage. *Journal of Biological Chemistry,* **256,** 6134–6139

NAKAJIN, S., SHINODA, M. and HALL, P. F. (1983) Purification and properties of 17α-hydroxylase from microsomes of pig adrenal: a second C_{21} side-chain cleavage system. *Biochemical and Biophysical Research Communications,* **111,** 512–517

NAKAMURA, Y., OTSUKA, H. and TAMAOKI, B. (1969) Requirement of a new flavoprotein and a non-heme iron containing protein in the steroid 11β and 18-hydroxylase system. *Biochimica et Biophysica Acta,* **122,** 34–42

NEHER, R., MILANI, A., SOLANO, A. R. and PODESTA, E. J. (1982) Compartmentalization of corticotropin-dependent steroidogenic factors in adrenal cortex: evidence for a post-translational stimulation of the cholesterol side-chain split. *Proceedings of the National Academy of Sciences (USA),* **79,** 1727–1731

OHASHI, M., SIMPSON, E. R., KRAMER, R. E. and CARR, B. R. (1982) Regulation of low-density lipoprotein receptors in cultured bovine adrenocortical cells. *Archives of Biochemistry and Biophysics,* **215,** 199–205

OMURA, T. (1979) Biosynthesis and drug-induced increase of microsomal enzymes. In *The Induction of Drug Metabolism,* edited by R. W. Estabrook and E. Lindenlaub, pp. 161–176. New York: F. K. Shaltauer Verlag

OMURA, T., ITO, A., OKADA, Y., SAGARA, Y. and ONO, H. (1983) Proteolytic processing of enzyme precursors by liver and adrenal cortex mitochondria.. In *Proteinase Inhibitors: Medical and Biological Aspects,* edited by N. Katanuma, pp. 307–312. Tokyo: Japan Science Society Press

OMURA, T., SANDERS, E., ESTABROOK, R. W., COOPER, D. Y. and ROSENTHAL, O. (1966) Isolation from adrenal cortex of a nonheme iron protein and a flavoprotein functional as a triphosphopyridine nucleotide-cytochrome P-450 reductase. *Archives of Biochemistry and Biophysics,* **117,** 660–673

PAUL, D. P., GALLANT, S., ORME-JOHNSON, N. R., ORME-JOHNSON, W. H. and BROWNIE, A. C. (1976) Temperature dependence of cholesterol binding to cytochrome $P\text{-}450_{SCC}$ of the rat adrenal. *Journal of Biological Chemistry,* **251,** 7120–7126

PEDERSON, R. C. and BROWNIE, A. C. (1980) Adrenocortical response to corticotropin is potentiated by part of the amino-terminal region of pro-corticotropin/endorphin. *Proceedings of the National Academy of Sciences (USA),* **77,** 2239–2243

PEDERSON, R. C. and BROWNIE, A. C. (1983) Cholesterol side-chain cleavage in the rat adrenal cortex: isolation of a cycloheximide-sensitive activator peptide. *Proceedings of the National Academy of Sciences (USA),* **80,** 1882–1886

PEDERSON, R. C., BROWNIE, A. C. and PING, N. (1980) Pro-adrenocorticotropin/endorphin-derived peptides: coordinate action on adrenal steroidogenesis. *Science,* **208,** 1044–1046

PERON, F. G. and MCCARTHY, J. L. (1968) Corticosteroidogenesis in the rat adrenal gland. In *Functions of the Adrenal Cortex,* Vol. **1,** edited by K. McKerns, pp. 261–322. New York: Appleton-Century-Crofts

PFEIFFER, D. R., CHU, J. W., KUO, T. H., CHAN, S. W., KIMURA, T. and TCHEN, T. T. (1972) Changes in some biochemical parameters including cytochrome P-450 after hypophysectomy and their restoration by ACTH administration in rats four months post hypophysectomy. *Biochemical and Biophysical Research Communications,* **48,** 486–490

PONTICORVO, L., GREENFIELD, N., WOLFSON, A., CHASALOW, F. and LIEBERMAN, S. (1980) The nature of the cytosolic activators of the adrenal steroid 21-hydroxylation system. *Archives of Biochemistry and Biophysics,* **200,** 223–231

POZNANSKY, M. J. and LANGE, Y. (1978) Transbilayer movement of cholesterol in phospholipid vesicles under equilibrium and non-equilibrium conditions. *Biochimica et Biophysica Acta,* **506,** 256–264

PRIVALLE, C. T., CRIVELLO, J. F. and JEFCOATE, C. R. (1983) Regulation of intramitochondrial cholesterol transfer to side-chain cleavage cytochrome P-450 in rat adrenal gland. *Proceedings of the National Academy of Sciences (USA),* **80,** 702–706

PURVIS, J. L., CANICK, J. A., MASON, J. I., ESTABROOK, R. W. and MCCARTHY, J. L. (1973) Lifetime of adrenal cytochrome P-450 as influenced by ACTH. *Annals of the New York Academy of Sciences,* **212,** 319–342

RAY, P. and STROTT, C. A. (1981) Cytosol stimulation of pregnenolone synthesis by isolated adrenal mitochondria. *Life Sciences,* **28,** 1529–1534

ROSENTHAL, O. and COOPER, D. Y. (1967) Methods of determining the photochemical action spectrum. *Methods in Enzymology,* **10,** 616–629

RYBAK, S. M. and RAMACHANDRAN, J. (1982) Mechanism of induction of Δ^5-3β-hydroxysteroid dehydrogenase-isomerase activity in rat adrenocortical cells by corticotropin. *Endocrinology,* **111,** 427–433

SALA, G. B., HAYASHI, K., CATT, K. J. and DUFAU, M. L. (1979) Adrenocorticotropin action in isolated adrenal cells. *Journal of Biological Chemistry,* **254,** 3861–3865

SARTRE, M., VIGNAIS, P. V. and IDELMAN, S. (1969) Distribution of the steroid 11β-hydroxylase and the cytochrome P-450 in membranes of beef adrenal cortex mitochondria. *FEBS Letters,* **5,** 135–140

SCHULSTER, D. (1976) Control of glucocorticoid synthesis by ACTH. In *Molecular Endocrinology of the Steroid Hormones,* edited by D. Schulster, S. Burstein and B. Cooke, pp. 167–207. New York: Wiley

SEYBERT, D. W. (1983) Lipid modulation of cytochrome P-450 activity. *Federation Proceedings,* **42,** 2065

SEYBERT, D. W., LANCASTER, J. R. JNR, LAMBETH, J. D. and KAMIN, H. (1979) Participation of the membrane in the side chain cleavage of cholesterol. *Journal of Biological Chemistry,* **254,** 12088–12098

SHIKITA, M. and HALL, P. F. (1974) The stoichiometry of the conversion of cholesterol and hydroxycholesterols to pregnenolone (3β-hydroxypregn-5-en-20-one) catalysed by adrenal cytochrome P-450. *Proceedings of the National Academy of Sciences (USA),* **71,** 1441–1445

SIMPSON, E. R. and BOYD, G. S. (1967a) The cholesterol side-chain cleavage system of bovine adrenal cortex. *European Journal of Biochemistry,* **2,** 275–285

SIMPSON, E. R. and BOYD, G. S. (1967b) Partial resolution of the mixed-function oxidase involved in the cholesterol side-chain cleavage reaction of bovine adrenal mitochondria. *Biochemical and Biophysical Research Communications,* **28,** 945–950

SIMPSON, E. R. and WATERMAN, M. R. (1983) Regulation by ACTH of steroid hormone bio-synthesis in the adrenal cortex. *Canadian Journal of Biochemistry and Cell Biology,* **61,** 692–707

SIMPSON, E. R., JEFCOATE, C. R., BROWNIE, A. C. and BOYD, G. S. (1972) The effect of ether anaesthesia stress on cholesterol side-chain cleavage and cytochrome P-450 in rat-adrenal mitochondria. *European Journal of Biochemistry,* **28,** 442–450

SIMPSON, E. R., MCCARTHY, J. L. and PETERSON, J. A. (1978) Evidence that the cycloheximide-sensitive site of adrenocorticotropic hormone action is in the mitochondrion. *Journal of Biological Chemistry,* **253,** 3135–3139

SIMPSON, E. R., WATERS, J. and WILLIAMS-SMITH, D. (1975) Effect of calcium on pregnenolone formation and cytochrome P-450 in rat adrenal mitochondria. *Journal of Steroid Biochemistry,* **6,** 395–400

SIMPSON, E. R. and WILLIAMS-SMITH, D. (1975) Effect of Ca^{++} uptake on pregnenolone formation and spectral properties of cytochrome P-450. *Biochimica et Biophysica Acta,* **404,** 309–320

SIMPSON, E. R. and WILLIAMS-SMITH, D. L. (1976). Electron paramagnetic resonance spectra of mitochondrial and microsomal cytochrome P-450 from the rat adrenal. *Biochimica et Biophysica Acta,* **449,** 59–71

STONE, D. and HECHTER, O. (1954) Studies on ACTH action in perfused bovine adrenals: the site of action of ACTH in corticosteroidogenesis. *Archives of Biochemistry and Biophysics,* **51,** 457–469

TAKEMOTO, C., NAKANO, H., SATO, H. and TAMAOKI, B. (1968) Fate of molecular oxygen required by endocrine enzymes for the side-chain cleavage of cholesterol. *Biochimica et Biophysica Acta,* **152,** 749–757

TAKEMORI, S., SUHARA, K., HASHIMOTO, S., HASHIMOTO, M., SATO, H., GOMI, T. and KATAGIRI, M. (1975a) Purification of cytochrome P-450 from bovine adrenocortical mitochondrial by an aniline-sepharose and the properties. *Biochemical and Biophysical Research Communications,* **63,** 588–593

TAKEMORI, S., SATO, H., GOMI, T., SUHARA, K. and KATAGIRI, M. (1975b) Purification of cytochrome $P\text{-}450_{11\beta}$ from adrenocortical mitochondria. *Biochemical and Biophysical Research Communications,* **67,** 1151–1157

TILLEY, B. E., WATANUKI, M. and HALL, P. F. (1977) Preparation and properties of side-chain cleavage cytochrome P-450 from bovine adrenal cortex by affinity chromatography with pregnenolone as ligand. *Biochimica et Biophysica Acta,* **493,** 260–271

TRZECIAK, W. H. and BOYD, G. S. (1973) The effect of stress induced by ether anaesthesia on cholesterol content and cholesteryl-esterase activity in rat-adrenal cortex. *European Journal of Biochemistry,* **37,** 327–333

TUCKEY, R. H., NEBERT, D. W. and NEGISHI, M.. (1981) Structural gene product of the [Ah] complex. Journal of Biological Chemistry, **256,** 6969–6974

VAHOUNY, G. V., CHANDERBHAN, R., NOLAND, B. J., IRWIN, D., DENNIS, P., LAMBETH, J. D. and SCALLEN, T. J. (1983) Sterol carrier protein. *Journal of Biological Chemistry,* **258,** 11731–11737

VAN DER WUSSE, G. J., KALKMAN, M. L., VAN WINSEN, M. P. T. and VAN DER MOLEN, J. J. (1975) Cholesterol exchange between microsomal, mitochondrial and erythrocyte membranes and its enhancement by cytosol. *Biochimica et Biophysica Acta,* **398,** 18–38

VAN DER WUSSE, G. S., KALKMAN, M. L., VAN WINSEN, M. P. I. and VAN DER MOLEN, H. J. (1976) Effect of Ca^{2+} ruthenium red and ageing on pregnenolone production by mitochondrial fractions from normal and luteinizing hormone treated rat testes. *Biochimica et Biophysica Acta,* **428,** 420–431

WANG, H.-P. and KIMURA, T. (1976) Purification and characterization of adrenal cortex mitochondrial cytochrome P-450 specific for cholesterol side chain cleavage activity. *Journal of Biological Chemistry,* **251,** 6068–6074

WARBURG, O. (1949) *Heavy Metal Prosthetic Groups and Enzyme Action.* Oxford: Clarendon

WATERMAN, M. R. (1982) ACTH-mediated induction of synthesis and activity of cytochrome P-450s and related enzymes in cultured bovine adrenocortical cells. *Xenobiotica,* **12,** 773–786

WILLIAMS-SMITH, D. L., SIMPSON, E. R., BARLOW, S. M. and MORRISON, P. J. (1976) Electron paramagnetic resonance studies of cytochrome P-450 and adrenal ferredoxin in single whole rat adrenal glands. Effect of corticotrophin. *Biochimica et Biophysica Acta,* **449,** 72–83

WILSON, L. O. and HARDING, B. W. (1970) Studies on adrenal cortical cytochrome P-450. III. Effect of carbon monoxide and light on steroid 11β-hydroxylation. *Biochemistry,* **9,** 1615–1620

YAGO, N. KOBAYASHI, S., SEKIYAMA, S., KUROKAWA, H., IWAI, V., SUZUKI, I. and ICHII, S. (1970) Further studies on the submitochondrial localization of cholesterol side-chain cleaving enzyme system in hog adrenal cortex by sonic treatment. *Journal of Biochemistry,* **68,** 775–783

4

On mineralocorticoid and glucocorticoid receptors

John W. Funder

INTRODUCTION

Over the past five years there have been a number of excellent reviews of glucocorticoid and mineralocorticoid hormone action (Baxter and Rousseau, 1979; Marver and Lombard, 1981; Marver, 1984). This chapter is intended as an essay, and in no sense meant to be a comparable, comprehensive review. It sets out to give an overview, rather than to detail steroid effects on the cells, tissues and organs of the various species studied; its focus is whole-body, rather than subcellular. It is a personal viewpoint and thus inevitably selective; it is discursive and on occasions speculative. The point of the essay is very much to stimulate questions, not to answer them.

Although hundreds of different steroids are produced in the body, there are currently only six recognized major classes of mammalian steroid hormones – mineralocorticoid, glucocorticoid, androgenic, oestrogenic, progestational, and the cholecalciferol (vitamin D) derivatives. These classes were originally defined in terms of their observed effects *in vivo*, for example alterations in urinary sodium and potassium ion concentrations, hepatic glycogen deposition and comb growth; subsequently this classification has been buttressed by the demonstration of high-affinity intracellular binding sites (receptors) for each of these classes of steroid, and characteristic 'domains' of action in target tissue cells at a biochemical level.

Inevitably, with our expanding appreciation of hormone action at the subcellular level, these definitions have been narrowed down from the whole animal to the receptor/effector level. Although this may be a comfortably reductionist approach to definition, there are several aspects of 'definition by receptor' that should not be ignored. Firstly, in terms of specificity, steroid receptors can be divided into two distinct classes. Receptors for oestradiol and 1,25-dihydroxycholecalciferol have very high affinity ($K_d \leqslant 0.1$ nmol/l) for their physiological ligands but bind other steroids, particularly those of other effector classes, less well, by several orders of magnitude. Even in these relatively stringent signal recognition systems, specificity is not absolute; for example, the androgen 5-androstene-3β-17β-diol has a relatively high affinity ($\simeq 2$ nmol/l) for oestrogen receptors, and may occupy them under certain circumstances *in vivo* and produce oestrogenic effects (Seymour-Munn and Adams, 1983).

Parenthetically, at a prereceptor level, the classification of androgens and oestrogens becomes even more complex. For example, except in ovulating women and pregnancy, the majority of circulating oestrogen derives from peripheral aromatization of adrenal androgens. In the rat, where α-fetoprotein (FP) binds oestradiol avidly, neonatal masculinization in brain and liver appears to be the result of circulating androgen (not bound to α-FP) undergoing aromatization in target tissue cells and acting via oestrogen receptors in the same cells (McEwen, 1976).

In contrast to receptors for oestradiol and 1,25-dihydroxycholecalciferol, those for the other four recognized classes of mammalian steroids are almost disquietingly non-specific. Historically, mineralocorticoid (type I) receptors were thought to have a 10 times higher affinity ($K_d \simeq 0.5$ nmol/l) for aldosterone than for corticosterone, the physiological glucocorticoid in the rat. More recent studies on isolated cells (Lan *et al.,* 1981; Meyer and Nichols, 1981) and on preparations free of plasma protein (Krozowski and Funder, 1983) have shown that such receptors have equivalent affinity for aldosterone and corticosterone, i.e. no intrinsic specificity for one or other class of adrenal steroid. Classical glucocorticoid (type II) receptors have a very modest affinity (20–50 nmol/l) for physiological glucocorticoids, and almost equivalent affinity for progesterone, deoxycorticosterone and aldosterone; of a variety of corticosteroids tested, only the potent synthetic steroids dexamethasone and triamcinolone have higher affinity for 'glucocorticoid' receptors than for 'mineralocorticoid' receptors (Lan *et al.,* 1982).

Certain steroids, e.g. progesterone and deoxycorticosterone, have relatively high affinity for all four classes of steroid receptor; the affinity of progesterone for 'mineralocorticoid' receptors, for example, is higher than that for 'progesterone' receptors in progesterone target tissues, such as the breast or uterus (Moguilewsky and Raynaud, 1980). The first caution against too simple a receptor-based steroid classification is therefore clear, namely that for at least four of the classes of steroid hormone the receptors are relatively non-specific; and that, for example, progesterone and cortisol/corticosterone bind with higher affinity to 'mineralocorticoid' receptors than to their 'own' receptors.

The second problem in the transition from a physiological to a receptor-based classification of steroids occurs when definition is based on a synthetic steroid used as a radiolabelled probe in receptor studies. This has been nowhere as marked as in the case of glucocorticoids, where tritiated dexamethasone has been very extensively used as a receptor probe and has *de facto* become the benchmark glucocorticoid – what dexamethasone does in any given system is *ipso facto* considered the glucocorticoid effect. The historical reasons for the use of dexamethasone were that it mimicked the effects of naturally occurring glucocorticoids in the bioassay systems examined, that it bound to a readily identifiable class of intracellular sites with higher affinity than cortisol or corticosterone, and that, unlike cortisol and corticosterone, it had very low affinity for plasma transcortin, which contaminates many receptor preparations and thus potentially confounds studies using cortisol or corticosterone as probes.

Although the use of dexamethasone as a radiolabelled probe has allowed extensive characterization of the receptor to which it binds, and of the intracellular events following receptor binding, it has artificially narrowed the focus of glucocorticoid physiology by diverting attention from other intracellular steroid receptors which are predominantly occupied by cortisol/corticosterone *in vivo* but have a low affinity for dexamethasone. If the predominant ligand for a particular receptor is the naturally occurring glucocorticoid, it would appear not unreasonable to class such receptors as glucocorticoid receptors and to explore physiological effects of glucocorticoid occupancy of such sites. Much of the following discussion centres on this fact, i.e. the existence of two distinct classes of glucocorticoid receptor; in this physiological context the use of the synthetic steroid dexamethasone as a benchmark glucocorticoid is clearly both inappropriate and potentially misleading.

TWO CLASSES OF GLUCOCORTICOID RECEPTORS

In the mid-1970s, Bruce McEwen and his colleagues in New York showed that adrenalectomized rats bound very much more injected tritiated corticosterone in the nuclei of cells from the septum and hippocampus than rats injected with equivalent doses of tritiated dexamethasone; rats injected with tritiated progesterone or tritiated deoxycorticosterone showed no nuclear uptake in brain areas (McEwen, De Kloet and Wallach, 1976). These 'corticosterone-preferring' sites were subsequently shown to be neuronal, in contradistinction to the glial dexamethasone-binding sites; their potential receptor role was suggested by studies in which administration of corticosterone, but not dexamethasone, to adrenalectomized rats increased levels of protein I (synapsin), a phosphoprotein of a relative molecular mass (M_r) of 80 000 (Nestler *et al.,* 1982).

Moguilewsky and Raynaud, (1980) subsequently reported studies on the brain binding of tritiated aldosterone in the presence of RU26988, a potent synthetic glucocorticoid with negligible affinity for renal mineralocorticoid receptors (Teutsch *et al.,* 1981). Since aldosterone has a moderate affinity for dexamethasone-binding glucocorticoid receptors (Funder, Feldman and Edelman, 1973b), the use of concurrent excess RU26988 converts tritiated aldosterone from a bivalent probe to one which uniquely labels mineralocorticoid receptors in studies on the kidney. Moguilewsky and Raynaud showed that, of all the brain regions studied, the hippocampus had the highest concentration of binding sites for tritiated aldosterone in the presence of excess RU26988. On the basis of competition studies, such sites had a higher affinity for the synthetic steroid 9α-fluorocortisol than for aldosterone; affinity for progesterone was equivalent to that for aldosterone, and that for dexamethasone was much lower. Because of the high affinity for aldosterone and 9α-fluorocortisol, and low affinity for dexamethasone, these authors classified the hippocampal sites as mineralocorticoid receptors. No studies using corticosterone as a competitor for tritiated aldosterone binding were reported.

These two lines of investigation have been reconciled and extended by recent work from the author's laboratory in which the hippocampal corticosterone-preferring binding site in the rat was shown to have the same intrinsic affinity for a range of natural and synthetic steroids as renal mineralocorticoid receptors (Krozowski and Funder, 1983). In both brain and kidney, the sites were shown to have high ($K_d \simeq 0.5$ nmol/l) and equal intrinsic affinity for corticosterone and aldosterone, and much lower affinity for dexamethasone (10% or less than that for the two endogenous corticoids). Two important corollaries follow this equivalence of affinity, considering the normally very much higher circulating free levels of corticosterone than aldosterone. Firstly, to be occupied by aldosterone rather than corticosterone, extrinsic specificity-conferring mechanisms must exist in aldosterone target tissues. Secondly, in the absence of such mechanisms, these sites will be predominantly occupied by corticosterone, and thus would be glucocorticoid rather than mineralocorticoid receptors.

MINERALOCORTICOID MECHANISMS

Cytosols prepared from very carefully perfused kidneys, with levels of haemoglobin below detection limits, retain very high levels of a binding species that behaves indistinguishably from transcortin (CBG). Using such cytosols, we have shown that in the presence of excess RU26988 tritiated aldosterone binds to a single class of sites, with an affinity characteristic of classical renal mineralocorticoid receptors, and an approximately 10-fold lower apparent affinity for corticosterone. If these receptors are separated from the transcortin-like corticosterone binding species by adsorption onto hydroxylapatite, washing and elution, their affinity for corticosterone increases approximately 10-fold, to equal that for aldosterone.

It would thus appear that renal cytosols contain large amounts of a transcortin-like binder with high affinity for corticosterone but not aldosterone. This transcortin or transcortin-like binder sequesters corticosterone *in vitro,* reducing its apparent affinity for renal tritiated aldosterone binding sites. In the absence of detectable levels of haemoglobin, the source of such transcortin-like activity is presumed to be extravasated transcortin. This extravasated transcortin appears to be very prominent in the inner medulla-papillary zone of the kidney, where interstitial concentrations four to eight times higher than in plasma have been demonstrated, with the concentration increasing towards the papillary tip (Stephenson, Krozowski and Funder, 1984).

The combination *in vivo* of the recurrent vasculature in this region and very high levels of interstitial transcortin allows the establishment of a passive, reverse counter-current mechanism. As blood flows down the descending limb of the vasa recta, plasma (and interstitial and intracellular) free corticosterone is progressively sequestered by the high interstitial levels of the corticosterone binder; as blood flows up the ascending limb, the interstitial bound corticosterone is progressively discharged back into the plasma. By contrast, the plasma concentrations of aldosterone are not reduced by sequestration; relative to corticosterone, therefore, the concentrations are locally increased so that aldosterone may preferentially occupy inherently non-specific type I receptors, for example in the medullary collecting tubule.

TYPE I GLUCOCORTICOID RECEPTORS

The second corollary of the equivalent intrinsic affinity of hippocampal corticosterone-preferring sites and renal type I receptors is that in the absence of such local specificity-conferring mechanisms, such sites will be occupied predominantly by corticosterone. In hippocampal cytosols, even from non-perfused rats, levels of transcortin-like binding sites are very low; in such cytosols, aldosterone and corticosterone (and deoxycorticosterone and 9α-fluorocortisol) show equivalent, very high affinity for these sites, whether the tritiated tracer is aldosterone, corticosterone or deoxycorticosterone (Krozowski and Funder, 1983). Under normal physiological circumstances, therefore, the predominant occupant of hippocampal type I receptors – (equivalent to the 'corticosterone-preferring sites' of McEwen (1976) or the 'mineralocorticoid receptors' of Moguilewsky and Raynaud (1980) – would be corticosterone, since its circulating free plasma level is much higher than that of aldosterone. If the principal occupant of such sites is corticosterone (in the rat), if such sites occupied by cortisterone translocate to the cell nuclei in the hippocampus, and if the administration of corticosterone (but not of equal doses of dexamethasone) is followed by demonstrable effects on synapsin, then these corticosterone-preferring (McEwen) or type I (Krozowski and Funder) hippocampal intracellular binding sites appear to fulfil the criteria appropriate for a glucocorticoid receptor.

Such sites have a modest affinity for dexamethasone, not dissimilar to that of dexamethasone for classical type II glucocorticoid receptors. The outstanding difference between type I and type II receptors lies in their relative affinity for the physiological glucocorticoids, for which type I receptors have an affinity approximately 100 times that of type II receptors. This difference will obviously be reflected in the level of occupancy at all but the extremes of plasma glucocorticoid concentration; the physiological consequences of this difference are discussed later in this chapter.

MINERALOCORTICOID?/GLUCOCORTICOID?

At the time when type I sites were considered synonymous with mineralocorticoid receptors, they were reported to occur in the following mammalian tissues: kidney (Rousseau *et al.,* 1972; Funder, Feldman and Edelman, 1973a), salivary gland (Funder, Feldman and Edelman, 1972) and gut (Pressley and Funder, 1975), all classical aldosterone target tissues. In heart (Funder, Duval and Meyer, 1973) and liver (Duval and Funder, 1974), levels of type I receptors were not detectable with the techniques available at the time; whether they can be demonstrated in molybdate-stabilized cytosol preparations remains to be examined.

Subsequently, high-affinity tritiated aldosterone-binding sites were described in cultured aortic cells (Meyer and Nichols, 1981) and in a pituitary tumor (GH) cell line (Lan *et al.,* 1981). In both these studies, corticosterone as well as deoxycorticosterone were found to have the same high affinity for the tritiated aldosterone-binding sites as did aldosterone itself; the authors were accordingly careful to distinguish these sites from classical renal mineralocorticoid receptors by dubbing them 'corticoid receptors' or 'mineralocorticoid-like receptors'.

Most recently, similar sites, with equivalent affinity for corticosterone and aldosterone, have been described in the rat mammary gland (Quirk, Gannell and Funder, 1983), where their occupancy by corticosterone has been shown to affect α-lactalbumin synthesis (Quirk *et al.,* 1984). Such type I sites, however, are not ubiquitous; in rat thymus, even in molybdate-stabilized cytosol preparations, tritiated corticosterone binding is completely displaced by RU26988, with a higher potency than non-radioactive corticosterone itself, evidence for binding of tracer uniquely to type II glucocorticoid receptors (Krozowski and Funder, 1983).

It remains to be determined whether such type I sites (in renal cortex, mammary gland, pituitary and gut) are physiological mineralocorticoid receptors occupied by aldosterone in response to Na^+ deficit, or glucocorticoid receptors occupied by corticosterone/cortisol in response to ACTH. If the sequestration of corticosterone by extravascular transcortin in the inner medulla-papillary zone of the kidney favours aldosterone occupancy of type sites in this area, similar mechanisms may operate in other physiological target issues.. Although this is an area which has not been extensively investigated, there are a number of observations consistent with such a possibility. Whether sweat glands are invested locally with extravascular transcortin remains to be established; they do, however, have an appropriate vascular supply, with a central afferent limb and a coiled recurrent efferent limb. Both the pituitary and the breast have been known for some time to contain relatively large amounts of extravasated transcortin; whether this is disposed around recurrent capillary loops in the lobular architecture of the breast, or in the anterior pituitary, remains to be investigated.

Beyond the question whether a type I receptor is occupied by aldosterone or corticosterone/cortisol lies the question whether such occupancy is agonist, antagonist, partial agonist/antagonist (procuring less than 100% of a given full agonist effect) or selectively agonist (effecting some agonist responses, but not others). This question is of particular importance for type I receptors, given their very high affinity for both mineralocorticoids and glucocorticoids. As an illustration, there is excellent, if incomplete, evidence from infusion and micropuncture studies in adrenalectomized rats that aldosterone at physiological concentrations stimulates sodium ion transport in cortical collecting tubules (unprotected by extravascular transcortin), whereas corticosterone at physiological concentrations does not. The study is incomplete in that the effect of concurrent infusion of corticosterone and aldosterone has not been investigated; based on previous experiments, it is to be expected that corticosterone will prove to be an aldosterone antagonist in the cortical collecting tubule.

The very high affinity of rat type I receptors for corticosterone means that such sites will be essentially fully occupied at the diurnal (unstressed) peak – and essentially unoccupied at the diurnal nadir. This has important implications for type I receptor occupancy by both aldosterone and corticosterone. In sodium replete states, plasma aldosterone shows a pattern of diurnal variation paralleling that of the glucocorticoids; under such conditions, therefore, aldosterone occupancy of type I sites – possibly even those protected by extravascular transcortin – may be minimal given the large difference in free plasma levels. In sodium deficiency, however, the nexus of diurnal parallelism is broken and aldosterone levels remain elevated during the trough of diurnal plasma glucocorticoid levels; in extreme cases, i.e. if corticosterone were totally absent from the plasma, aldosterone would occupy type I receptors (in kidney, hippocampus, and other tissues) to an extent determined by its plasma concentration. Transcortin sequestration of corticosterone in the kidney thus represents a mechanism which renders the medullary collecting tubule more sensitive to aldosterone in the presence of modest levels of corticosterone, shifting the aldosterone/corticosterone curve to the left.

TYPE I RECEPTORS AND THE GLUCOCORTICOID '*ZEITGEBER*'

The high affinity of type I receptors for corticosterone, and their being filled or empty over the diurnal range in plasma glucocorticoid levels, may be of far wider physiological significance than the aldosterone occupancy of type I receptors in sodium deficiency described above. Although the physiological roles of glucocorticoid occupancy remain almost entirely unknown, the existence of type I receptors in tissues as diverse as lung (Krozowski and Funder, 1981a), brain (Krozowski and Funder, 1983), kidney (Krozowski and Funder, 1983), mammary gland (Quirk, Gannell and Funder, 1983), pituitary (Lan *et al.*, 1981; Krozowski and Funder, 1981b), vascular wall (Meyer and Nichols, 1981) and gut (Pressley and Funder, 1975), taken with their being occupied at peak plasma levels but not at the nadir, strongly suggest a role for glucocorticoids as a '*zeitgeber*', setting in train a gamut of metabolic responses evoked coordinately in a multitude of organs and systems. Doubtless, the onset and offset of such responses will be found to vary within and between target tissue cells; and minor, for instance meal-related, excursions in ACTH and glucocorticoid concentrations may be found to be of limited physiological significance compared to their sustained nadir during, and often before, sleep, and the subsequent abrupt peak in concentration. Given the commonly observed effects of high-dose steroid administration on mood, sleep/wake patterns and appetite, an action via hippocampal type I receptors, for example, seems worthy of investigation. Similarly, the depression of steroid withdrawal (or the '3 am blues') may be a reflection of empty type I receptors in the hippocampus or other brain areas concerned with affect.

TYPE II RECEPTORS: THE ENDOCRINE EQUIVALENT OF THE SYMPATHETIC NERVOUS SYSTEM

In contrast to type I receptors, classical tritiated dexamethasone-binding type II receptors have a much lower affinity for physiological glucocorticoids (K_d=30–50 nmol/l). At peak plasma concentrations they will be occupied to some extent, but since their sensitivity is only approximately 1% that of type I receptors, they are much less well placed to monitor diurnal changes in unstressed glucocorticoid levels. They are, however, much better placed – because of their relatively low affinity – to monitor and respond to elevated glucocorticoid levels following stress-induced changes in ACTH. The essentially ubiquitous distribution of type II glucocorticoid receptors suggests that they may serve to integrate a whole-body response to stress.

A wide variety of stressors are known to increase ACTH and glucocorticoid concentrations: pyrexia increases free glucocorticoid levels even more than total levels, reflecting the very marked temperature dependence of glucocorticoid binding to transcortin. The historical association of glucocorticoids with carbohydrate metabolism, and the potent stimulatory effect of hypoglycaemia on ACTH secretion have been interpreted as indicating that the primordial role of the ACTH–adrenal axis was energy conservation (Rousseau and Baxter, 1979). That this may also involve redistribution of resource utilization is suggested by the effect of glucocorticoids on the key glycolytic enzyme phosphoenol pyruvate carboxykinase, levels of which are raised, lowered or unaltered depending on the tissue examined (Feldman, Funder and Loose, 1978).

Transcortin shows a remarkable temperature dependence in its binding of naturally occurring glucocorticoids; affinity at 4 °C is 12–25 times higher than at 37 °C (Westphal, 1971). There would appear to be various physiological processes in which this phenomenon may be involved. One, for example, is that of fractionating the distribution of free glucocorticoid to different tissues on the basis of temperature: core organs, for example, will be exposed to larger amounts of free steroid than skin, which in turn will see much larger amounts than the feet of wading birds (which have an efficient vascular heat exchange counter-current system). Similarly, the 'calor' of the local inflammatory response serves to increase local concentrations of free glucocorticoids. In terms of the whole body, temperature control is obviously intimately related to energy balance; the extent to which the ACTH–adrenal axis – in either the diurnal or the stress mode – monitors and modulates temperature independent of the glycaemic state remains to be investigated.

Finally, glucocorticoids have long been implicated in development; since in many of the experimental studies dexamethasone has been shown to be a potent effector, these actions would appear to be mediated via type II receptors. For most species, including man and the rat, the outstandingly stressful life event is surely birth, entailing the challenges of breathing and oral alimentation, to name but two. In both these processes glucocorticoids have been shown to play key roles – on surfactant production, on the transition from fetal to adult forms of haemoglobin, and on the expression of enzymes in the gut and liver. In this situation, glucocorticoids may be functioning as a 'zeitgeber' via type II receptors, not in the diurnal mode, as for type I receptors, but to coordinate and integrate an enormous variety of cells, tissues and organs for the spectacular transition from intrauterine to extrauterine life.

SUMMARY

Over the 30 years since the isolation of aldosterone, there has emerged a coherent physiological picture of mineralocorticoid hormones. In contrast, glucocorticoid hormones, isolated more than 40 years ago, have remained a physiological enigma, involved both in an array of 'pleiotypic' and/or 'permissive' effects and in an ill-defined response to 'stress'. The term glucocorticoid was first used in a classic bioassay, that of mouse liver glycogen deposition. Glucocorticoids are secreted on demand from the adrenal cortex under immediate control of pituitary ACTH, and probably other products of pro-opiomelanocortin in the longer term. ACTH has two patterns of secretion: circadian, with a major peak around waking, minor meal-related peaks, and troughs of very low levels; and episodic, in response to stress (e.g. hypoglycaemia, fever, trauma). In most species, including man, cortisol (hydrocortisone, compound F) is the predominant glucocorticoid, while in the rat and mouse it is corticosterone (compound B). Almost all of these glucocorticoids (more than 90%) circulate bound to protein, predominantly to a high-affinity globulin – transcortin, or corticosteroid-binding globulin (CBG). Their specificity is not absolute; high therapeutic doses of hydrocortisone, for example, overlap into the mineralocorticoid (aldosterone) domain, and result in fluid and electrolyte retention.

Recent studies have shown quite clearly that there are two classes of glucocorticoid receptors. Classical glucocorticoid (type II) receptors have been described over the past 15 years using tritiated dexamethasone ([^{3}H]DM) as a probe, since this synthetic steroid binds minimally to CBG which may contaminate cytosol preparations. The affinity of steroids for these receptors parallels their anti-inflammatory potency; it is moderate for dexamethasone (K_d=3–10 nmol/l), and even lower for the naturally occurring glucocorticoids (B, F).

Very recently, a series of receptors with very high affinity for corticosterone ($K_d \simeq 0.3$ nmol/l) and much lower affinity for DM ($K_d \simeq 10$ nmol/l) have been found in some rat tissues (hippocampus, kidney, mammary gland) but not in others (thymus). These type I sites have equivalent affinity for corticosterone and aldosterone; in certain tissues, e.g. kidney, there are mechanisms favouring receptor occupancy by aldosterone rather than corticosterone, allowing a mineralocorticoid receptor role for such binding sites. However, in the hippocampus, for example, such sites would be largely occupied (at the peak of diurnal secretion) or unoccupied (at the troughs) by the glucocorticoid corticosterone; thus, they are best defined as glucocorticoid receptors in these tissues.

It therefore seems that glucocorticoids operate in two modes. High-affinity receptors, occupied or empty during the diurnal variations in glucocorticoid levels, are ideally placed to mediate, for example, glucocorticoid effects on mood, appetite and sleep/wake patterns (hippocampus), or permissive effects on production of milk proteins (mammary gland). Classical glucocorticoid receptors have much lower affinity for natural glucocorticoids; these sites are substantially occupied only by levels of glucocorticoid which are a response to supranormal levels of ACTH following stress.

In general terms, therefore, glucocorticoids may play two physiological roles. The first is a in a time dimension, serving as a diurnal '*zeitgeber*' for various neural and secretory processes via type I receptors. The other is via type II receptors, as the endocrine equivalent of the sympathetic nervous system, serving to integrate cellular and organ adaptations to stress, probably primordially in response to changes in energy balance (substrate/temperature). Another crucial role currently under investigation is developmental, in anticipation of the stress of birth.

References

BAXTER, J. D. and ROUSSEAU, G. G. (Eds) (1979) Glucocorticoid hormone action. In *Monographs on Endocrinology,* pp. 1–24. New York: Springer-Verlag

DUVAL, D. and FUNDER, J. W. (1974) The binding of tritiated aldosterone in the rat liver cytosol. *Endocrinology,* **94,** 575–579

FELDMAN, D., FUNDER, J. W. and LOOSE, D. (1978) Is the glucocorticoid receptor identical in various target organs? *Journal of Steroid Biochemistry,* **9,** 141–145

FUNDER, J. W., DUVAL, D. and MEYER, P. (1973) Cardiac glucocorticoid receptors: the binding of tritiated dexamethasone in rat and dog heart. *Endocrinology,* **93,** 1300–1308

FUNDER, J. W., FELDMAN, D. and EDELMAN, I. S. (1972) Specific aldosterone binding in rat kidney and parotid. *Journal of Steroid Biochemistry,* **3,** 209–218

FUNDER, J. W., FELDMAN, D. and EDELMAN, I. S. (1973a) The roles of plasma binding and receptor specificity in the mineralocorticoid action of aldosterone. *Endocrinology,* **92,** 994–1004

FUNDER, J. W., FELDMAN, D. and EDELMAN, I. S. (1973b) Glucocorticoid receptors in the rat kidney: the binding of tritiated dexamethasone. *Endocrinology,* **92,** 1005–1013

KROZOWSKI, Z. and FUNDER, J. W. (1981a) Mineralocorticoid receptors in the lung. *Endocrinology,* **109,** 1811–1813

KROZOWSKI, Z. and FUNDER, J. W. (1981b) Mineralocorticoid receptors in rat anterior pituitary; towards a redefinition of 'mineralocorticoid' hormone. *Endocrinology,* **109,** 1221–1224

KROZOWSKI, Z. S. and FUNDER, J. W. (1983) Renal mineralocorticoid receptors and hippocampal corticosterone-binding species have identical intrinsic steroid specificity. *Proceedings of the National Academmy of Sciences (USA),* **80,** 6056–6060

LAN, N. C., GRAHAM, B., BARTTER, F. C. and BAXTER, J. D. (1982) Binding of steroids to mineralocorticoid receptors: implications for *in vivo* occupancy by glucocorticoids. *Journal of Clinical Endocrinology and Metabolism,* **54,** 332–342

LAN, N. C., MATULICH, D. T., MORRIS, J. A. and BAXTER, J. D. (1981) Mineralocorticoid receptor-like aldosterone-binding protein in cell culture. *Endocrinology,* **109,** 1963–1970

MARVER, D. (1984) Evidences of corticosteroid action along the nephron. *American Journal of Physiology* (in press)

MARVER, D. and LOMBARD, W. E. (1981) Localization of aldosterone target sites in rabbit renal medulla. *Kidney International,* **19,** 248

MCEWEN, B. S. (1976) Interactions between hormones and nerve tissue. *Scientific American,* **235,** 48–58

MCEWEN, B. S., DE KLOET, R. and WALLACH, G. (1976) Interactions *in vivo* and *in vitro* of corticoids and progesterone with cell nuclei and soluble macromolecules from rat brain regions and pituitary. *Brain Research,* **105,** 129–136

MEYER, W. J. and NICHOLS, N. R. (1981) Mineralocorticoid binding in cultured smooth muscle cells and fibroblasts from rat aorta. *Journal of Steroid Biochemistry,* **14,** 1157–1168

MOGUILEWSKY, M. and RAYNAUD, J. P.. (1981) Evidence for a specific mineralocorticoid receptor in rat pituitary and brain. *Journal of Steroid Biochemistry,* **12,** 309–314

NESTLER, E. J., RAINBOW, T. C., MCEWEN, B. S. and GREENGARD, P. (1982) Corticosterone increases the amount of protein I, neurone-specific phosphoprotein in rat-hippocampus. *Science,* **212,** 1162–1164

PRESSLEY, L. A. and FUNDER, J. W. (1975) Glucocorticoid and mineralocorticoid receptors in gut mucosa. *Endocrinology,* **97,** 588–596

QUIRK, S. J., GANNELL, J. E. and FUNDER, J. W. (1983) Aldosterone-binding sites in the pregnant and lactating rat mammary gland. *Endocrinology,* **113,** 1812–1817

QUIRK, S. J., FULLERTON, M. J., GANNELL, J. E. and FUNDER, J. W. (1984) Effects of adrenal steroids on α-lactalbumin production in the rat mammary gland. In *Proceedings of the 7th International Congress of Endocrinology* (Quebec, 1984), p. 1880. Amsterdam: Elsevier

ROUSSEAU, G. G. and BAXTER, J. D. (Eds) (1979) Glucocorticoids and the metabolic code. In *Glucocorticoid Hormone Action,* pp. 613–629. New York: Springer-Verlag

ROUSSEAU, G., BAXTER, J. D., FUNDER, J. W., EDELMAN, I. S. and TOMKINS, G. M. (1972) Glucocorticoid and mineralocorticoid receptors for aldosterone. *Journal of Steroid Biochemistry,* **3,** 219–227

SEYMOUR-MUNN, K. and ADAMS, J. (1983) Estrogenic effects of 5-androstene-3β,17β-diol at physiological concentrations and its possible implication in the etiology of breast cancer. *Endocrinology,* **112,** 486–491

STEPHENSON, G., KROZOWSKI, Z. and FUNDER, J. W. (1984) Extravascular CBG-like sites in rat kidney and mineralocorticoid receptor specificity. *American Journal of Physiology,* **247,** F665–F671

TEUTSCH, G., COSTEROUSSE, G., DERAEDT, R., BENZONI, J., FORTIN, M. and PHILBERT, D. (1981) 17α-Alkany-11β,17-dihydroxyandrosterone derivative: a new class of potent glucocorticoids. *Steroids,* **38,** 651–655

WESTPHAL, U. (1971) *Steroid-Protein Interactions,* edited by F. Gross., A. Labhart., T. Mann., L. T. Samuels and J. Zander. *Monographs on Endocrinology,* **4.** New York: Springer-Verlag

5
The adrenarche and adrenal hirsutism

P. Dewis and D. C. Anderson

THE ADRENARCHE

That the adrenal gland may have a part to play in normal sexual maturation was first considered by Albright, Smith and Fraser (1942), who observed pubic and axillary hair development in patients with gonadal dysgenesis. Albright (1947) subsequently named this process the 'adrenarche'. More recently, the development of sensitive hormone assay methods has enabled many workers to demonstrate that there is a progressive rise in the circulating concentrations of the adrenal C19-steroids dehydroepiandrosterone (DHEA) and dehydroepiandrosterone sulphate (DHEA-S), adrenal in origin, from the age of about 7 years to the middle of the second decade. Plasma levels of these two steroids rise at least tenfold during this period (Rosenfield and Eberlein, 1969; Hopper and Yen, 1975; Lee *et al.*, 1975; Sizonenko and Paunier, 1975; Ducharme *et al.*, 1976; Korth-Schutz, Levine and New, 1976; Reiter, Fuldaner and Root, 1977; Genazzani *et al.*, 1978; Grumbach *et al.*, 1978; de Peretti and Forest, 1978). Meanwhile, plasma concentrations and secretion rates (corrected for body weight) of cortisol remain constant (Kenny, Preeyasombat and Migeon, 1966; Forest, 1978); the actual secretion rate of cortisol must, of course, increase. These changes occur in parallel with the progressive development of the zona reticularis, which usually begins between the age of 3 and 7 years (Dohm, 1973), and is preceded by loss of the fibrous stroma left after degeneration of the fetal zone (the medullary capsule).

Albright's observation that the adrenarche occurred independently of puberty (the 'gonadarche') has been confirmed by studies on non-human primates. While the chimpanzee shows a similar temporal relationship between adrenarche and gonadarche to man, in other primates, e.g. *Macaca nemestrina*, the adrenarche actually follows the onset of puberty (Cutler *et al.*, 1978; Smail *et al.*, 1982). Furthermore, in man, timing of the onset of adrenarche is not disturbed in either patients with precocious puberty or those with hypogonadism (Lee *et al.*, 1975; Copeland, Paunier and Sizonenko, 1977; Grumbach *et al.*, 1978; Sklar, Kaplan and Grumbach, 1980).

The dissociation between the adrenarche and gonadarche is further emphasized by studies on premature adrenarche. This syndrome is more frequently seen in girls; pubic and axillary hair develops prematurely without any of the other features of puberty. The syndrome is associated with premature elevation of the adrenal androgens DHEA and DHEA-S, and a less marked but significant elevation of androstenedione (Korth-Schutz, Levine and New, 1976; Lee and Gareis, 1976; Reiter, Fuldaner and Root, 1977; Warne *et al.*, 1978; Sklar, Kaplan and Grumbach, 1981). Warne *et al.* (1978) have also reported premature elevation of oestradiol levels, in contrast to the findings of the other studies (Jenner *et al.*, 1972; Lee and Gareis, 1976; Sklar, Kaplan and Grumbach, 1981). Although the cause of this syndrome is unknown it is clear that it is unrelated to puberty.

THE ADRENOPAUSE

Serum cortisol levels also remain constant during later life (West *et al.*, 1961), with only a small decrease in production rates in old age (Romanoff *et al.*, 1961; Migeon, Green and Eckert, 1963). In contrast, serum concentrations of the adrenal androgens DHEA and DHEA-S fall markedly in old age (to about 20% of young adult levels) in both sexes (Migeon *et al.*, 1957; Abraham *et al.*, 1973; Cattaneo *et al.*, 1975). This appears to be due to a marked fall in production rates (Vermeulen, 1980). Similarly, while the cortisol response to ACTH stimulation is maintained with ageing, the response of the adrenal androgens is reduced (Parker *et al.*, 1981).

A recent study has confirmed the reduced ACTH response for the Δ^5-steroids (DHEA, pregnenolone, 17-hydroxypregnenolone), but has found that the response of the Δ^4-steroids (androstenedione, progesterone and 17-hydroxyprogesterone) was maintained in elderly subjects (Vermeulen *et al.*, 1982).

THE CONTROL OF ADRENAL ANDROGEN SECRETION

The mechanisms for the dissociation between adrenal androgens and cortisol secretion remain controversial.

Control by an adrenal androgen secreting hormone?

Some have argued that a hormonal factor distinct from ACTH must control adrenal androgen secretion (*for review see* Parker and Odell, 1980). Such a hormone would have to act, through a change in relative activities of the enzymes 3β-hydroxysteroid dehydrogenase or 17,20-desmolase, on part or all of the ACTH-responsive zones of the adrenal. Many candidates for such a hormone have been proposed and subsequently eliminated; for instance, their plasma levels do not change in parallel with the adrenarche. These include oestrogens (Albright, 1942, 1947; Warne *et al.*, 1978), prolactin (Bassi *et al.*, 1977; Carter *et al.*, 1977; Vermeulen, Suy and Rubens, 1977), gonadotropins (Saez and Bertrand, 1968; Herrera-Justiano *et al.*, 1979), growth hormone (Lim and Dingman, 1964; Brown, Ginz and Oakey, 1978) and angiotensin (Parker *et al.*, 1983). Other workers have suggested that some uncharacterized pituitary factor may control adrenal androgen secretion (Mills, 1956, 1968; Prunty, 1956; Mills, Brooks and Prunty, 1962; Parker and Odell, 1979). Further ACTH-related peptides and prostaglandins have also been postulated to be involved, but so far with little evidence.

Parker, Lifak and Odell (1983) have recently reported the existence of a glycopeptide with a molecular weight of about 60 000 in extracts of bovine pituitary powder. This substance apparently selectively stimulated DHEA production by collagenase-dispersed dog adrenal cells, without affecting cortisol production. (Parenthetically, neither cow nor dog show development of the zona reticularis.) ACTH, in contrast, stimulated production of both DHEA and cortisol. This peptide seems to be unusually large for a tropic hormone, and no controls for other tissues were reported. We do not accept that it is necessarily a tropic hormone. For example, an enzyme such as 17,20-desmolase might have the same apparent effect since it would convert 17-hydroxypregnenolone selectively to DHEA. In the above experiments the activity was low, with the production of DHEA stimulated by the unidentified substance 200 times less than that of cortisol stimulated by ACTH.

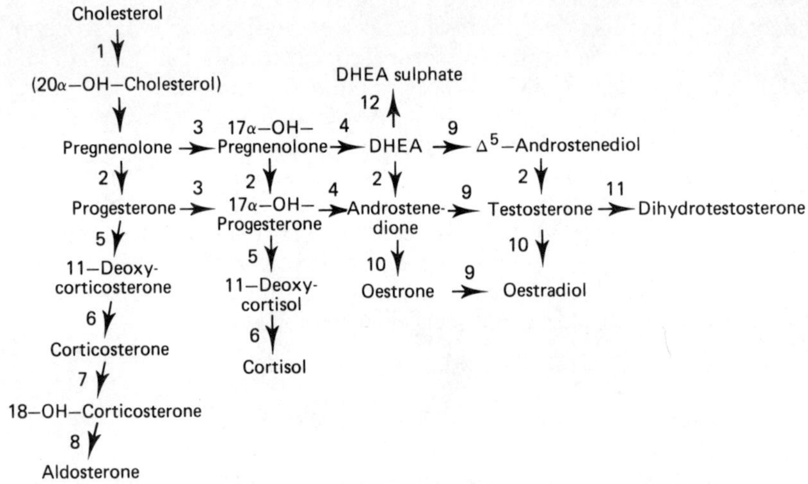

Figure 5.1 The major steroid pathways. Numbers indicate the major enzymes involved in steroid synthesis: (1) cholesterol desmolase complex; (2) 3β-hydroxysteroid dehydrogenase/isomerase complex; (3) 17α-hydroxylase; (4) 17,20-desmolase; (5) 21-hydroxylase; (6) 11β-hydroxylase; (7) 18-hydroxylase; (8) 11-hydroxysteroid dehydrogenase; (9) 17β-hydroxysteroid oxidoreductase; (10) aromatase; (11) 5α-reductase; (12) sulphate conjugase

The factor has been named the adrenal androgen stimulating hormone (AASH) by some (Grumbach *et al.*, 1978; Sklar, Kaplan and Grumbach, 1980) and the cortical androgen stimulating hormone (CASH) by others (Parker and Odell, 1979). The proponents do not deny the validity of the various studies which have demonstrated that ACTH modulates adrenal androgen secretion (Rosenfeld *et al.*, 1971; Irvine *et al.*, 1974; Reiter, Fuldauer and Root, 1977), but propose that a further hormone of pituitary origin in some way modifies the response to ACTH so that the ratio of adrenal androgen to cortisol secretion is increased.

A number of studies have demonstrated various changes in adrenal enzyme activity from childhood to adulthood which would account for the increased androgen secretion. The adrenal steroidogenic pathways and major enzymes are shown in *Figure 5.1*. Rich *et al.* (1981) studied the response of steroid intermediates to injection of ACTH from a dexamethasone-suppressed base in prepubertal children, girls with premature adrenarche and adults in the follicular phase of the menstrual cycle. Their results suggest that the adrenarche is associated with an increased 17,20-desmolase activity and a reduced 3β-hydroxysteroid dehydrogenase activity. Kelnar and Brook (1983) measured the steroid metabolites in 24-hour urine specimens in boys of various ages. The observed age-related differences were consistent with a decrease in 3β-hydroxysteroid dehydrogenase, a slight rise in 17α-hydroxylase, a marked rise in 17,20-desmolase activity and a fall in 11β-hydroxylase activity. This would result in a reduced ability to synthesize cortisol, which in turn would lead to an increase in ACTH secretion to maintain cortisol secretion, and a consequent rise in adrenal androgen secretion. A similar conclusion had tentatively been drawn by Anderson *et al.* (1976), who looked at differences in the plasma levels of multiple steroid intermediates between prepubertal children and adult females.

Schiebinger *et al.* (1981) measured the activities of microsomal 3β-hydroxysteroid dehydrogenase, 17,20-desmolase, 17-hydroxylase and 21-hydroxylase *in vitro* in human adrenal glands obtained from individuals between 3 months and 60 years of age. From birth to early childhood there was a rise in 3β-hydroxysteroid dehydrogenase activity along with a fall in 17,20-desmolase and 17α-hydroxylase activity, while the adrenarche was associated with a rise in 17,20-desmolase and 17α-hydroxylase activity. While it is possible that these enzyme changes could be brought about by the action of an androgen-secreting hormone, an alternative is that they occur mainly as a result of intra-adrenal mechanisms.

Intra-adrenal control of adrenal androgens?

(*See also* Chapter 1). This theory as originally propounded maintains that the maturation of the zona reticularis and the consequent increase in adrenal androgen secretion is dependent on intra-adrenal levels of cortisol (and other steroids) (Anderson, 1980). Blood flow in the adrenal is from the cortex to the medulla (Dobbie, MacKay and Symington, 1968); blood vessels in the outer cortex are straight, and those in the inner cortex lattice-like. Thus, conditions in the reticularis are right for the pooling of blood, and a concentration gradient of cortisol and other steroids is to be expected from cortex to medulla. Cells in the outer fasciculata will be exposed to relatively little cortisol after responding to a burst of ACTH secretion because it is rapidly flushed out towards the inner zones of the cortex. It is proposed that the inner cells begin to respond to the (episodically) very high levels of cortisol by undergoing morphological changes (to become zona reticularis cells) and also by changing the enzyme activities which result in increased androgen production.

Various observations might be explained by this theory. Androgen levels from the adrenals are suppressed for far longer than plasma cortisol levels in patients treated with glucocorticoids (Ettinger *et al.*, 1973; Cutler *et al.*, 1979). Glucocorticoid therapy causes suppression of ACTH, and cortisol levels within the adrenal will be too low to maintain the structural and/or functional identity of the zona reticularis. When glucocorticoids are stopped the adrenal responds to endogenous ACTH like that of a young child. According to the intra-adrenal theory, time is required for the inner cells to differentiate into reticularis cells, and respond again as such cells, once an intra-adrenal cortisol gradient is restored.

In congenital adrenal hyperplasia (CAH) due to 21-hydroxylase deficiency (*see* Chapter 6) there is a massive increase in production and intra-adrenal levels of the major substrate for 21-hydroxylase, 17α-hydroxyprogesterone, within the zona fasciculata and reticularis. The major androgen to be secreted in excess from the adrenals is Δ^4-androstenedione; one might well expect a marked increase in secretion of the Δ^5-steroids associated with the adrenarche. However, even in untreated cases the rise in serum DHEA concentrations is relatively modest (tenfold) considering the 100-fold or more increase in 17α-hydroxyprogesterone. In patients treated by ACTH suppression with corticosteroids, the serum DHEA levels are low and the increment after injection of ACTH is only 4% of normal

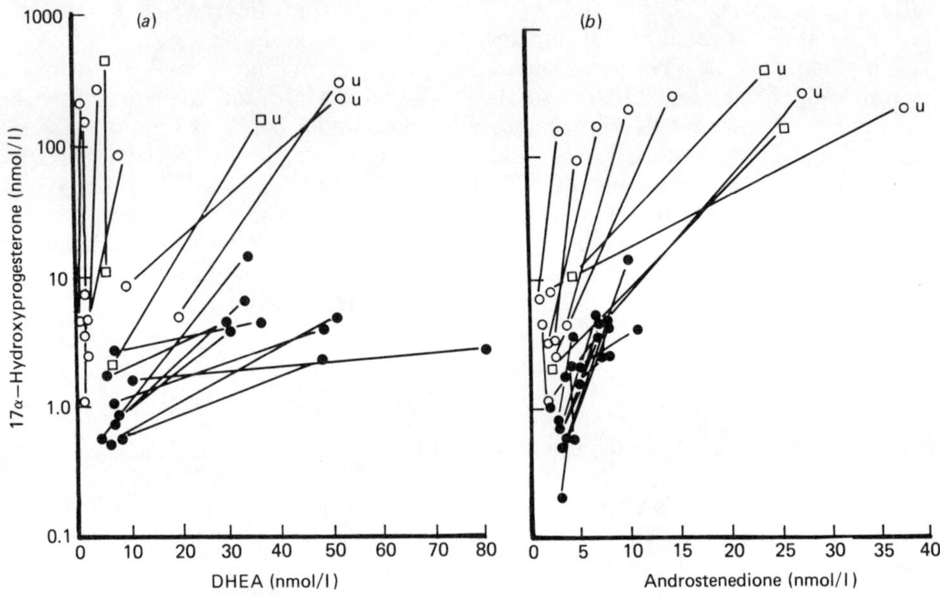

Figure 5.2 Serum steroid levels after overnight dexamethasone before and 60 min after 1-24 ACTH (250 μg i.m.) in female control subjects (●) and males (□) and females (○) with 21-hydroxylase deficiency (congenital adrenal hyperplasia). Levels for each individual are connected by a line. All patients were treated except those which are marked 'u' on the graphs. (*a*) 17-Hydroxyprogesterone (log scale) plotted against DHEA (linear scale). (*b*) 17-Hydroxyprogesterone (log scale) plotted against androstenedione. (From Conway, Anderson and Bu'Lock, 1982, Courtesy of the Editor and Publishers, *Clinical Endocrinology*)

(Conway, Anderson and Bu'Lock, 1982; *Figure 5.2*), yet the rise in 17α-OH-progesterone is 60 times greater than normal. These observations could be explained if an intra-adrenal gradient of cortisol or other steroids is important in producing the changes in inner cortical cells that are reflected in the adrenarche. Intra-adrenal cortisol levels will be low in untreated and very low in treated congenital adrenal hyperplasia.

It has also been demonstrated that serum DHEA-S levels are low in women with primary hypothyroidism (Bassi *et al.*, 1980). A possible explanation for this phenomenon is that cortisol clearance is reduced in thyroid hormone deficiency. A new steady state is achieved with lower ACTH and cortisol secretion; peripheral cortisol levels are normal, but intra-adrenal levels would be reduced.

Central to the theory is the development of the zona reticularis and its importance in secreting androgens. Development of the reticularis coincides with the onset of the adrenarche, regardless of whether or not it is precocious (Dhom, 1973). In primates that do not have an adrenarche the reticularis is absent (Levine *et al.*, 1982). It is apparent that there is a movement of adrenocortical cells from the outer to the inner zone (*see also* Chapter 1). In the rat, the zona glomerulosa loses 0.7% of its cells per hour, which is almost balanced by the gain of 0.5% per hour in the zona reticularis (Wright, 1981). The latter also have a high content of pigmented lipids, which is attributed to their greater age (Reichel, 1968).

Glucocorticoid and androgen receptors have been demonstrated in the adrenal cortex (Calandra *et al.*, 1978; Rifka *et al.*, 1978; Do, Loose and Feldman, 1979), and there is now evidence that the activity of various enzymes in cultured suspensions of fasciculata–reticularis cells can be affected by the addition of various steroids to the medium (Hornsby, 1980, 1982; *see also* Chapter 1).

The studies described by Hornsby were performed mainly on bovine adrenocortical cells. The human differs from the cow in producing much more DHEA and DHEA-S. This appears to be due to a much lower 3β-hydroxysteroid dehydrogenase (3β-OHSD) activity – apparently a particular characteristic of all human adrenocortical cells (Hornsby and Crivello, 1983). It has also been shown that its activity is lower in the zona reticularis than in the fasciculata (Cavallero and Chappino, 1962; Dawson, Pryse-Davis and Snape, 1962). Inactivity of 3β-OHSD, together with high levels of ACTH, is a feature of the fetal zone – here it is likely that the very high oestrogen levels produce direct inhibition (*see* Chapter 2). In contrast, the levels of oestrogen seen in the non-pregnant state were shown not to affect 3β-OHSD activity in short- or long-term experiments in ovariectomized women (Anderson and Yen, 1976).

The properties of isolated zona reticularis and zona fasciculata cells have been studied *in vitro*. The guinea pig is an excellent model; it has a well-developed zona reticularis, secretes cortisol (rather than corticosterone) as well as DHEA and DHEA sulphate, and, like the human, cannot synthesize ascorbic acid. Strott, Goff and Lyons (1981) demonstrated that the concentration of cortisol was 30 times greater in the cytosol of the outer zone of the guinea-pig adrenal cortex than in the inner zone. Tait and his colleagues have done extensive work on isolated dispersed zona reticularis and zona fasciculata cells. Working with the rat, they found that the contribution of the reticularis to steroidogenesis was relatively small, i.e. 20% of the deoxycorticosterone and 10% of the corticosterone output in the basal state and even less after ACTH stimulation (Bell *et al.*, 1979). Basal androstenedione outputs were smaller for both cell types, and ACTH had very little effect, but did produce a slightly greater increase in androstenedione secretion in fasciculata than in reticularis cells. In the basal state the ratio of deoxycorticosterone to corticosterone was higher in the reticularis than in the fasciculata, indicating that this zone is relatively deficient in 11β-hydroxylase. Similar results were obtained when their work was repeated on the guinea pig adrenal cortex (Hyatt *et al.*, 1983). They concluded that a distinct role for the zona reticularis could not be established. Nishikawa and Strott (1984) have shown that cells from the guinea pig fasciculata produced much more cortisol than those from the reticularis, both basally and in response to ACTH. They did not, however, study androgen secretion.

Recently, with W. R. Robertson and colleagues (Davison *et al.*, 1983), we have studied steroid production from the zona reticularis and zona glomerulosa/fasciculata separated by means of microdissection (*Figure 5.3*). Under basal conditions the fasciculata produced nine times more cortisol than the reticularis, while the latter produced more androgen. Applying planimetry to assess relative volumes, we found that the reticularis was apparently responsible for 7% of the cortisol and 66% of the androgen secreted in the basal state. We were not able to demonstrate an increase in DHEA or cortisol secretion in response to ACTH in these cells. In contrast, zona fasciculata cells in dispersed preparations respond briskly with an increase in cortisol and adrenal androgen secretion. It may be

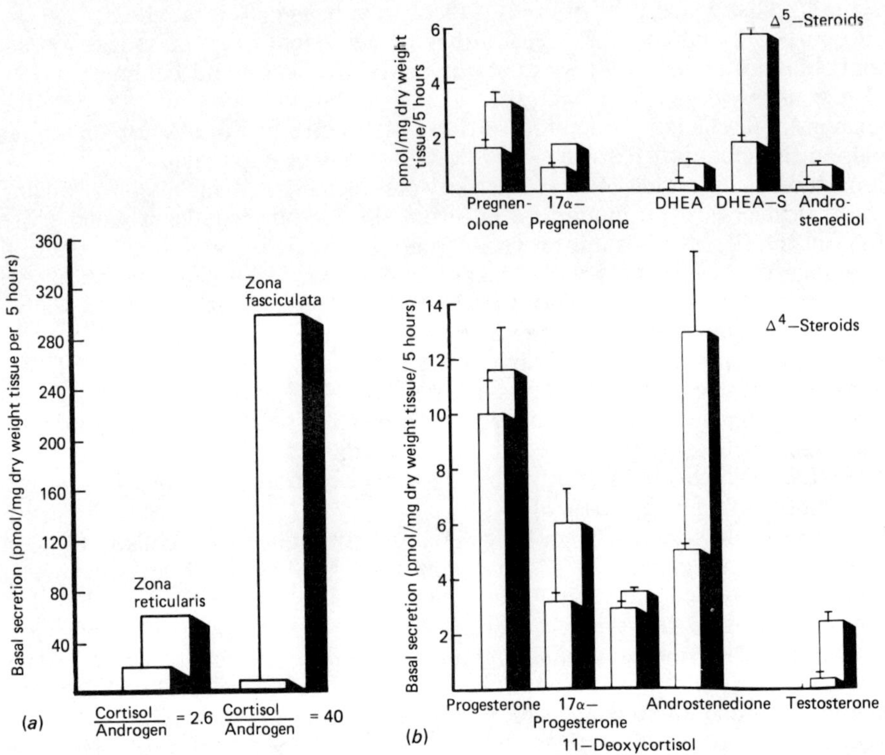

Figure 5.3 Basal secretion of different steroids from microdissected guinea-pig zona fasciculata (ZF) and zona reticularis (ZR) maintained in organ culture. The results are expressed as the mean output ± s.d. A minimum of six experiments was performed for each analysis. (*a*) Cortisol (rear columns) and total androgen (front columns) secretion in ZR and ZF. Total androgen equals the sum of levels of DHEA, DHEA sulphate, androstenediol, androstenedione and testosterone. (*b*) Secretion of other individual steroids from ZR (rear columns) and ZR (front columns). (Modified from Davidson *et al.*, 1983)

therefore that the zona reticularis is secreting predominantly androgens in the basal state, and that the outer fasciculata makes most cortisol and the inner fasciculata transition zone most DHEA and DHEA sulphate in response to ACTH. The work of Jones and Griffiths (1968) with the guinea pig and of O'Hare, Nice and Neville (1980) with human adrenals also suggests that the zona reticularis has an important role to play in androgen, particularly DHEA-S, secretion. Another important observation by Neville and O'Hare (1979) and O'Hare, Nice and Neville (1980) was that while zona fasciculata and zona reticularis in culture secreted different products initially (the former producing glucocorticoids and the latter glucocorticoids and androgens), after a few days the cells became indistinguishable in terms of the steroids secreted, both cell types behaving as fasciculata cells. This observation is in keeping with the functional changes seen in the zona reticularis maintained by local environmental factors. The role of the zona reticularis in synthesizing and sulphating androgens may not be its sole function, since other enzymes, such as ethylmorphine demethylase and Δ^4-hydrogenase, have been found in high concentrations there (Martin and Black, 1982, 1983).

Conclusion: the adrenarche

The adrenarche is an ill-defined and gradual process that contributes to the development of secondary sexual hair, and is associated with a rise in plasma adrenal androgens and the development of the zona reticularis. There are two main theories concerning the control of adrenal androgens and thus underlying the adrenarche. One states that an as yet unidentified hormone is responsible for controlling adrenal androgen synthesis or at least for altering the response of the adrenal cortex to ACTH in favour of androgens. The other states that the increase in adrenal androgen synthesis is an inevitable result of the changing intra-adrenal steroid environment. Although the dilemma has yet to be resolved, the latest *in vitro* data appear to support the latter theory (*see also* Chapter 1).

ADRENAL HIRSUTISM

Androgens in the normal adult woman

In the adult woman, ovarian androgens are obligate precursors of oestrogen synthesis. The major circulating androgens in the adult female are testosterone (normal range about 0.5–2.0 nmol/l), androstenedione (normal range 2–10 nmol/l), DHEA (normal range 12–40 nmol/l), and DHEA-S (normal range 2–10 μmol/l) (Osborne and Yannone, 1971). Dihydrotestosterone (DHT), the active androgen in many target tissues, is also present within the circulation (normal range 0.5–1.0 nmol/l). DHEA and DHEA-S are very weak androgens in comparison to testosterone and DHT, but they may act as pre-hormones for the synthesis of more active compounds (Kirschner *et al.*, 1973). In the normal woman, between 5 and 20% of the circulating testosterone comes directly from the ovaries and between 1 and 30% from the adrenal cortex (Kirschner and Bardin, 1972). About 50% of circulating testosterone is derived from the extraglandular conversion of androstenedione (Horton and Tait, 1966; Bardin and Lipsett, 1967). About 50% of circulating androstenedione is secreted by the adrenal cortex and 50% by the ovary (except at mid-cycle, where twice as much comes from the ovary as from the adrenal cortex) (Abraham, 1974). The adrenal cortex is the predominant source of DHEA and DHEA-S (Abraham, 1974). Circadian variations in the serum concentrations of androstenedione and DHEA synchronous with those of cortisol have been demonstrated in normal women (Kirschner, Lipsett and Collins, 1965; Tunbridge, Rippon and James, 1973; Abraham, 1974; Givens *et al.*, 1975). Rich *et al.* (1981) have shown that despite its adrenal origin DHEA-S does not respond acutely to ACTH with a rise in its peripheral level because of its long half-life; thus a circadian rhythm is not seen. When very frequent samples are taken, a small circadian rhythm for testosterone synchronous with cortisol has also been found (Lachelin *et al.*, 1979, 1982), although this is not apparent with a longer sampling interval (Tunbridge, Rippon and James, 1973; West *et al.*, 1973). While serum testosterone and androstenedione show cyclical variations throughout the menstrual cycle (Judd and Yen, 1973; Abraham, 1974; Vermeulen and Verdonck, 1976), serum DHEA and DHEA-S concentrations do not (Abraham, 1974).

Hirsutism

Hirsutism has been defined by Vermeulen (1981) as 'excessive terminal androgen-dependent hair with a typical male distribution'. The most commonly affected areas are the chin, upper lip, side of the face, lower abdomen and thighs, but in more severe cases the whole body is involved. The diagnosis is often arbitrary, with a large overlap between the 'hirsute' and the 'normal' population. Whether a woman complains of being hirsute will depend on a number of factors, such as psychological state and race. Caucasian women are generally more hirsute than Japanese women, regardless of androgen levels (Hamilton, 1958); the same overall greater development of body hair is evident in Caucasian men. In European women hirsutism is a common clinical problem. Thomas and Ferriman (1957) estimated that 3% of women between the age of 15 and 44 had hair on the upper lip, 9% on the chin and 6% on the sides of the face.

The contribution of the adrenal cortex to hirsutism

Any condition which increases circulating serum androgen levels is likely to lead to hirsutism.

Congenital adrenal hyperplasia

This is discussed in detail in Chapter 6; here, comments will be confined to androgen secretion and hirsutism. In the female, congenital adrenal hyperplasia classically presents with ambiguous genitalia in infancy, but occasionally presents for the first time in adult life with hirsutism (Brooks and Prunty, 1960; Lipsett and Riter, 1961). Often these patients have other signs of virilization, such as frontal baldness, clitoral enlargement, breast atrophy and amenorrhoea, although in some cases late onset congenital adrenal hyperplasia may be indistinguishable from the milder forms of hirsutism discussed below (Gabrilove, Sharma and Dorfman, 1965; Rosenwaks *et al.*, 1979; Blankstein *et al.*, 1980; Cathelineau *et al.*, 1980; Lobo and Goebelsmann, 1980; Migeon *et al.*, 1980; Chrousos *et al.*, 1982). Patients with CAH constitute only a small proportion of those presenting to any endocrine clinic and are easily identified by the demonstration of high levels of serum 17α-hydroxyprogesterone, which is under ACTH control. The reason for their mild clinical symptoms appears to be a relatively mild enzyme deficiency; perhaps some have a low cortisol requirement, and so a lower obligatory ACTH drive.

Polycystic ovary syndrome and idiopathic hirsutism

DEFINITIONS

Most hirsute women have either the polycystic ovary syndrome or idiopathic hirsutism. In 1935, Stein and Leventhal described the association between enlarged polycystic ovaries, amenorrhoea, hirsutism and obesity. The underlying cause is unknown and indeed it may be a spectrum of disorders of differing aetiology. The strictness of the diagnostic criteria varies between workers. For example, Yen and his colleagues insist that most patients with this syndrome have all the above features (Yen, Chaney and Judd, 1976; Yen, 1980). Others will accept patients of normal body weight and without hirsutism (Goldzieher, 1968) and even those with normal-sized ovaries (Smith, Steinberger and Perloff, 1965; Vejlsted and Albrechtsen, 1976). An example of this clinical diversity can be seen in Japan, where women with polycystic ovaries are rarely hirsute, for reasons discussed above (Kasuga, 1980).

It is normal in studies of hirsute women to divide patients into those with idiopathic hirsutism and those with polycystic ovary syndrome; the above discussion, however, shows that this distinction may be purely arbitrary. A variety of criteria have been used to distinguish these patients; for example, some insist on the presence of the histological changes of polycystic ovaries (Bardin and Lipsett, 1967; Kirschner *et al.*, 1973), others on the presence of palpably enlarged ovaries (Ettinger *et al.*, 1973; Meikle *et al.*, 1976), and yet others on the absence of regular menstrual cycles regardless of ovarian size or histology (Raj *et al.*, 1982; Moore *et al.*, 1983). This means that different workers may well not be talking about the same conditions, having quite different criteria for distinguishing between these two conditions.

THE ROLE OF ADRENAL ANDROGENS IN POLYCYSTIC OVARIES AND IDIOPATHIC HIRSUTISM

Serum androgen levels and blood production rates. Stein and Leventhal (1935) suggested that an endocrine disturbance may produce both the ovarian pathology and the clinical picture. However, little useful information emerged until the introduction of competitive protein-binding and radioimmunoassay techniques. These provided convenient and accurate methods for measuring serum androgen levels and established that while mean plasma testosterone levels were high in hirsute women there was substantial overlap with normal women. This was true regardless of menstrual regularity or ovarian size and histology (Forchielli *et al.*, 1963; Hudson *et al.*, 1963; Burger, Kent and Kellie, 1964; Dignam *et al.*, 1964; Korenman, Kirschner and Lipsett, 1965; Lloyd *et al.*, 1966). A similar situation was found to exist for other androgens, including those of predominantly adrenal origin – androstenedione (Bardin and Lipsett, 1967; Givens *et al.*, 1975), DHEA (Kirschner *et al.*, 1973; Abraham and Chakmakjian, 1974) and DHEA-S (Abraham and Chakmakjian, 1974; Abraham *et al.*, 1975).

Bardin and Lipsett (1967) measured blood production rates of testosterone and androstenedione and found the former to be higher than normal in all the hirsute women studied and the latter higher in all but one of the hirsute women. There was no difference between the two groups of hirsute women. Kirschner *et al.* (1973) later demonstrated that blood production rates of DHEA and Δ^5-androstenediol were also elevated in hirsute women. The discrepancy between serum levels and blood production rates of androgens was most marked for testosterone; increased androgen production leads to decreased levels of serum sex-hormone binding globulin (SHBG), which has a much higher affinity for testosterone than for androstenedione and DHEA. The fall in SHBG thus obscures the rise in unbound testosterone, which reflects the production rate much better than the total testosterone level. The possible role of SHBG in hirsutism is outside the scope of this chapter and has been discussed elsewhere (Anderson, 1976). Having established the occurrence of increased androgen production it was important to establish whether the adrenal cortex or the ovary was the major source.

Stimulation–suppression data. In early studies the finding of elevated urinary 17-ketosteroids suppressible by glucocorticoids (Prunty, 1966) was held as evidence of an important adrenal role in the production of androgen excess. In addition, it was well recognized that many women with polycystic ovaries menstruate regularly when treated with glucocorticoids (Smith, Steinberger and Perloff, 1965). With the introduction of convenient methods for measuring circulating serum androgen levels have come many attempts to identify the relative adrenal and ovarian contributions more precisely.

For the adrenal gland, the most obvious way is to suppress its activity acutely with dexamethasone and subsequently to stimulate it with ACTH, and measure the changes in plasma levels. Givens *et al.* (1975) studied 10 normal and 20 hirsute women by looking at diurnal rhythm of plasma levels of cortisol, androstenedione, testosterone, 17α-hydroxyprogesterone, follicle stimulating hormone (FSH) and luteinizing hormone (LH). Dexamethasone 1 mg was then given to seven normal and 17 hirsute women and plasma hormone levels were assessed at 8.00 hours the following morning. All subjects then received an intravenous injection of ACTH at a dose calculated to reproduce the normal morning cortisol peak. All subjects had a diurnal rhythm for cortisol and an androstenedione rhythm synchronous with that of cortisol. The hirsute women could be divided into two groups: those with a normal (group II) and those with an exaggerated androstenedione rhythm (group I). There was no significant diurnal rhythm for testosterone, FSH or LH in any of the subjects. In response to dexamethasone, the hirsute women in group I showed a greater suppression of the two androgens than those in group II, but still not the same as normal individuals. In response to ACTH, group I hirsute women had an exaggerated androstenedione response, and seven of the nine had an exaggerated 17-hydroxyprogesterone but normal cortisol response.

The general conclusions drawn from this study were that women in group I had a substantial ACTH-dependent (and presumably adrenal) component to their androgen excess, although dexamethasone data suggested that they still had an ACTH-independent (and therefore presumably ovarian) component. In group II women the androgen excess was largely ACTH-independent. In only a very small proportion was androgen secretion wholly ACTH-dependent (6%). The ACTH data, and particularly the 17-hydroxyprogesterone response, suggested that the excess androgen production may be due to a mild deficiency of the enzyme 21-hydroxylase. Group I women with a normal response of this steroid were considered to have a minor defect of 3β-hydroxysteroid dehydrogenase. However, the degree of deficiency would be minimal in comparison to the classical forms of congenital adrenal hyperplasia; it is even possible that the minor adrenal enzyme abnormality could be secondary to the already high androgen levels. Androgens can inhibit 21- and 11-hydroxylase enzymes *in vitro* (Morrow, Burrow and Mulrow, 1967; Sharma, Forchielli and Dorfman, 1973; Hornsby, 1980, 1982). That this might be the case in these women is further suggested by another study from Givens' group (Wild *et al.*, 1982). Similar tests were performed and repeated after treatment with a combined oral contraceptive for 21 days. This led to an increase in the mean plasma cortisol level over 24 hours and an exaggeration of the normal rhythm (bigger difference between peak and trough values), while at the same time mean levels of androstenedione fell and its rhythm was suppressed. In addition, the androstenedione response to ACTH was suppressed. Thus, these authors inferred that suppressing ovarian androgens also brought about a reduction in the adrenal contribution to this androgen excess. From these studies it is possible that most hirsute women have a primary excess of ovarian androgen production, which in some brings about suppression of adrenal enzyme activities, with a consequent secondary increase in adrenal androgen production. Most of the women in Givens' studies had irregular periods. The presence or absence of enlarged ovaries did not seem to affect whether they fell into group I or group II.

A similar study has been performed in nine women with menstrual irregularities and polycystic ovaries demonstrated at either laparotomy or laparoscopy (Lachelin *et al.*, 1979). Similar responses to dexamethasone and low doses of ACTH were found in patients and normal subjects. However, with an ACTH infusion designed to produce a maximal adrenal response, there were significantly greater increases in circulating levels of DHEA, 17α-hydroxypregnenolone, progesterone and 17-hydroxyprogesterone, and significantly smaller increases in androstenediol, androstenedione and testosterone in those with polycystic ovaries. Again, these results were considered to be incompatible with a classical enzyme block and it was suggested that they could be brought about by increased androgen levels modulating adrenal enzyme activities.

Kandeel *et al.* (1980) also used large doses of ACTH, but did not start the stimulation from a dexamethasone-suppressed base. They, again, found differences between normal subjects and those with polycystic ovaries which would fit with the above. Oestrone levels were raised in certain patients and, indeed, oestrogens in high concentrations have been shown to inhibit 3β-hydroxysteroid dehydrogenase (Yates and Deshpande, 1974; Winter *et al.* 1980), although Anderson and Yen (1976) have shown that acute infusion of oestradiol or chronic administration of ethinyl oestradiol to ovariectomized women has no such effect *in vivo.* They also found that maximal adrenal stimulation with ACTH causes a disproportionately greater rise in each Δ^5-3β-hydroxysteroid than its Δ^4 counterpart, something which has been ignored in many studies.

Other workers have suggested that abnormalities of 11-hydroxylase (Gibson *et al.*, 1980) and 3β-hydroxysteroid dehydrogenase (Abraham *et al.*, 1975; Gibson *et al.*, 1980) are present in a substantial proportion of patients with either idiopathic hirsutism or polycystic ovaries. However, these abnormalities are much less marked than those seen in the rare patients who first present in adult life with deficiencies of 21-hydroxylase (Blankstein *et al.*, 1980; Lobo and Goebelsmann, 1980; Chrousos *et al.*, 1982) and 11-hydroxylase (Gabrilove, Sharma and Dorfman, 1965; Cathelineau *et al.* 1980).

In our laboratory, Anderson, Child and Bu'Lock (1978) found basal Δ^5-steroid levels in hirsute women that might have been in keeping with a mild defect in adrenal 3β-hydroxysteroid dehydrogenase. However, after dexamethasone suppression and stimulation with ACTH there was no difference between the hirsute patients and a group of normal subjects. The difference in basal levels was attributed to increased ACTH drive at the time of sampling, possibly due to stress (the normal controls were all hospital staff and medical students and thus less likely to be worried by venepuncture). Consequently, it was concluded that there was no evidence of reduced adrenal enzyme activity in these hirsute patients (furthermore, no differences were found between those with regular and those with irregular periods).

Moore *et al.* (1983), in a group of hirsute women with regular ovulatory cycles, found exaggerated plasma responses to ACTH for androstenedione, DHEA and cortisol. All these abnormalities resolved after treatment with dexamethasone 0.5 mg each night for three months. These findings were interpreted as evidence for an exclusively adrenal source for the androgen excess in this subgroup of hirsute women. However, since the initial testing was not performed for a dexamethasone-suppressed base, while the final testing was, the results may be no different from those discussed above.

In a more recent study from our laboratory, Child, Bu'Lock and Anderson (1980) found normal responses of Po and 17Po to ACTH in seven of 10 hirsute women after dexamethasone suppression, while three showed an exaggerated response similar to that seen in heterozygotes for 21-hydroxylase deficiency (Child, Bu'Lock and Anderson, 1979a) and a certain proportion of normal women (Child, Bu'Lock and Anderson, 1979b). This suggested that some hirsute women have a minor deficiency of 21-hydroxylase. However, we have since studied the incidence of such a hyper-response in normal women and in those with a variety of gynaecological disorders, including polycystic ovaries and idiopathic hirsutism (Conway *et al.,* 1983), and conclude that it is due to a common minor variant of 21-hydroxylase, which is not apparently of clinical significance.

Abraham *et al.* (1981) have claimed that a predominantly adrenal source of excess androgens can only be demonstrated by suppression with high doses of dexamethasone over a prolonged period (2 mg/day for 14 days). However, Lachelin *et al.* (1982) have shown that DHEA can be suppressed to a nadir after two days of treatment with dexamethasone 0.5 mg at night. Because of its long half-life, DHEA sulphate requires much longer before it is maximally suppressed.

While these studies indicate that there is an adrenal source of androgen excess in some hirsute women, they do not allow any definite conclusions about its extent and nature. One of their disadvantages is that the effects of dexamethasone and ACTH may not be confined to the adrenal glands. Thus, corticosteroids have been shown to inhibit the LH surge in the face of a rapidly rising oestrogen level in rats, baboons and humans (Cunningham *et al.,* 1978). Dexamethasone has been shown to suppress plasma levels of LH and FSH in patients with elevated LH levels, while it increases LH and decreases FSH levels in women with normal LH levels (Givens *et al.,* 1975). Moreover, there is evidence to suggest that dexamethasone suppresses ovarian vein androgen levels (Kirschner and Jacobs, 1971). ACTH has been shown to have a stimulatory effect on ovarian theca cells *in vitro* (Wilson *et al.,* 1979) and may also affect plasma LH levels (Givens *et al.,* 1975, 1976).

Following reports that 'medical oophorectomy' could be achieved using a long-acting gonadotropin releasing hormone (GnRH) agonist (Meldrum *et al.,* 1982), such an agent has recently been used to suppress ovarian function in patients with polycystic ovaries (PCO) in an attempt to elucidate the relative importance of adrenal and ovarian androgen excess in this condition (Chang *et al.,* 1983). Six women were studied, and the results were compared to those from six regularly ovulating women with endometriosis and eight who had undergone bilateral oophorectomy for benign gynaecological diseases. The GnRH agonist was given daily by subcutaneous injection for 28 days. Plasma LH levels in patients with polycystic ovaries showed a sharp rise, which was maintained at a level about half that in oophorectomized subjects until the end of the study. In contrast, a small fall in plasma FSH levels was seen in both normal and PCO patients. The amplitude of LH and FSH pulses was reduced in both groups. Plasma oestradiol and oestrone levels showed a sharp rise, followed by a decline to castrate levels in normal subjects, while in PCO patients there was a steady decline to levels identical to those in oophorectomized subjects.

Plasma levels of testosterone and androstenedione declined progressively to castrate levels both in normal and PCO patients. On the other hand, androgens of a predominantly adrenal origin, DHEA and DHEA-S, were unchanged, remaining elevated in patients with polycystic ovaries. Moreover, their diurnal rhythm and response to ACTH infusion was unchanged. From changes in gonadotropin levels the mechanism of action of this GnRH analogue is unclear. It may act at both pituitary and ovarian levels (Hseuh and Jones, 1981), but from the lack of change in DHEA and DHEA-S in these subjects it does not appear to be affecting the adrenal cortex. Oestradiol, oestrone, testosterone and androstenedione were all suppressed to levels identical to those in castrate women, suggesting that ovarian secretion of these steroids has been abolished. The fact that these two androgens were suppressed to similar levels in normal and PCO subjects suggests that the ovary is largely responsible for their excess. There is obviously scope for further studies along these lines.

Venous catheterization data. The other main method used to determine the site of excess androgen production in hirsute women has been to catheterize selectively the adrenal and ovarian veins and to measure androgen levels in the venous effluent from each gland (Rivarola, Singleton and Migeon, 1967; Kirschner and Jacobs, 1971; Kirschner *et al.,* 1973; Kirschner, Zucker and Jespersen, 1976).

Kirschner's patients were said to have idiopathic hirsutism – none had palpably enlarged ovaries, but most had oligomenorrhoea. Metabolic clearance rates and blood production rates were calculated as described by Horton and Tait (1966). Simultaneous peripheral and catheter venous samples were taken for calculation of concentration gradients for each of the steroids studied. Significant gradients were found across both ovarian and adrenal veins for testosterone, androstenedione, androstenediol and DHEA, while for cortisol a gradient was found only across the adrenal vein. The gradients for the former three were greater across the ovarian vein while that for DHEA was much greater across the adrenal vein. The adrenal contribution to the total androgen production rates was also estimated, using the equation

$$\frac{\text{Adrenal gradient for androgen X}}{\text{adrenal secretion rate for androgen X}} = \frac{\text{Adrenal cortisol gradient}}{\text{cortisol secretion rate}}$$

and using literature figures for the cortisol secretion rate. It was concluded that the ovary was by far the major source of excess testosterone production, while DHEA was largely adrenal in origin. They also found that their calculations of the site of androgen excess bore no relationship to the ability of the peripheral androgen levels to be suppressed by dexamethasone.

While it is possible to conclude from these data that the ovary is an important site of excess androgen production in these patients, these authors may have underestimated the contribution of the adrenal gland since they assumed that there is a constant relationship between androgen and cortisol secretion by the adrenal. Their studies were performed at a time when the patients were likely to be stressed, and under such conditions it is likely that the ratio of cortisol to androgen secretion by the adrenal would be at its lowest. Indeed, Kirschner himself (Kirschner, Zucker and Jesperson, 1976) has shown that the percentage change in adrenal vein testosterone in response to ACTH is about half that for cortisol.

Although the number of such studies is small, others have suggested a more important role for the adrenal cortex (Stakl, Teeshink and Greenblatt, 1973; Oake *et al.,* 1974). It could be argued that such studies are not justified for the investigation of relatively benign conditions such as idiopathic hirsutism and PCO. Laatikainen *et al.* (1980), however, have examined ovarian and peripheral venous blood samples from PCO patients undergoing laparotomy for wedge resection and have compared the results with those of normal subjects undergoing surgery for tubal ligation. Ovarian vein concentrations of androstenedione, testosterone and DHEA were markedly raised in PCO patients compared to normal subjects, while those of dihydrotestosterone and androsterone were elevated to a lesser extent. Ovarian vein oestradiol concentrations were similar in the PCO and control groups. Similar results have also been obtained in Japanese patients with PCO (Kasuga, 1980). Although these studies did not assess the adrenal contribution to the androgen excess they did indicate that the ovary was important in producing excessive amounts of androgens.

We may conclude that there is an important adrenal source of androgen excess in perhaps half of all hirsute women, but that in most the ovary is a more important source. Except in those rare patients where adult onset congenital adrenal hyperplasia mimics PCO or idiopathic hirsutism, there is no evidence for a primary adrenal abnormality, and the adrenal cortex may merely be responding to the prevailing hormonal environment. Because of problems of definition and the drawbacks involved in suppression/stimulation studies it has not been possible to differentiate the syndromes of idiopathic hirsutism and PCO on the basis of the relative contribution of the adrenal glands and the ovaries to the androgen excess. From a practical point of view there seems to be little point in performing suppression or stimulation tests in order to try to define those patients in whom the adrenal is contributing significantly to the androgen excess. A single measurement of plasma DHEA-S gives equally reliable information (Wild *et al.,* 1982).

The role of glucocorticoids in the treatment of idiopathic hirsutism and PCO

Suppression of adrenal androgen production can be achieved by giving small doses of glucocorticoids, the most popular regimen being dexamethasone 0.5 mg at night, sometimes with 0.25 mg in the morning. There have been several clinical studies on the use of glucocorticoids in hirsutism (Perloff *et al.,* 1958; Jeffries and Levy, 1959; Casey *et al.,* 1966; Ettinger *et al.,* 1973; Casey, 1975; Abraham *et al.,* 1981; Moore *et al.,* 1983).

Abraham *et al.* (1981) selected their patients by means of a prolonged dexamethasone suppression test and treated those showing significant suppression of plasma androgens, regardless of menstrual history. They reported that 73%

showed moderate to marked improvement in hirsutism. Notwithstanding the objections to this dexamethasone suppression test their results appear to be much better than those reported by other groups. Ettinger *et al.* (1973) reported that only one-third of patients with either idiopathic hirsutism or PCO reported an improvement in hirsutism after three months on dexamethasone, although the serum testosterone level was suppressed in 80% – sometimes for several months after stopping treatment. In the same group, 50% reported improvement on subsequent treatment with medroxyprogesterone acetate and ethinyl oestradiol. Moore *et al.* (1983) reported that 58% of hirsute women with regular periods showed improvement of hirsutism with dexamethasone. Both these groups found that the response to treatment could not be predicted from the results of short-term dynamic tests. Casey (1975) found that betamethasone was as effective as a combination of ethinyl oestradiol and medroxyprogesterone acetate in reducing testosterone blood production rates, but was less effective clinically. A combination of the two forms of therapy was the most effective both biochemically and clinically, with 60% of patients having moderate success in reducing hair growth.

In these studies (as in most studies on the treatment of hirsutism) there is a lack of objective data on changes in hair growth. The general impression is that glucocorticoids are less effective then either combined oral contraceptive agents or a combination of ethinyl oestradiol and cyproterone acetate in the treatment of hirsutism.

In those who are infertile and wish to conceive, glucocorticoids are the only well recognized endocrine treatment for hirsutism which will not interfere with fertility. Indeed, there is evidence that glucocorticoids may improve menstrual regularity and cause a return to ovulatory menstrual cycles (Smith, Steinberger and Perloff, 1965; Casey *et al.*, 1966; Sciarra *et al.*, 1976; Rodriguez-Rigau *et al.*, 1979; Abraham *et al.*, 1981).

Adrenal tumours

Functioning adrenal tumours may be either adenomas or carcinomas and are a rare cause of hirsutism. The most recent review of virilizing adrenal adenomas is that of Gabrilove *et al.* (1981). All 34 cases of androgen-secreting adenomas and the eight androgen-secreting carcinomas in their series were in women, possibly because they are easier to detect in females. Indeed, apparently non-functioning adrenal carcinomas are more common in men (Rapaport *et al.*, 1952; McFarlane, 1958; Lipsett, Hertz and Ross, 1963; Hutter and Kayhoe, 1966). The age range at presentation for virilizing adenoma was 15–76 years, with eight patients being over 50 years. Of those with virilizing adrenal carcinoma, seven presented at 28–49 years of age and one at 16 years of age. For adenomas the duration of symptoms ranged from 6 months to 14 years. Marked hirsutism was the most prominent symptom, with amenorrhoea, enlargement of the clitoris and deepening of the voice being other common features. The tumour size varied from 1 × 1 cm to 16 cm in diameter. The clinical features of patients with virilizing adrenal carcinomas do not differ from those with virilizing adenomas. However, functioning adrenal carcinomas often present with Cushing's syndrome as well as virilization (Bertagna and Orth, 1981). Virilizing adenomas in children are rare while carcinomas are a little more common.

The majority of patients with virilizing adrenal tumours have markedly elevated serum testosterone levels. Suppression and stimulation tests are unreliable in diagnosing virilizing adenomas and are no longer necessary since the vast majority of tumours should be picked up by computerized tomography or ultrasonography. If these methods fail to reveal the diagnosis, then selective venous catheterization studies should be employed.

Where adrenal androgen DHEA and DHEA-S have been measured in cases of virilizing adenoma they do not appear to have been of value in localizing the problem to the adrenal. DHEA was normal in three out of six cases, as was the DHEA-S. While most virilizing tumours will suppress gonadotropin secretion, gonadotropin-responsive virilizing adrenal adenomas have been described (Werk, Sholiton and Kalejs, 1973; Givens *et al.*, 1974; Larson *et al.*, 1975; De Lange, Pratt and Doorenbos, 1980). This indicates that reliance on the response to human chorionic gonadotropin (HCG) stimulation tests to distinguish ovarian from adrenal androgen production may be misleading.

Although virilizing adrenal tumours are rare they should be suspected in any woman with a recent onset of hirsutism or other signs of virilization (e.g. amenorrhoea, clitoral enlargement). A marked elevation of serum testosterone (above 5 nmol/l) also indicates a virilizing tumour. Diagnosis and distinction from ovarian tumours relies on computerized tomography (CT), ultrasonography and, if necessary, selective venous catheterization.

Cushing's syndrome

Although hirsutism may be present in Cushing's syndrome it is usually mild and other symptoms predominate. The presence of severe hirsutism in Cushing's syndrome, particularly is associated with major virilization, should suggest the diagnosis of adrenocortical carcinoma.

Very recently, Hauffa, Kaplan and Grumbach (1984) have studied adrenal androgen levels in children with Cushing's disease and found that Δ^5-androgens are not elevated above what would be expected for the patient's age. They argue that this speaks against a role of intra-adrenal cortisol gradients or ACTH in maintaining the zona reticularis/adrenarche. The dynamics of ACTH and cortisol secretion in Cushing's disease are clearly quite abnormal, so these findings do not exclude a physiological effect which requires cortisol to be released episodically.

CONCLUSIONS

Evidently there is still much that we do not understand about the adrenarche, adrenal androgen secretion and the role of the adrenal glands in hirsutism. Intra-adrenal factors can no longer be ignored, and undoubtedly have a role in adrenal androgen production and development of the zona reticularis (*see also* Chapter 1). In hirsutism, the problem is compounded by variations in skin responsiveness to androgens, by differences of definition in different series, and by the fallibility of dynamic tests and catheterization studies. In most cases, the adrenals appear to be a less major source of biologically active androgen than the ovaries. Overt enzyme deficiencies or virilizing tumours are rare causes of hirsutism that can usually be confirmed easily by the appropriate biochemical tests and CT scanning.

References

ABRAHAM, G. E. (1974) Ovarian and adrenal contribution to the peripheral androgens during the menstrual cycle. *Journal of Clinical Endocrinology and Metabolism,* **39,** 340–346

ABRAHAM, G. E., BUSTER, J., KYLE, F., CORRALES, P. C. and TELLER, R. C. (1973) Radioimmunoassay of plasma pregnenolone, 17-hydroxypregnenolone and DHA under various physiological conditions. *Journal of Clinical Endocrinology and Metabolism,* **37,** 140–144

ABRAHAM, G. E. and CHAKMAKJIAN, Z. H. (1974) Plasma steroids in hirsutism. *Obstetrics and Gynecology,* **44,** 171–175

ABRAHAM, G. E., CHAKMAKJIAN, Z. H., BUSTER, J. E. and MARSHALL, J. R. (1975) Ovarian and adrenal contributions to peripheral androgens in hirsute women. *Obstetrics and Gynecology,* **46,** 169–173

ABRAHAM, G. E., MAROULIS, G. B., BOYES, S. P., BUSTER, J. E., HAGYAR, D. M. and ELSNER, C. W. (1981) Dexamethasone suppression test in the management of hyperandrogenised patients. *Obstetrics and Gynecology,* **57,** 158–165

ALBRIGHT, F. (1947) Osteoporosis. *Annals of Internal Medicine,* **27,** 861–882

ALBRIGHT, F., SMITH, P. H. and FRASER, R. (1942) A syndrome characterized by primary ovarian insufficiency and decreased stature: report of 11 cases with a digression on hormonal control of axillary and pubic hair. *American Journal of Medical Science,* **204,** 625–648

ANDERSON, D. C. (1976) The role of sex hormone binding globulin in health and disease. In *The Endocrine Function of the Human Ovary,* edited by V. H. T. James, M. Serio and G. Giusti, pp. 141–158. London: Academic Press

ANDERSON, D. C. (1980) The adrenal androgen stimulating hormone does not exist. *Lancet,* **2,** 8192, 454–456

ANDERSON, D. C., CHILD, D. F. and BU'LOCK, D. E. (1978) Adrenal steroids in hirsutism and related conditions. In *The Endocrine Function of the Human Ovary,* edited by V. H. T. James, M. Serio, G. Giusti and L. Martini, pp. 301–323. London: Academic Press

ANDERSON, D. C., HOPPER, B. R., LASLEY, B. L. and YEN, S. S. C. (1976) A simple method for the assay of eight steroids in small volumes of plasma. *Steroids,* **28,** 179–195

ANDERSON, D. C. and YEN, S. S. C. (1976) Effects of estrogens on adrenal 3β-hydroxysteroid dehydrogenase in ovariectomised women. *Journal of Clinical Endocrinology and Metabolism,* **43,** 561–570

BARDIN, C. W. and LIPSETT, M. B. (1967) Testosterone and androstenedione blood production rates in normal women and women with idiopathic hirsutism or polycystic ovaries. *Journal of Clinical Investigation,* **46,** 891–902

BASSI, F., GIUSTI, G., BORSI, L. *et al.* (1977) Plasma androgens in women with hyperprolactinaemic amenorrhoea. *Clinical Endocrinology,* **6,** 5–10

BASSI, F., PUPI, A., GIANNOTT, P. *et al.* (1980) Plasma dehydroepiandrosterone sulphate in hypothyroid premenopausal women. *Clinical Endocrinology,* **13,** 111–113

BELL, J. B. G., GOULD, R. P., HYATT, P. J., TAIT, J. F. and TAIT, S. A. S. (1979) Properties of rat adrenal zona reticularis cells: production and stimulation of certain steroids. *Journal of Endocrinology,* **83,** 435–447

BERTAGNA, C. and ORTH, D. N. (1981) Clinical and laboratory findings and results of therapy in 58 patients with adrenocortical tumors admitted to a single medical center (1951 to 1978). *American Journal of Medicine,* **71,** 855–874

BLANKSTEIN, J., FAIMAN, C., REYES, F. J., SCHROEDER, M. L. and WINTER, J. S. D. (1980) Adult onset familial adrenal 21-hydroxylase deficiency. *American Journal of Medicine,* **68,** 441–448

BROOKS, R. B. and PRUNTY, F. T. G. (1960) Patterns of steroid secretion in three types of post-pubertal hirsutism. *Journal of Endocrinology,* **21,** 263–276

BROWN, T., GINZ, B. and OAKEY, R. (1978) Effects of polypeptide hormone preparations on steroid production by the human fetal adrenal. *Journal of Endocrinology,* **79,** 60P–61P

BURGER, H. G., KENT, J. R. and KELLIE, A. E. (1964) Determination of testosterone in human peripheral and adrenal venous plasma. *Journal of Clinical Endocrinology and Metabolism,* **24,** 432–441

CALANDRA, R. S., PURVIS, K., NAESS, O., ATTRAMADAL, A., DJOSELAND, O. and HANSSON, V. (1978) Androgen receptors in the rat adrenal gland. *Journal of Steroid Biochemistry,* **9,** 1009–1015

CARTER, J. N., TYSON, J. E., WARNE, G. L., MCNEILLY, A. S., FAIMAN, C. and FRIESEN, H. G. (1977) Adrenocortical function in hyperprolactinaemic women. *Journal of Clinical Endocrinology and Metabolism,* **45,** 973–980

CASEY, J. H. (1975) Chronic treatment regimes for hirsutism in women: effect on blood production rates of testosterone and on hair growth. *Clinical Endocrinology,* **4,** 313–325

CASEY, J. H., BURGER, H. G., KENT, J. R. *et al.* (1966) Treatment of hirsutism by adrenal and ovarian suppression. *Journal of Clinical Endocrinology and Metabolism,* **26,** 1370–1374

CATHELINEAU, G., BRERAULT, J.-L., FIET, J., JULIEN, R., DREUX, C. and CANIVET, J. (1980) Adrenocortical 11β-hydroxylation defect in adult women with post-menarchial onset of symptoms. *Journal of Clinical Endocrinology and Metabolism,* **51,** 287–291

CATTANEO, S., FORTI, G., FIORELLI, G., BARBERIA, U. and SERIO, M. (1975) A rapid radioimmunoassay for determination of DHAS in human plasma. *Clinical Endocrinology,* **4,** 505–514

CAVALLERO, C. and CHAPPINO, G. (1962) Histochemistry of steroid-3β-ol-dehydrogenase in human adrenal cortex. *Experientia,* **18,** 119–120

CHANG, R. J., LAUTER, L. R., MELDRUM, D. R. *et al.* (1983) Steroid secretion in polycystic ovarian disease after ovarian suppression by a long-acting gonadotrophin releasing hormone agonist. *Journal of Clinical Endocrinology and Metabolism,* **56,** 897–903

CHILD, D. F., BU'LOCK, D. E. and ANDERSON, D. C. (1979a) Adrenal steroidogenesis in heterozygotes for 21-hydroxylase deficiency. *Clinical Endocrinology,* **11,** 391–398

CHILD, D. F., BU'LOCK, D. E. and ANDERSON, D. C. (1979b) Heterogeneity in adrenal steroidogenesis in normal men and women. *Clinical Endocrinology,* **11,** 383–389

CHILD, D. F., BU'LOCK, D. E. and ANDERSON, D. C. (1980) Adrenal steroidogenesis in hirsute women. *Clinical Endocrinology,* **12,** 595–601

CHROUSOS, G. P., LORIAUX, D. L., MANN, D. L. and CUTLER, G. B. (1982) Late onset 21-hydroxylase deficiency mimicking idiopathic hirsutism or polycystic ovarian disease. *Annals of Internal Medicine,* **96,** 143–148

CONWAY, D. I., ANDERSON, D. C. and BU'LOCK, D. E. (1982) The steroid response to controlled adrenal stimulation in congenital adrenal hyperplasia. *Clinical Endocrinology,* **16,** 215–226

CONWAY, D. I., ANDERSON, D. C., GORDON, M. T., BU'LOCK, D. E. and HILLIER, V. F. (1983) Adrenal progesterone and 17α-hydroxyprogesterone and cortisol responses to Synathen in normal women and women with various gynaecological disorders. *Clinical Endocrinology,* **19,** 77–85

COPELAND, K. C., PAUNIER, L. and SIZONENKO, P. C. (1977) The secretion of adrenal androgens and growth patterns of patients with hypogonadotrophic hypogonadism and idiopathic delayed puberty. *Journal of Pediatrics,* **91,** 985–990

CUNNINGHAM, G. R., GOLDZIEHER, J. W., DE LA PENA, A. and OLIVER, M. (1978) The mechanism of ovulation induction by triamcinolone acetonide. *Journal of Clinical Endocrinology and Metabolism,* **46,** 8–14

CUTLER, G. B. JR, DAVIS, S. E., JOHNSONBOUGH, R. E. and LORIAUX, D. L. (1979) Dissociation of cortisol and androgen secretion in patients with secondary adrenal insufficiency. *Journal of Clinical Endocrinology and Metabolism,* **49,** 604–609

CUTLER, G. B. JR, GLENN, M., BUSH, M., HODGEN, G. D., GRAHAM, G. E. and LORIAUK, D. C. (1978) Adrenarche: a survey of rodents, domestic animals and primates. *Endocrinology,* **103,** 2112–2118

DAVISON, B., LARGE, D. M., ANDERSON, D. C. and ROBERTSON, W. R. (1983) Basal steroid production by the zona reticularis of the guinea-pig adrenal cortex. *Journal of Steroid Biochemistry,* **18,** 285–290

DAWSON, I. M. P., PRYSE-DAVIS, J. and SNAPE, I. M. (1962) The distribution of six enzymes and of lipid in the human and rat adrenal cortex before and after administration of steroid and ACTH with comments on the distribution in human fetuses and in some natural disease conditions. *Journal of Pathology and Bacteriology,* **81,** 181–195

DELANGE, W. E., PRATT, J. J. and DOORENBOS, H. (1980) A gonadotrophin-responsive testosterone producing adrenocortical adenoma and high gonadotrophin levels in an elderly woman. *Clinical Endocrinology,* **12,** 21–28

DE PERETTI, E. and FOREST, M. (1978) Pattern of plasma DHAS levels in humans from birth to adulthood: evidence for testicular production. *Journal of Clinical Endocrinology and Metabolism,* **47,** 572–580

DHOM, G. (1973) The prepubertal and pubertal growth of the adrenal. *Beitrage zur Pathologie,* **150,** 357–377

DIGNAM, W. J., PION, R. J., LAMB, E. J. and SIMMER, H. H. (1964) Plasma androgens in women. II. Patients with polycystic ovaries and hirsutism. *Acta Endocrinologica,* **45,** 254–271

DO, S. T., LOOSE, D. S. and FELDMAN, D. (1979) Heterogeneity of glucocorticoid binders: a unique and a classical dexamethasone binding site in bovine tissues. *Endocrinology,* **105,** 1055–1063

DOBBIE, J. W., MACKAY, A. M. and SYMINGTON, T. (1968) The structure and functional zonation of the human adrenal cortex. In *The Investigation of Hypothalamic–Pituitary–Adrenal Function,* edited by V. H. T. James and J. Landon. *Memoirs of the Society for Endocrinology,* **17,** 103–111

DUCHARME, J. R., FOREST, M. G., DE PERETTI, E., SEMPE, M., COLLU, R. and BETRAND, J. (1976) Plasma adrenal and gonadal sex steroids in human pubertal development. *Journal of Clinical Endocrinology and Metabolism,* **42,** 468–476

ETTINGER, B., GOLDFIELD, E. B., BURVILL, K. C., VON WERDER, K. and FORSHAM, P. H. (1973) Plasma testosterone stimulation–suppression dynamics in hirsute women. Correlation with long-term therapy. *American Journal of Medicine,* **54,** 195–200

FORCHIELLI, E., SORCINI, G., NIGHTINGALE, M. S. *et al.* (1963) Testosterone in human plasma. *Analytical Biochemistry,* **5,** 146–421

FOREST, M. G. (1978) Age-related response to plasma testosterone, Δ^4-androstenedione and cortisol to adrenocorticotrophin in infants, children and adults. *Journal of Clinical Endocrinology and Metabolism,* **47,** 931–937

GABRILOVE, J. L., SAMAN, A. T., SABET, R., MITTY, H. A. and NICOLIS, G. L. (1981) Virilising adrenal adenoma with studies on the adrenal venous effluent and a review of the literature. *Endocrine Reviews,* **2,** 462–470

GABRILOVE, J. L., SHARMA, D. C. and DORFMAN, R. I. (1965) Adrenocortical 11β-hydroxylase deficiency and virilism first manifest in the adult woman. *New England Journal of Medicine,* **272,** 1189–1194

GENAZZANI, A. R., PINTO, C., FACCHINETTI, F., CARBONI, G., PELOSI, V. and CORDA, R. (1978) Adrenal and gonadal steroids in girls during sexual maturation. *Clinical Endocrinology,* **8,** 15–25

GIBSON, M., LACKRITZ, R., SCHIFF, I. and TULCHINSKY, D. (1980) Abnormal adrenal responses to adrenocorticotrophic hormone in hyperandrogenic women. *Fertility and Sterility,* **33,** 43–48

GIVENS, J. R., ANDERSON, R. N., WISER, W. L., COLEMAN, S. A. and FISH, S. A. (1974) A gonadotrophin-responsive adrenocortical adenoma. *Journal of Clinical Endocrinology and Metabolism,* **38,** 126–133

GIVENS, J. R., ANDERSON, R. N., RAGLAND, J. B., WISER, W. L. and UMSTOT, E. S. (1975) Adrenal function in hirsutism. 1. Diurnal change and response of plasma A, T, 17Po, F, LH and FSH to dexamethasone and ½ unit of ACTH. *Journal of Clinical Endocrinology and Metabolism,* **40,** 988–1000

GIVENS, J. R., ANDERSON, R. N., UMSTOT, E. S. and WISER, W. L. (1976) Clinical findings and hormone responses in patients with polycystic ovarian disease with normal versus elevated LH levels. *Obstetrics and Gynecology,* **47,** 388–394

GOLDZIEHER, J. W. (1968) Polycystic ovarian disease. In *Progress in Infertility,* edited by S. J. Behrman and R. W. Kistner, p. 351. Boston: Little, Brown & Co.

GRUMBACH, M. M., RICHARDS, G. E., CONTE, F. A. and KAPLAN, S. L. (1978) Clinical disorders of adrenal function and puberty: an assessment of the role of the adrenal cortex in normal and abnormal puberty in man and evidence for an ACTH-like pituitary adrenal androgen stimulating hormone. In *The Endocrine Function of the Human Adrenal Cortex,* edited by V. H. T. James, M. Serio and G. Giusti. London: Academic Press

HAMILTON, J. B. (1958) Age, sex and genetic factors in the regulation of hair growth in man: a comparison of Caucasian and Japanese populations. In *The Biology of Hair Growth,* edited by W. Montagna and R. R. Ellis, pp. 399–433. New York and London: Academic Press

HAUFFA, B. P., KAPLAN, S. and GRUMBACH, M. M. (1984) Dissociation between plasma adrenal androgens and cortisol in Cushing's disease and ectopic ACTH-producing tumour; relation to adrenarche. *Lancet,* **1,** 1373–1375

HERRERA-JUSTINIANO, E., GALVEZ, M., AZNAR, A., GOMEZ, S., SANDON, P., ZURITA, A., MALAGON, C. and AZNAR, R. (1979) Changes in the plasma levels of androstenedione, DHA and cortisol after stimulation with ACTH and HCG and suppression with dexamethasone during male puberty. *Acta Endocrinologica,* **90,** 113–121

HOPPER, B. R. and YEN, S. S. C. (1975) Circulating concentrations of dehydroepiandrosterone and dehydroepiandrosterone sulfate during puberty. *Journal of Clinical Endocrinology and Metabolism,* **40,** 458–461

HORNSBY, P. J. (1980) Regulation of cytochrome P-450-supported 11β-hydroxylation of deoxycortisol by steroids, oxygen and antioxidants in adrenocortical cell cultures. *Journal of Biological Chemistry,* **255,** 4020–4027

HORNSBY, P. J. (1982) Regulation of 21-hydroxylase activity by steroids in cultured bovine adrenocortical cells: possible significance of adrenocortical androgen synthesis. *Endocrinology,* **111,** 1092–1101

HORNSBY, P. J. and CRIVELLO, J. F. (1983) The role of lipid peroxidation and biological antioxidants in the function of the adrenal cortex. Part II. *Molecular and Cellular Endocrinology,* **30,** 123–148

HORTON, R. and TAIT, J. F. (1966) Androstenedione production and interconversion rates measured in peripheral blood and studies on the possible site of its conversion to testosterone. *Journal of Clinical Investigation,* **45,** 301–313

HSEUH, A. J. W. and JONES, P. B. C. (1981) Extrapituitary actions of gonadotrophin releasing hormones. *Endocrine Reviews,* **2,** 437–461

HUDSON, B., COGHLAND, J., DULMANIS, A., WINTOUR, M. and EKKEL, I. (1963) The estimation of testosterone in biological fluids. I. Testosterone in plasma. *Australian Journal of Experimental Biology and Medical Science,* **41,** 235–246

HUTTER, A. M. and KAYHOE, D. E. (1966) Adrenal cortical carcinoma. Clinical features of 138 patients. *American Journal of Medicine,* **41,** 572–580

HYATT, P. J., BELL, J. B. G., BHATT, K. and TAIT, J. F. (1983) Preparation and steroidogenic properties of purified zona fasciculata and zona reticularis cells from the guinea-pig adrenal gland. *Journal of Endocrinology,* **96,** 1–14

IRVINE, W., TOFT, A., WILSON, K., FRASER, R., WILSON, A., YOUNG, J., HUNTER, W., ISMAIL, A. and BURGER, P. (1974) The effect of synthetic corticotropin on adrenocortical anterior pituitary and testicular function. *Journal of Clinical Endocrinology and Metabolism,* **39,** 522–529

JEFFRIES, W. MCK. and LEVY, R. P. (1959) Treatment of ovarian dysfunction with small doses of cortisone or hydrocortisone. *Journal of Clinical Endocrinology and Metabolism,* **19,** 1069–1079

JENNER, M. R., KELCH, R. P., KAPLAN, S. L. and GRUMBACH, M. M. (1972) Hormonal changes in puberty. IV. Plasma estradiol, LH and FSH in prepubertal children, pubertal females and in precocious puberty, premature thelarche, hypogonadism and in a child with a feminising ovarian tumour. *Journal of Clinical Endocrinology and Metabolism,* **34,** 521–530

JONES, T. and GRIFFITHS, K. (1968) Ultramicrochemical studies on the site of formation of dehydroepiandrosterone sulphate in the adrenal cortex of the guinea-pig. *Journal of Endocrinology,* **42,** 559–565

JUDD, H. L. and YEN, S. S. C. (1973) Serum androstenedione and testosterone levels during the menstrual cycle. *Journal of Clinical Endocrinology and Metabolism,* **36,** 475–477

KANDEEL, F. R., LONDON, D. R., BUTT, W. R. *et al.* (1980) Adrenal function in subgroups of the polycystic ovary synchome assessed by a long ACTH test. *Clinical Endocrinology,* **13,** 601–612

KASUGA, Y. (1980) Ovarian steroidogenesis in Japanese patients with polycystic ovary syndrome. *Endocrinolica Japonica,* **27,** 541–550

KELNAR, C. J. H. and BROOK, C. G. D. (1983) A mixed longitudinal study of adrenal steroid excretion in childhood and the mechanism of adrenarche. *Clinical Endocrinology,* **19,** 117–129

KENNY, F. M., PREEYASOMBAT, C. and MIGEON, C. J. (1966) Cortisol production rate. II. Normal infants, children and adults. *Pediatrics,* **37,** 34–42

KIRSCHNER, M. A. and BARDIN, C. W. (1972) Androgen production and metabolism in normal and virilised women. *Metabolism,* **21,** 667–688

KIRSCHNER, M. A. and JACOBS, J. B. (1971) Combined ovarian and adrenal vein catheterisation to determine the site of androgen overproduction in hirsute women. *Journal of Clinical Endocrinology and Metabolism,* **33,** 199–209

KIRSCHNER, M. A., LIPSETT, M. B. and COLLINS, D. R. (1965) Plasma ketosteroids and testosterone in man. *Journal of Clinical Investigation,* **44,** 657–665

KIRSCHNER, M. A., SINKAMAHAPATRA, S., ZUCKER, I. R., LORIAUX, L. and NIESCHLAG, E. (1973) The production, origin and role of dehydroepiandrosterone and 5-androstenediol as androgen pre-hormones in hirsute women. *Journal of Clinical Endocrinology and Metabolism,* **37,** 183–189

KIRSCHNER, M. A., ZUCKER, I. R. and JESPERSEN, D. L. (1976) Ovarian and adrenal vein catheterisation studies in women with idiopathic hirsutism. In *The Endocrine Function of the Human Ovary,* edited by V. H. T. James, M. Serio and G. Giusti, pp. 443–456. London: Academic Press

KORENMAN, S. G., KIRSCHNER, M. A. and LIPSETT, M. B. (1965) Testosterone production in normal and virilised women and in women with the Stein-Steventhal syndrome or idiopathic hirsutism. *Journal of Clinical Endocrinology and Metabolism,* **25,** 798–803

KORTH-SCHUTZ, S., LEVINE, L. S. and NEW, M. I. (1976) Serum androgens in normal prepubertal and pubertal children and in children with precocious adrenarche. *Journal of Clinical Endocrinology and Metabolism,* **42,** 117–124

LAATIKAINEN, T. J., APTER, D. L., PAAVONEN, J. A. and WAHLSTROM, T. R. (1980) Steroids in ovarian and peripheral venous blood in polycystic ovarian disease. *Clinical Endocrinology,* **13,** 125–134

LACHELIN, G. C. L., BARNETT, M., HOPPER, B. R., BRINK, G. and YEN, S. S. C. (1979) Adrenal function in normal women and women with the polycystic ovary syndrome. *Journal of Clinical Endocrinology and Metabolism,* **49,** 892–898

LACHELIN, G. C. L., JUDD, H. L., SWANSON, S. C., HAUCK, M. E., PARKER, D. C. and YEN, S. S. C. (1982) Long-term effects of nightly dexamethasone administration in patients with polycystic ovarian disease. *Journal of Clinical Endocrinology and Metabolism,* **55,** 768–773

LARSON, B. A., VANDERLAAN, W. P., JUDD, H. L. and MCCULLOUGH, D. (1975) A testosterone-producing adrenal cortical adenoma in an elderly woman. *Journal of Clinical Endocrinology and Metabolism,* **42,** 882–887

LEE, P. A. and GAREIS, F. J. (1976) Gonadotrophin and sex steroid response to luteinising hormone-releasing hormone in patients with premature adrenarche. *Journal of Clinical Endocrinology and Metabolism,* **43,** 195–197

LEE, P. A., KOWARSKI, A., MIGEON, C. J. and BLIZZARD, R. M. (1975) Lack of correlation between gonadotrophin and adrenal androgen levels in agonadal children. *Journal of Clinical Endocrinology and Metabolism,* **40,** 664–669

LEVINE, J., WOLFE, L. G., SCHIEBINGER, R. J., LORIAUX, D. L. and CUTLER, G. B. JR (1982) Rapid regression of fetal adrenal zone and absence of adrenal reticular zone in the marmoset. *Endocrinology,* **111,** 1797–1802

LIM, N. and DINGMAN, J. (1964) Androgenic adrenal hyperfunction in acromegaly. *New England Journal of Medicine,* **271,** 1189–1194

LIPSETT, M. B., HERTZ, R. and ROSS, G. T. (1963) Clinical and pathophysiological aspects of adrenocortical carcinoma. *American Journal of Medicine,* **35,** 374–383

LIPSETT, M. B. and RITER, B. D. (1961) Urinary steroids in postnatal adrenal hyperplasia with virilism. *Acta Endocrinologica,* **38,** 481–489

LLOYD, C. W., LOBOTSKY, J., SERGRE, E. J., KOBAYASHI, T., TAYMOR, M. L. and BATT, R. E. (1966) Plasma testosterone and urinary 17-ketosteroids in women with hirsutism and polycystic ovaries. *Journal of Clinical Endocrinology and Metabolism,* **26,** 314–324

LOBO, R. A. and GOEBELSMANN, U. (1980) Adult manifestation of congenital adrenal hyperplasia due to incomplete 21-hydroxylase deficiency. *American Journal of Obstetrics and Gynecology,* **138,** 720–726

LOBO, R. A., GRANGER, L., GOEBELSMANN, U. and MISCHELL, J. R. (1981) Elevations in unbound serum estradiol as a possible mechanism for inappropriate gonadotrophin secretion in women with PCO. *Journal of Clinical Endocrinology and Metabolism,* **52,** 156–158

MARTIN, K. O. and BLACK, V. H. (1982) Δ^4-Hydrogenase in guinea-pig adrenal; evidence of localisation in zona reticularis and age-related change. *Endocrinology,* **110,** 1749–1757

MARTIN, K. O. and BLACK, V. H. (1983) Effects of age and adrenocorticotrophin on microsomal enzymes in guinea-pig adrenal inner and outer cortices. *Endocrinology,* **112,** 573–579

MCFARLANE, D. A. (1958) Cancer of the adrenal cortex. The natural history, prognosis and treatment in a study of fifty-five cases. *Annals of the Royal College of Surgeons of England,* **23,** 155–186

MEIKLE, A. W., STRINGHAM, J. D., DOHMAN, L. I., LAGERQUIST, L. G. and TYLER, F. H. (1976) Plasma 5α-androstanediol, an androgen marker of hirsutism in women. *Transactions of the Association of American Physicians,* **89,** 133–143

MELDRUM, D. R., CHANG, R. J., LU, J., VALE, W., RIVIER, J. and JUDD, H. L. (1982) 'Medical oophorectomy' using a long-acting GnRH agonist – a possible new approach to the treatment of endometriosis. *Journal of Clinical Endocrinology and Metabolism,* **54,** 1081–1083

MIGEON, C., GREEN, O. and ECKERT, J. (1963) Study of adrenocortical function in obesity. *Metabolism,* **12,** 718–729

MIGEON, C., KELLER, A., LAWRENCE, B. *et al.* (1957) DHA and androstene levels in human plasma. Effect of age and sex; day to day and diurnal variations. *Journal of Clinical Endocrinology and Metabolism,* **17,** 1051–1062

MIGEON, C., ROSENWAKS, Z., LEE, P. A., URBAR, M. P. and BIAS, W. B. (1980) The attenuated form of congenital adrenal hyperplasia as an allelic form of 21-hydroxylase deficiency. *Journal of Clinical Endocrinology and Metabolism,* **51,** 647–649

MILLS, I. H. (1956) Adrenal steroid metabolism studies in pregnancy in relation to the general problem. *MD Thesis,* Cambridge University

MILLS, I. H. (1968) The control of adrenal precursors of 17-oxosteroids. In *The Investigation of Hypothalamo–Pituitary–Adrenal Function,* edited by V. H. T. James and J. Landon. *Memoirs of the Society for Endocrinology,* **17,** 83–100

MILLS, I. H., BROOKS, R. V. and PRUNTY, F. T. G. (1962) The relationship between the production of cortisol and of androgen in the human adrenal. In *The Human Adrenal Cortex,* edited by A. R. Currie, R. Symington and J. K. Grant, pp. 204–229. London: Churchill Livingstone

MOORE, A., MAGEE, F., CUNNINGHAM, S., CULLITON, M. and MCKENNA, T. J. (1983) Adrenal abnormalities in idiopathic hirsutism. *Clinical Endocrinology,* **18,** 391–399

MORROW, L. B., BURROW, G. N. and MULROW, P. J. (1967) Inhibition of adrenal protein synthesis by steroids *in vitro. Endocrinology,* **80,** 883–888

NEVILLE, A. M. and O'HARE, M. J. (1979) Aspects of structure, function and pathology. In *The Adrenal Gland,* edited by V. H. T. James, pp. 1–65. New York: Raven Press

NISHIKAWA, T. and STROTT, C. A. (1984) Cortisol production by cells from the outer and inner zones of the adrenal zones of the guinea-pig. *Endocrinology,* **114,** 486–491

OAKE, R. J., DAVIES, S. J., MCLACHLAN, M. S. F. and THOMAS, J. P. (1974) Plasma testosterone in adrenal and ovarian vein blood of hirsute women. *Quarterly Journal of Medicine,* **43,** 603–613

O'HARE, M. J., NICE, E. C. and NEVILLE, A. M. (1980) Regulation of androgen secretion and sulpho-conjugation in adult human adrenal cortex: studies with primary monolayer cultures. In *Adrenal Androgens,* edited by A. R. Genazzani, J. H. H. Thijsson and P. K. Siiteri, pp. 7–25. New York: Raven Press

OSBORN and YANNONE, M. E. (1971) Plasma androgens in the normal and androgenic female – a review. *Obstetrical and Gynecological Survey,* **26,** 195–228

PARKER, L. N., GRAL, T., PERRIGO, V. and SHOWSKY, R. (1981) Decreased adrenal androgen sensitivity to ACTH during aging. *Metabolism,* **30,** 601–604

PARKER, L. N., LIFAK, E. T., KAURAHARA, C. K., GEDULD, S. I. and KOZBUR, X. M. (1983) Angiotensin II potentiates ACTH stimulated adrenal androgen secretion. *Journal of Steroid Biochemistry,* **18,** 205–209

PARKER, L. N., LIFAK, E. T. and ODELL, W. D. (1983) A 60 000 molecular weight human pituitary glycopeptide stimulates adrenal androgen secretion. *Endocrinology,* **113,** 2092–2096

PARKER, L. N. and ODELL, W. D. (1979) Evidence for existence of cortical androgen stimulating hormone. *American Journal of Physiology,* **236,** 616–620

PARKER, L. N. and ODELL, W. D. (1980) Control of adrenal androgen secretion. *Endocrine Reviews,* **1,** 392–410

PERLOFF, W. H., CHANMICK, B. J., SUPLICK, B. and CARRINGTON, E. R. (1958) Clinical management of idiopathic hirsutism (adrenal virilism). *Journal of the American Medical Association,* **167,** 2041–2047

PRUNTY, F. T. G. (1956) Chemical and clinical problems of the adrenal cortex. *British Medical Journal,* **2,** 615–622, 673–686

PRUNTY, F. T. G. (1966) Androgen metabolism in man – some current concepts. *British Medical Journal,* **2,** 605–613

RAJ, S. G., RAJ, M. H. G., TALBERT, L. M., SLOAN, C. S. and HICKS, B. (1982) Normalisation of testosterone levels using a low estrogen-containing oral contraceptive in women with polycystic ovary syndrome. *Obstetrics and Gynecology,* **60,** 15–19

RAPAPORT, E., GOLDBERG, M. B., GORDAN, G. S. and HINMAN, F. (1952) Mortality in surgically treated adrenocortical tumours. *Postgraduate Medicine,* **11,** 325–353

REICHEL, W. (1968) Lipofuscin pigment accumulation and distribution in five rat organs as a function of age. *Journal of Gerontology,* **23,** 145–153

REITER, E. O., FULDAUER, V. G. and ROOT, A. W. (1977) Secretion of the adrenal androgen, dehydroepiandrosterone sulfate, during normal infancy, childhood and adolescence, in sick infants, and in children with endocrinological abnormalities. *Journal of Pediatrics,* **90,** 766–770

RICH, B. H., ROSENFELD, R. L., LUCKY, A. W., HELKE, J. C. and OTTO, P. (1981) Adrenarche: changing adrenal response to adrenocorticotrophin. *Journal of Clinical Endocrinology and Metabolism,* **52,** 1129–1136

RIFKA, S., CUTLER, G., SAUER, M. and LORIAUX, D. (1978) Rat adrenal androgen receptor: a possible mediator of androgen-induced decrease in rat adrenal weight. *Endocrinology,* **103,** 1103–1110

RIVAROLO, M. A., SINGLETON, R. T. and MIGEON, C. L. (1967) Splanchnic extraction and interconversion of testosterone and androstenedione in man. *Journal of Clinical Investigation,* **46,** 2095–2100

RODRIQUEZ-RIGAU, L. J., SMITH, K. D., TCHOLAKIAN, R. K. and STEINBERGER, E. (1979) Effects of prednisone on plasma testosterone levels and on duration of phases of the menstrual cycle in hyperandrogenic women. *Fertility and Sterility,* **32,** 408–413

ROMANOFF, L., MORRIS, C., WELCH, P., RODRIGUEZ, R. M. and PINCUS, G. (1961) The metabolism of cortisol-4-C^{14} in young and elderly men. *Journal of Clinical Endocrinology and Metabolism,* **21,** 1413–1425

ROSENFIELD, R. L. and EBERLEIN, W. R. (1969) Plasma 17-ketosteroid levels during adolescence. *Journal of Pediatrics,* **74,** 932–936

ROSENFIELD, R., HELLMAN, L., ROFFWARG, H., WEITZMAN, E., FUKUSHIMA, D. and GALLAGHER, T. (1971) Dehydroepiandrosterone is secreted episodically and synchronously with cortisol by normal men. *Journal of Clinical Endocrinology and Metabolism,* **33,** 87–92

ROSENWAKS, Z., LEE, P. A., JONES, G. S., MIGEON, C. L. and WENTZ, A. C. (1979) An attenuated form of congenital virilising adrenal hyperplasia. *Journal of Clinical Endocrinology and Metabolism,* **59,** 335–339

SAEZ, J. and BERTRAND, J. (1968) Studies on testicular function in children: plasma concentrations of testosterone, DHA and DHAS before and after HCG stimulation. *Steroids,* **12,** 749–761

SCIARRA, F., TOSCANO, V., CONCOLINO, G., SORCINI, G., MAROCCHI, A. and CONTI, C. (1976) Simultaneous estimation of four plasma androgens before and after dynamic tests in women with hirsutism: correlation with long-term therapy. *Hormone Research,* **7,** 16–27

SCHIEBINGER, R. J., ALBERTSON, B. D., CASSORLA, F. G. *et al.* (1981) The developmental changes in plasma adrenal androgens during infancy and adrenarche are associated with changing activities of adrenal microsomal 17-hydroxylase and 17-20 desmolase. *Journal of Clinical Investigation,* **67,** 1177–1182

SHARMA, D. C., FORCHIELLI, E. and DORFMAN, R. I. (1973) Inhibition of enzymatic steroid 11β-hydroxylation by androgens. *Journal of Biological Chemistry,* **238,** 572–575

SIZONENKO, P. C. and PAUNIER, L. (1975) Hormonal changes in puberty. III. Correlation of plasma dehydroepiandrosterone, testosterone, FSH and LH with stages of puberty and bone age in normal boys and girls and in patients with Addison's disease or hypogonadism or with premature or late adrenarche. *Journal of Clinical Endocrinology and Metabolism,* **41,** 894–904

SKLAR, C. A., KAPLAN, S. L. and GRUMBACH, M. M. (1980) Evidence for dissociation between adrenarche and gonadarche. Studies in patients with idiopathic precocious puberty, gonadal dysgenesis, isolated gonadotrophin deficiency and constitutionally delayed growth and adolescence. *Journal of Clinical Endocrinology and Metabolism,* **53,** 548–556

SKLAR, C. A., KAPLAN, S. L. and GRUMBACH, M. M. (1981) Lack of effect of oestrogens on adrenal androgen secretion in children and adolescents with a comment on oestrogens and pubic hair growth. *Clinical Endocrinology,* **14,** 311–320

SMAIL, P. J., FAIMAN, C., HOBSON, W. C., FULLER, G. B. and WINTER, J. S. D. (1982) Further studies on adrenarche in non-human primates. *Endocrinology,* **111,** 844–848

SMITH, K. D., STEINBERGER, E. and PERLOFF, W. H. (1965) Polycystic ovarian disease. *American Journal of Obstetrics and Gynecology,* **93,** 994–999

STAKL, N. L., TEESHINK, C. R. and GREENBLATT, R. B. (1973) Ovarian, adrenal and peripheral testosterone levels in the polycystic ovary syndrome. *American Journal of Obstetrics and Gynecology,* **117,** 194–200

STEIN, I. F. and LEVENTHAL, M. L. (1935) Amenorrhoea associated with bilateral polycystic ovaries. *American Journal of Obstetrics and Gynecology,* **29,** 181–186

STROTT, C. A., GOFF, A. K. and LYONS, C. D. (1981) Functional differences between the inner and outer zone of the guinea-pig adrenal cortex. *Endocrinology,* **109,** 2249–2251

THOMAS, P. K. and FERRIMAN, D. G. (1957) Variations in facial and pubic hair growth in white women. *American Journal of Physical Anthropology,* **15,** 171–180

TUNBRIDGE, D., RIPPON, A. E. and JAMES, V. H. T. (1973) Circadian relationships of plasma androgen and cortisol levels. *Journal of Endocrinology,* **59,** 21

VEJLSTED, H. and ALBRECHTSEN, R. (1976) Biochemical and clinical effects of ovarian wedge resection in the polycystic ovary syndrome. *Obstetrics and Gynecology,* **47,** 575–580

VERMEULEN, A. (1980) Adrenal androgens and aging. In *Adrenal Androgens,* edited by A. Genazzani, J. Thijssen and P. Siiteri, pp. 207–217. New York: Raven Press

VERMEULEN, A. (1981) Hirsutism and virilism. *Medicine International,* **1,** 297–301

VERMEULEN, A., DESLYPERE, J. P., SCHELFHOUT, W., VERDONCK, L. and RUBENS, P. (1982) Adrenocortical function in old age: response to acute adrenocorticotrophin stimulation. *Journal of Clinical Endocrinology and Metabolism,* **54,** 187–191

VERMEULEN, A., SUY, E. and RUBENS, R. (1977) Effect of prolactin on plasma DHEA(S) levels. *Journal of Clinical Endocrinology and Metabolism,* **44,** 1222–1227

VERMEULEN, A. and VERDONCK, L. (1976) Plasma androgen levels during the menstrual cycle. *American Journal of Obstetrics and Gynecology,* **125,** 491–494

WARNE, G. L., CARTER, J. N., FAIMAN, C., REYES, F. I. and WINTER, J. S. D. (1978) Hormonal changes in girls with precocious adrenarche: a possible role for estradiol or prolactin. *Journal of Pediatrics,* **92,** 743–747

WERK, E., SHOLITON, L. J. and KALEJS, L. (1973) Testosterone-secreting adrenal adenoma under gonadotrophin control. *New England Journal of Medicine,* **289,** 767–770

WEST, C., BROWN, H., SIMONS, E., CARTER, D. B., KUMAGAI, L. F. and ENGLERT, E. (1961) Adrenocortical function and cortisol metabolism in old age. *Journal of Clinical Endocrinology and Metabolism,* **21,** 1197–1207

WEST, C. D., MAHAJAN, D. K., CHAVRE, V. J., NABORS, C. J. and TYLER, F. H. (1973) Simultaneous measurement of multiple plasma steroids by radio-immunoassay demonstrating episodic secretion. *Journal of Clinical Endocrinology and Metabolism,* **36,** 1230–1236

WILD, R. A., UMSTOT, E. S., ANDERSEN, R. N. and GIVENS, J. R. (1982) Adrenal function in hirsutism. II. Effects of an oral contraceptive. *Journal of Clinical Endocrinology and Metabolism,* **54,** 676–681

WILSON, E. A., ERIKSON, G. F., ZARUTSKI, P., FINN, A. E., TULCHINSKY, D. and RYAN, K. J. (1979) Endocrine studies of normal and polycystic ovarian tissues *in vitro. American Journal of Obstetrics and Gynecology,* **134,** 56–63

WINTER, J. S. D., FUJIEDA, K., FAIMAN, C., REYES, F. I. and THLIVERIS, J. (1980) In *Adrenal Androgens,* edited by A. R. Genazzani, J. H. H. Thijssen and P. K. Siiteri, pp. 55–65. New York: Raven Press

WRIGHT, N. A. (1981) In *Cell Death in Biology and Pathology,* edited by I. D. Bowen and R. A. Lockshin, pp. 171–200. London: Chapman and Hall

YATES, J. and DESHPANDE, N. (1974) Kinetic studies on the enzymes catalysing the conversion of 17α-hydroxyprogesterone to androstenedione in the human adrenal gland *in vitro. Journal of Endocrinology,* **60,** 27–35

YEN, S. S. C. (1980) The polycystic ovary syndrome. *Clinical Endocrinology,* **12,** 177–208

YEN, S. S. C., CHANEY, C. and JUDD, H. L. (1976) Functional aberrations of the hypothalamic pituitary system in polycystic ovary syndrome: a consideration of the pathogenesis. In *The Endocrine Function of the Human Ovary,* edited by V. H. T. James, M. Serio and G. Giusti, pp. 373–389. New York: Academic Press

6
Congenital enzymatic defects of the adrenal

Maria I. New

INTRODUCTION

Congenital adrenal hyperplasia (CAH) consists of a group of disorders of adrenal steroidogenesis, each of which results from an inherited deficiency of one of several enzymes necessary for normal steroid synthesis. The different enzyme deficiencies produce characteristic patterns of hormonal abnormalities; the clinical symptoms of the different forms of CAH depend on the particular hormones that are either deficient or produced in excess.

The adrenal synthesizes three classes of hormones: mineralocorticoids (17-deoxy pathway), glucocorticoids (17-hydroxy pathway) and sex steroids. A simplified scheme of adrenal steroidogenesis is shown in *Figure 6.1.* A more detailed discussion of these processes can be found in Chapter 2 and published reviews (Finkelstein and Shaefer, 1979; New *et al.,* 1982).

FETAL SEXUAL DEVELOPMENT

In order to understand the pathophysiology of CAH, a brief review of normal sexual differentiation is necessary (*Figure 6.2*). Normal differentiation of male genitalia depends on two functions of the fetal testes:

(1) The Leydig cells secrete testosterone, which stimulates the Wolffian ducts to develop into male internal genitalia (the epididymis, vas deferens, seminal vesicles, and ejaculatory ducts). Androgen from the testes as well as that produced by the adrenal in CAH causes masculinization of the external genitalia.
(2) The Sertoli cells secrete a glycoprotein substance (anti-Müllerian hormone) that inhibits development of the female internal genitalia (Josso, Picard and Tran, 1977).

The fetal ovary secretes neither testosterone nor anti-Müllerian hormone and thus does not play a determining role in sex differentiation. However, female

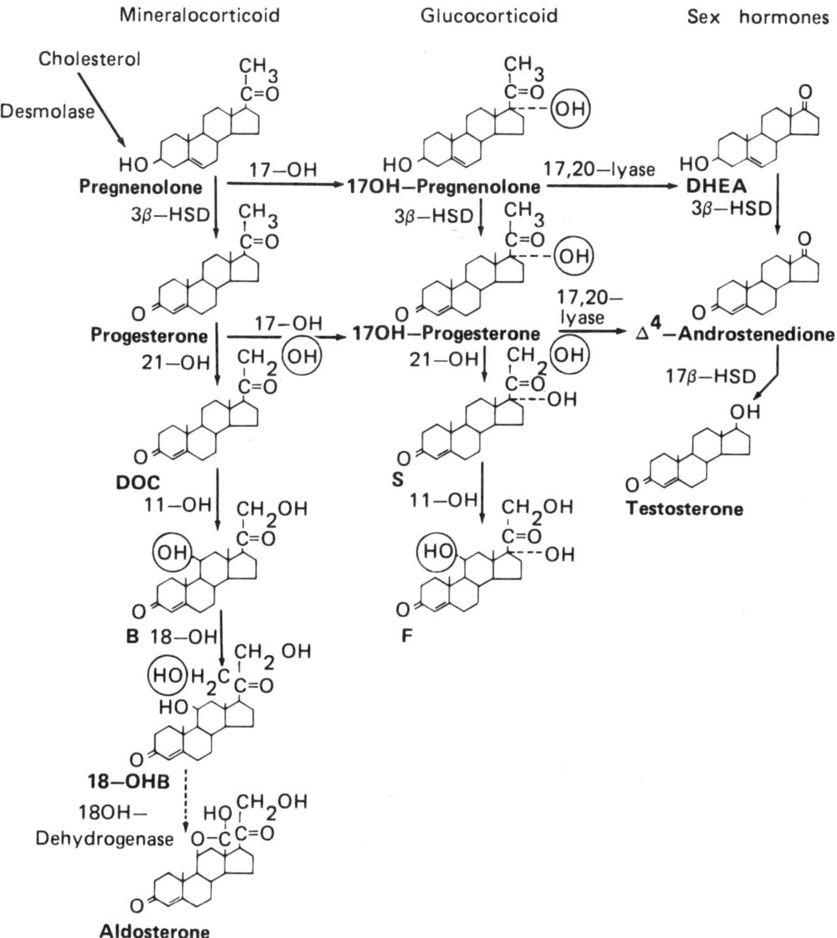

Figure 6.1 Simplified scheme of adrenal steroidogenesis. Each hydroxylation step is indicated, and the newly added hydroxyl group is circled. B = corticosterone; DOC = deoxycorticosterone; F = cortisol; DHEA = dehydroepiandrosterone; 3β-HSD = 3β-hydroxysteroid dehydrogenase; 18-OHB = 18-hydroxycorticosterone; S = 11-deoxycortisol. (From New and Levine, 1973, courtesy of the Publishers, *Advances in Human Genetics*)

fetuses exposed to high levels of androgen as a consequence of CAH, an androgen-producing tumor in the mother, or maternal ingestion of androgens, manifest virilization of the external genitalia; their internal genitalia remain female.

ENZYME DEFECTS IN CONGENITAL ADRENAL HYPERPLASIA

A summary of the clinical and biochemical features of the steroidogenic enzyme deficiencies is shown in *Table 6.1*. Adrenal hyperplasia occurs only in defects associated with cortisol deficiency. In general, the absolute level of hormone secretion in patients with errors of adrenal steroidogenesis is less informative than the relation of a hormonal level to a given stimulus, such as ACTH or angiotensin.

The best diagnostic measure for identifying the enzymatic defect is to examine the ratio of precursor to product when the tropic hormone is increased.

21-Hydroxylase deficiency

Impairment of 21-hydroxylation is the commonest enzymatic defect seen in CAH. Decreased cortisol synthesis induces increased ACTH secretion (Sydnor *et al.*, 1953; Binoux *et al.*, 1972; Fukushima *et al.*, 1975; Pham-Huu-Trung *et al.*, 1978; LaFranchi, 1980), and thus overproduction of cortisol precursors and sex steroids,

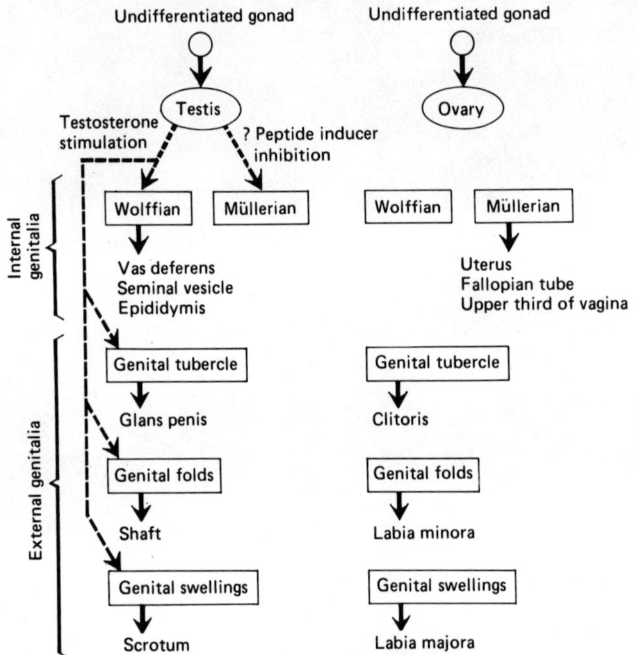

Figure 6.2 Fetal sex differentiation. (From New and Levine, 1973, courtesy of the Publishers, *Advances in Human Genetics*)

which do not require 21-hydroxylase for their biosynthesis. The diagnosis of 21-hydroxylase deficiency depends on the demonstration of elevated levels of the precursor 17OH-progesterone (17-OHP), with or without decreased levels of product (Pang *et al.*, 1979). An excessive rise of serum 17-OHP with ACTH administration is the best indicator of a 21-hydroxylase deficiency.

Virilization

The most prominent clinical feature of 21-hydroxylase deficiency is virilization. Prenatal virilization of the genetic female affects only the androgen-responsive external genitalia, while the internal genitalia develop normally. Affected females present with labioscrotal fusion and a urogenital sinus, in addition to an enlarged clitoris. Although the clitoris remains sensitive to the masculinizing effects of androgens throughout fetal life, and even postnatally, the vagina separates from

Table 6.1 Clinical and laboratory features of various disorders of adrenal steroidogenesis

Clinical features: Newborn with sexual ambiguity						Laboratory findings: Urinary excretion				Circulating hormones				
Female	*Male*	*Salt-wasting*	*Hyper-tension*	*Postnatal virilization*	*Enzyme deficiency*	*17-KS*	*17-OH*	*P'triol*	*Aldo*	*17-OHP*	*Δ^4-Steroids*	*DHEA*	*Testo-sterone*	*Renin*
					21-Hydroxylase	↑ ↑	*N or* ↓	↑ ↑	*N*	↑ ↑	↑ ↑	*N or* ↑	↑	*N or* ↑
+	0	0	0	+	simple virilizing							(DHEA/Δ^4 ↓)		
+	0	+	0	+	salt-wasting	↑ ↑	↓	↑ ↑	↓	↑ ↑	↑ ↑	N or ↑	↑	↑ ↑
0	0	0	0	+	'late onset'	↑	N	↑	N	↑	↑	N or ↑	N or ↑	N
+	0	0	+	+	11β-Hydroxylase	↑ ↑	↑ ↑*	↑	↓	↑	↑ ↑	↑	↑	↓ ↓
+	+	+	0	+	3β-HSD†	↑ §	↓ ↓	N or ↓	↓	N or ↑	N or ↑	↑ ↑ ↑	**	↑
0	+	0	+	0	17α-Hydroxylase	↓ ↓	↓ ↓	↓ ↓	↓	↓	↓	↓	↓	↓
0	+	+	0	0	Cholesterol desmolase	↓ ↓	↓ ↓	↓ ↓	↓ ↓	↓	↓	↓	↓	↑
0	0	+	0	0	18-Hydroxylase	N	N	N	↓ ↓	N	N	N	N	↑
0	0	+	0	0	Methyloxidase type II	N	↑ ††	N	↓ ↓	N	N	N	N	↑
?	+	–	–	+§§	17β-HSD***	N or ↑	N	N	N	N	↑ ↑	N or ↑	N or ↓ (Δ^4/T ↑ ↑)	N

* Mostly tetrahydro-11-deoxycortisol.
† The values presented apply to infants and very young children.
§ Mostly Δ^5-17-ketosteroids.
** ↓ or normal in male; ↑ or normal in female.
†† Largely 18-hydroxytetrahydro-aldosterone, which gives a Porter-Silber reaction.
§§ Only in males at puberty.
*** This defect may occur only in the testes.
N = normal.
17-OH = 17-hydroxysteroids; Aldo = aldosterone; DHEA = dehydroepiandrosterone; HSD = hydroxysteroid dehydrogenase; KS = ketosteroids (oxosteroids); 17-OHP = 17-hydroxyprogesterone; P'triol = pregnanetriol; T = testosterone; THA = tetrahydro-aldosterone.

the urogenital sinus before the twelfth week of gestation and then ceases to respond to the virilizing influence of androgens. Thus, maternal ingestion of androgens after the twelfth fetal week, while stimulating clitoral enlargement, will not cause labioscrotal fusion or formation of a urogenital sinus (Grumbach and Conte, 1981). The fetal adrenal cortex probably initiates steroidogenesis by the sixth week of gestation (*see* Chapter 2), but little is known about the timing of the onset of androgen secretion in the fetus with CAH.

A puzzling feature of the virilization is the development of the internal genital ducts in affected females. As expected, Müllerian duct development is normal because of the absence of testes, which are required for the secretion of Müllerian inhibiting factor. It is surprising that Wolffian duct development does not occur in these virilized females, despite elevated androgen levels. It has been suggested that higher local concentrations of testosterone may be required for Wolffian stimulation than for virilization of external genitalia and that the local presence of a testis is therefore required for the former (Grumbach and Conte, 1981). Federman (1968) has proposed that in 21-hydroxylase deficiency circulating levels of adrenal androgens do not achieve sufficient concentration in the Wolffian duct region. However, Wilson and Walsh (1979) have pointed out that in animal experiments androgens administered to the mother can stimulate Wolffian development in female fetuses. In CAH the major androgen secreted by the fetal adrenal is Δ^4-androstenedione (Villee, 1973), and it is likely that this androgen is not sufficiently potent to induce Wolffian development. Wilson and Walsh (1979) suggested that it is timing rather than degree or locality of androgen concentration that is responsible for the lack of Wolffian development in CAH. Yet Wolffian development is believed to take place during the same stage of fetal development as labioscrotal differentiation, making it difficult to account for the occurrence of one of these developmental events without the other solely on the basis of a timing factor.

Clarification of the process of virilization of females *in utero* must await more detailed studies of fetal adrenal function and fetal sexual differentiation. There has been a report of a female with 21-hydroxylase deficiency born with labioscrotal fusion but without clitoromegaly (Wolff *et al.,* 1977), an unexpected variation of the virilization caused by this disorder.

Males with 21-hydroxylase deficiency do not have genital abnormalities at birth. Postnatally, in both untreated males and untreated females, continued excessive androgen production results in rapid somatic growth, advanced epiphyseal maturation, progressive penile or clitoral enlargement, early appearance of facial, axillary and pubic hair, and acne. Without treatment, early epiphyseal closure and eventual short stature result.

Salt-wasting

Some infants with 21-hydroxylase deficiency develop signs of adrenal insufficiency, with low serum sodium and high serum potassium concentrations and vascular collapse. Life-threatening crises may occur within the first few weeks of life and early recognition of this disorder is therefore essential. It seems likely that the presence or absence of salt-wasting depends on both the level of aldosterone production and the anti-mineralocorticoid effect on the kidney of steroids such as 17OH-progesterone and progesterone (Janoski, 1977), which are present in very high concentrations.

11β-Hydroxylase deficiency

A defect, with excess production of 11-deoxycorticosterone (DOC) and 11-deoxycortisol, causes this hypertensive form of CAH. Markedly increased serum levels of these steroids, which increase further with ACTH administration, together with suppressed plasma renin activity (PRA), distinguish 11β-hydroxylase deficiency from 21-hydroxylase deficiency (Levine *et al.,* 1980a). Aldosterone values are usually within the normal range (Lim, Mimica and Dingman, 1969). Prepubertal gynecomastia is sometimes seen in genetic males with this disorder (Zachmann and Prader, 1975; Rösler and Leiberman, 1984) and may be a result of the elevated DOC levels (Lawrence, 1943).

Hypertension of varying severity is a common finding in patients with this form of CAH, but normotensive patients have been described. Cardiovascular complications are a function of the duration and severity of hypertension rather than of the age of the patient. There appears to be little correlation between severity of hypertension and degree of virilization; both mildly virilized females with severe hypertension leading to vascular accidents and normotensive, completely masculinized females have been reported (Rösler and Leiberman, 1980). The hypertension is thought to reflect excessive production of DOC, an aldosterone precursor with mineralocorticoid activity, but it seems likely that sodium intake is another variable, since normotensive patients with elevated DOC levels (Blunck and Bierich, 1968), as well as hypertensive patients with normal DOC levels (Glenthøj *et al.,* 1980), have been reported.

There have been reports of a 'late-onset' form of 11β-hydroxylase deficiency, with postmenarchial onset of menstrual disturbances and hirsutism (Cathelineau *et al.,* 1980). This clinical heterogeneity may be due to allelism with varying degrees of enzymatic deficiency, as has been demonstrated in 21-hydroxylase deficiency (*see below*).

3β-Hydroxysteroid dehydrogenase deficiency

Deficiency of 3β-hydroxysteroid dehydrogenase (3β-HSD) results in decreased synthesis of all three classes of adrenal steroids. In its most severe form there is impaired secretion of aldosterone, cortisol and testosterone, which leads to male pseudohermaphroditism and life-threatening salt-wasting in infancy. A surprising finding in female newborns with 3β-HSD deficiency is slight clitoral enlargement; this may be due to the extremely high levels of dehydroepiandrosterone (DHEA), a weak androgen which is secreted in very large amounts. Masculinization of the external genitalia in the male is incomplete, although sporadic cases of almost complete masculinization have been reported (Parks *et al.,* 1971). Salt-wasting symptoms secondary to aldosterone deficiency occur in varying degrees (Jänne, Perheentupa and Vihko, 1970).

Markedly elevated plasma levels of Δ^5-steroids, e.g. 17OH-pregnenolone and DHEA, and excessive urinary excretion of pregnenetriol, are diagnostic of 3β-HSD deficiency. Urinary metabolites of Δ^4-steroids, e.g. pregnanetriol, may also be elevated, possibly due to peripheral conversion from Δ^5-steroids. This may be a potential source of confusion with 21-hydroxylase deficiency if the overall pattern of urinary steroids is not carefully examined for the increased ratio of Δ^5/Δ^4 steroids.

A non-salt-losing form of 3β-HSD deficiency has been described in two siblings (Pang *et al.*, 1983). These children never had a crisis due to adrenal hormone deficiency, even under stress (including major surgery). Both manifested signs of mild androgen excess in mid-childhood, evidenced by premature pubarche, advanced linear growth and, in the girl, an enlarged clitoris, acne, and advanced skeletal maturation.

A 'late-onset' form has been described in older women presenting with menstrual irregularity and hirsutism in adult life (Bongiovanni, 1981). Lobo and Goebelsmann (1980) reported increased serum pregnenolone concentrations and an increased ratio of serum pregnenolone to progesterone in affected subjects; exogenous ACTH stimulation further increased this ratio. Treatment with corticoids suppressed the elevated levels of adrenal products and allowed normal, apparently ovulatory, menstrual cycles to occur.

Rosenfield *et al.* (1980) reported an abnormally high ratio of 17-hydroxypregnenolone to 17-hydroxyprogesterone in response to ACTH stimulation in the mother of a patient with late-onset 3β-HSD deficiency, while the father's response was within the normal range. The hormone response to ACTH of the mother of the patients reported by Pang *et al.* (1983) was also normal (unpublished observation). Thus it is not clear whether heterozygous carriers of this disorder can be identified by ACTH stimulation.

17α-Hydroxylase deficiency

A defect in 17α-hydroxylase results in diminished secretion of cortisol and sex steroids, and increased secretion of precursor steroids, especially the weak glucocorticoid corticosterone. Hypertension and hypokalemia can be attributed to excessive secretion of DOC. Sexual infantilism occurs in untreated females with 17α-hydroxylase deficiency. In males the androgen deficiency causes male pseudohermaphroditism, since differentiation of the androgen-dependent external genitalia is incomplete.

Laboratory tests show decreased serum androgen levels and decreased urinary excretion of 17-ketosteroids and 17-hydroxycorticoids. In many cases the aldosterone levels are also low. This is not a direct effect of the enzyme deficiency, since 17α-hydroxylation is not required for aldosterone synthesis, but rather the result of increased plasma volume, with secondary suppression of renin production and aldosterone synthesis (New, 1970). This effect disappears when excess DOC secretion is suppressed by dexamethasone (New, 1970; Kater *et al.*, 1982).

A PROVOCATIVE CONCEPTUAL APPROACH TO CONGENITAL ADRENAL HYPERPLASIA: THE FASCICULATA AND GLOMERULOSA AS TWO FUNCTIONALLY DISCRETE GLANDS

11β-Hydroxylase deficiency

When proposing that the adrenal zona fasciculata and zona glomerulosa could be viewed as two distinct glands under different regulatory control, New and Seaman (1970) pointed out that 11β-hydroxylase deficiency can be expressed to a greater degree in the fasciculata, with less effect on aldosterone secretion. This would

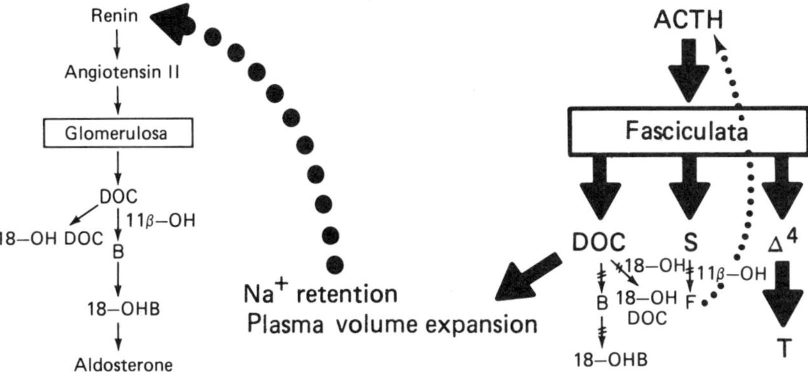

Figure 6.3 Regulation of adrenocortical steroidogenesis in untreated 11β-hydroxylase deficiency considering the fasciculata and glomerulosa as two separate glands. Dotted arrows indicate negative feedback. Elevated or diminished secretion levels are indicated by relative size of print. Note that the excessive secretion of deoxycorticosterone (DOC) due to an 11β-hydroxylase (11β-OH) defect impairing the conversion of DOC to corticosterone (B) by the fasciculata results in suppression of the glomerulosa. F = cortisol; Δ^4 = Δ^4-androstenedione; 18-OH = 18-hydroxylase; 18-OHB = 18-hydroxycorticosterone; S = 11-deoxycortisol; T = testosterone. (From New *et al.*, 1982, courtesy of the Publishers, *The Metabolic Basis of Inherited Disease*)

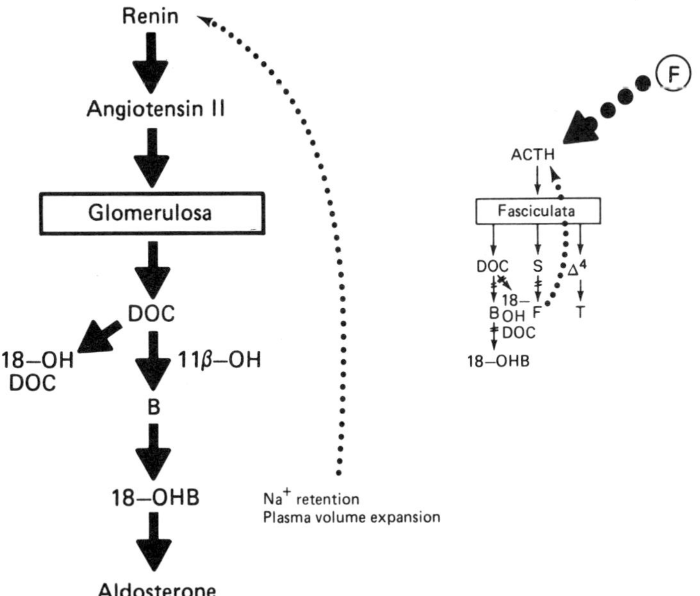

Figure 6.4 Regulation of adrenocortical steroidogenesis in treated 11β-hydroxylase deficiency. Dotted arrows indicate negative feedback. Elevated or diminished synthesis is indicated by relative size of print. Ⓕ represents glucocorticoid administered in therapy, suppressing fasciculata secretion of deoxycorticosterone (DOC) and permitting renin–angiotensin stimulation of the glomerulosa, which is not affected with an 11β-hydroxylase defect. For key *see Figure 6.3*. (From New *et al.*, 1982, courtesy of the Publishers, *The Metabolic Basis of Inherited Disease*)

explain the rise in aldosterone secretion in this disorder after suppression of excessive DOC levels (*Figures 6.3* and *6.4*). Thus in the glomerulosa, fasciculata-derived 11β-hydroxylase may be not rate-limiting and sodium depletion may lead to an appropriate increase in aldosterone secretion.

This hypothesis has been supported by subsequent reports (Sizonenko *et al.*, 1972; Gregory and Gardner, 1976) and confirmed by the recent extensive study of four patients with 11β-hydroxylase deficiency (Levine *et al.*, 1980a), in which the fasciculata and glomerulosa were stimulated by ACTH and renin, respectively (*Figure 6.5*). In the untreated state, when renin was suppressed by excessive DOC

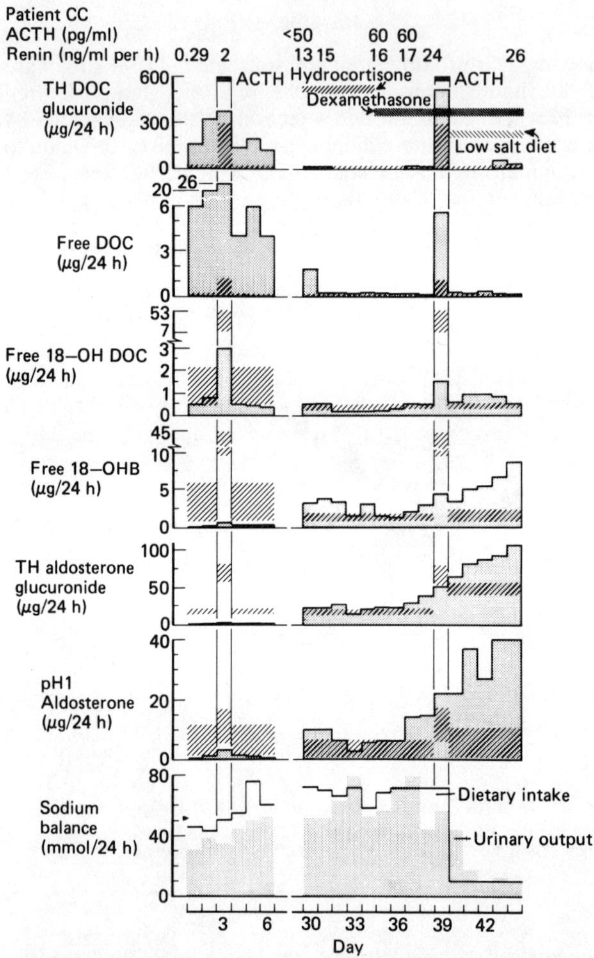

Figure 6.5 Metabolic balance and urinary hormone excretion in a prepubertal boy with 11β-hydroxylase deficiency during baseline, ACTH, dexamethasone, and dexamethasone and low salt periods. (▨) Patient's response; (▨) normal control response. THDOC = tetrahydrodeoxycorticosterone; 18-OHB = 18-hydroxycorticosterone. (From Levine *et al.*, 1980a, courtesy of the Editor and Publishers, *Journal of Clinical Endocrinology and Metabolism*)

secretion, ACTH stimulation resulted in a further increase in urinary excretion of 11-deoxysteroids (DOC and its metabolite tetrahydrodeoxycorticosterone) while the levels of 11-hydroxylated steroids (aldosterone and 18-hydroxycorticosterone) did not rise. This indicated impaired 11β-hydroxylation in the ACTH-responsive zona fasciculata. In contrast, when ACTH was suppressed by dexamethasone and renin stimulated by a low salt diet, the levels of aldosterone and 18-hydroxycorticosterone increased normally, suggesting that the renin-responsive zona glomerulosa was not absolutely defective in 11β-hydroxylation.

The hormonal data reported in this study also corroborate the report of Ulick (1976), who detected parallel deficiencies in both 11β- and 18-hydroxylase function in patients with the hypertensive form of CAH. He proposed that 11β- and 18-hydroxylation are functionally related and may involve the same enzyme protein and catalytic site. As can be seen in *Figure 6.5*, the failure of 18-hydroxydeoxycorticosterone (18-OH DOC) levels to rise in concert with DOC during ACTH stimulation suggests the presence of an 18-hydroxylase defect as well as an 11β-hydroxylase defect in the ACTH-responsive zona fasciculata. The normal rise of 18-hydroxycorticosterone levels with renin stimulation indicates that, in the glomerulosa, 18-hydroxylase as well as 11β-hydroxylase activity remains unimpaired.

It has been shown that 11β- and 18-hydroxylase functions can be inhibited by androgens (Molteni, Skelton and Brownie, 1970; McCall *et al.*, 1978), suggesting that the elevated androgen levels observed in the hypertensive form of CAH may be the major cause of the defective enzyme activity. Our findings argue against this proposal, in that persistent 11β- and 18-hydroxylase deficiencies could be demonstrated both when adrenal androgens were well suppressed and when they were increased by ACTH administration. Furthermore, in patients with 21-hydroxylase deficiency, aldosterone secretion, which requires both 11β-hydroxylation and 18-hydroxylation, may be normal despite extremely high levels of androgens (New and Seaman, 1970).

21-Hydroxylase deficiency

Additional evidence for the concept that the adrenal fasciculata and glomerulosa function as two separate glands may be gained from the study of patients with 21-hydroxylase deficiency. There has long been controversy over the biochemical basis for the variable degree of salt-wasting in this disorder. All patients show defective 21-hydroxylation of 17-hydroxysteroids, leading to elevation of serum 17OH-progesterone and diminished production of cortisol. The nature of the 21-hydroxylase defect in the 17-deoxy pathway remains less clear, although there is evidence that aldosterone secretion may be more deficient in those with the salt-wasting form. The hypotheses proposed by others to explain this difference invoke either a 'one-enzyme' or a 'two-enzyme' defect. Neither *in vivo* nor *in vitro* studies have been conclusive or have produced results compatible with either theory.

According to the one-enzyme theory, one 21-hydroxylase is capable of 21-hydroxylation in both the 17-hydroxy and 17-deoxy pathways, and the difference between salt-wasting and simple virilizing CAH is due either to different degrees of enzymatic deficiency (Kowarski *et al.*, 1965) or to different substrate affinities for the abnormal enzyme protein. The two-enzyme theory postulates that two different enzymes regulate the two pathways, and that simple virilizers have an enzymatic deficiency only in the 17-hydroxy pathway, whereas the 'salt-wasters' have deficiencies in both the 17-hydroxy and 17-deoxy pathways of adrenal steroidogenesis. The two-enzyme theory is supported by reports of normal or even elevated aldosterone secretion in simple virilizers in contrast to deficient aldosterone production in 'salt-wasters' (Degenhart *et al.*, 1965; Bartter, Henkin and Bryan, 1968; Simopoulos *et al.*, 1971; West *et al.*, 1979).

Recent studies from our laboratory (Kuhnle *et al.*, 1981a) have suggested a new hypothesis to explain the difference between salt-wasting and simple virilizing CAH. This hypothesis uses the two-gland model of the adrenal cortex and states that (1) in both simple virilizers and salt-wasters there is a fasciculata defect of

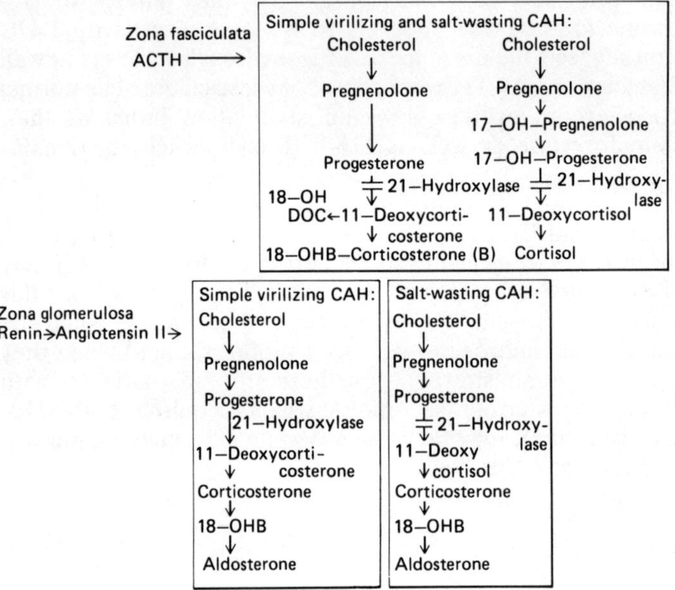

Figure 6.6 Pathway of adrenal steroidogenesis in the simple virilizing and salt-wasting forms of congenital adrenal hyperplasia (CAH) due to 21-hydroxylase deficiency. DOC = deoxycorticosterone; 18-OHB = 18-hydroxycorticosterone. (From Kuhnle *et al.*, 1981a, by permission of the *Endocrine Society*)

21-hydroxylation in both the 17-hydroxy and 17-deoxy pathways; and (2) in the salt-waster there is an additional defect in 21-hydroxylation in the glomerulosa, while in the simple virilizer the glomerulosa is spared this defect. This hypothesis is schematically presented in *Figure 6.6.* A sample of the data supporting this hypothesis is presented in *Figures 6.7* and *6.8.*

The marked rise in progesterone (P), together with the minimal response of DOC, corticosterone (compound B) and cortisol (compound F), to ACTH stimulation in both simple virilizers and salt-wasters indicates a 21-hydroxylase defect for both the 17-hydroxy and 17-deoxy pathways in the ACTH-responsive zona fasciculata (*Figure 6.7*). The 21-hydroxylase deficiency in the zona fasciculata may be more severe in the salt-wasters since the response to ACTH demonstrates greater deficiency in the products of 21-hydroxylation. After suppression of the

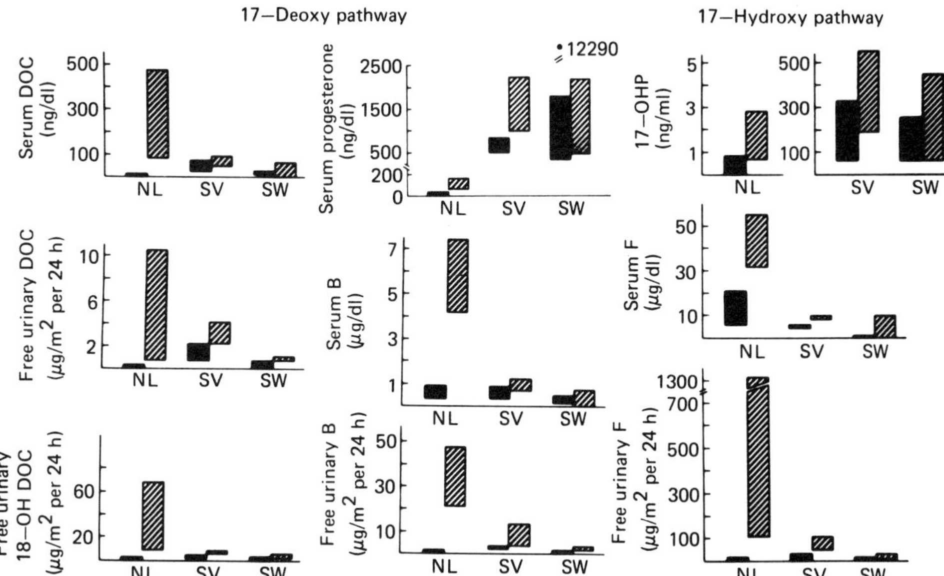

Figure 6.7 Hormonal response to ACTH stimulation of the zona fasciculata. NL = normal subjects; SV = patient with the simple virilizing form of 21-hydroxylase deficiency; SW = patient with the salt-wasting form of 21-hydroxylase deficiency. (■) Before ACTH stimulation; (▨) after ACTH stimulation. B = corticosterone; F = cortisol; DOC = deoxycorticosterone; 17-OHP = 17-hydroxyprogesterone. (From Kuhnle *et al.*, 1981a, by permission of the *Endocrine Society*)

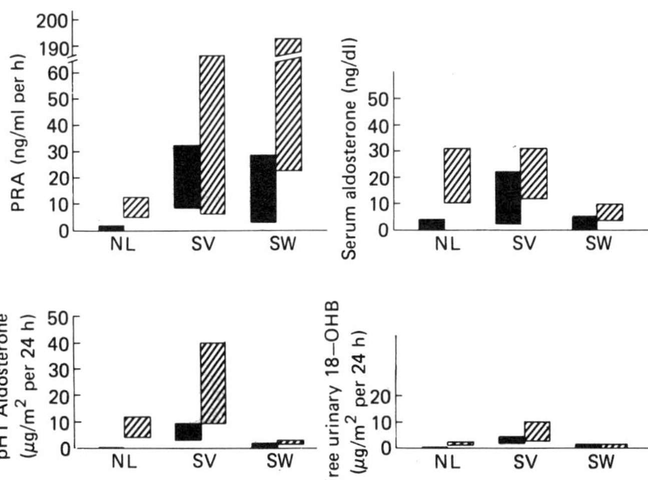

Figure 6.8 Hormonal response to sodium deprivation of the zona glomerulosa. For key *see Figure 6.7*. PRA = plasma renin activity; 18-OHB = 18-hydrocorticosterone. (■) High sodium and (▨) low sodium intake during dexamethasone suppression. (From Kuhnle *et al.*, 1981a, by permission of the *Endocrine Society*)

fasciculata with dexamethasone and stimulation of the glomerulosa by renin during low sodium intake (*Figure 6.8*), both normal subjects and simple virilizers demonstrate a rise in serum and urinary aldosterone and 18-OHB, indicating normal glomerulosa 21-hydroxylase function. In contrast, the salt-wasters show almost no increase in aldosterone and 18-OHB levels in response to renin stimulation, as is consistent with a deficiency of 21-hydroxylation in the glomerulosa.

These findings support the hypothesis that there is a 21-hydroxylase defect in the zona fasciculata of simple virilizers and salt-wasters, whereas the zona glomerulosa is defective only in salt-wasters. Naturally these *in vivo* studies do not indicate whether these functional differences reflect genetic differences or only secondary effects of changes in the adrenal microenvironment. However, they do indicate that there is one enzyme involved in the 21-hydroxylation of both the 17-hydroxy and 17-deoxy pathways of adrenal steroidogenesis in the zona fasciculata.

Reports of elevated levels of DOC, corticosterone and aldosterone in simple virilizing 21-hydroxylase deficiency remain perplexing. Proponents of the two-enzyme theory contend that these findings point to unimpaired 21-hydroxylation in the 17-deoxy pathway (West *et al.*, 1979). Others attribute the elevated DOC levels to inhibition of 11-hydroxylation by excessive androgens (Sharma, Forchielli and Dorfman, 1963) or to an intra-adrenal event caused by ACTH (Schambelan *et al.*, 1980). The frequent reports of increased PRA in simple virilizers (Simopoulos *et al.*, 1971, Edwin *et al.*, 1979) suggest instead that angiotensin is stimulating the 17-deoxy pathway of the glomerulosa. The basis for this high renin activity in simple virilizing CAH remains unclear since most of these patients appear to be capable of conserving sodium to the same degree as normal control subjects. It has been suggested that oversecretion of aldosterone secondary to increased plasma renin activity is a compensatory response to natriuretic hormones secreted by the adrenals in patients with 21-hydroxylase deficiency (Klein, 1960; Schaison *et al.*, 1980). The identity of such hormones has not been established, although a study by Kuhnle *et al.* (1981b) of patients with 21-hydroxylase deficiency provides additional evidence for the presence of mineralocorticoid antagonists in this disorder. Recently, there have been reports of extra-adrenal production of DOC from progesterone in patients with 21-hydroxylase deficiency (Winkel *et al.*, 1980; Antonipillai *et al.*, 1983), which may explain the increased DOC levels.

17α-Hydroxylase deficiency

Further evidence that the zonae fasciculata and glomerulosa function independently has been provided by the study of 17α-hydroxylase deficiency. In many of the reported cases aldosterone secretion is low (New, 1970; Biglieri, 1979) despite the fact that a 17α-hydroxylase defect does not impair aldosterone synthesis. Biglieri (1979) and Mantero *et al.* (1971) have proposed that this hypoaldosteronism could result from an additional defect in 18-hydroxylase. However, it is also possible that 17α-hydroxylase deficiency leads to overproduction of 17-deoxysteroids by the ACTH-responsive fasciculata, resulting in excessive secretion of DOC and corticosterone. Excessive amounts of DOC cause expansion of plasma volume, suppressed renin secretion and a consequent failure of aldosterone synthesis. Dexamethasone suppresses ACTH and diminishes DOC secretion, resulting in renin–angiotensin stimulation of the glomerulosa and a rise in aldosterone secretion, an effect which is similar to that observed in 11β-hydroxylase deficiency. Biglieri *et al.* (1979) have studied the dynamics of aldosterone secretion in treated and untreated 17α-hydroxylase deficiency and proposed that both suppressed renin activity and an inhibitory effect of chronic ACTH stimulation are responsible for the low aldosterone secretion in untreated patients.

3β-Hydroxysteroid dehydrogenase deficiency

The recent study, mentioned earlier, of two siblings with non-salt-losing 3β-HSD deficiency (Pang *et al.*, 1983) also lends support to the two-gland hypothesis. In both children levels of Δ^5-pregnenolone, Δ^5-17OH-pregnenolone, and Δ^5-3β-DHEA increased in response to ACTH, and decreased dramatically following dexamethasone treatment, suggesting that these steroids were products of the zona fasciculata and reticularis. Despite the normal or slightly elevated serum concentration of Δ^4-steroids, the markedly elevated precursor/product ratios of Δ^5/Δ^4 steroids of the 17-deoxysteroid, 17-hydroxysteroid and androgen pathways demonstrated a significant degree of 3β-HSD deficiency in all biosynthetic pathways in the zona fasciculata and reticularis in these children. However, aldosterone levels increased appropriately in response to salt depletion, indicating that 3β-HSD was not totally deficient in the zona glomerulosa.

Analysis of basal and ACTH-stimulated ratios of Δ^5/Δ^4 steroids suggests that the extent of the 3β-HSD defect is greater in the 17-hydroxysteroid pathway than in the 17-deoxysteroid pathway. This may be due to a variable degree of enzyme defect in the different steroidogenic pathways or to different substrate affinities (de Peretti *et al.*, 1980). It is also possible that high intra-adrenal levels of Δ^5-pregnenolone may inhibit 17,20-desmolase activity (Neher and Kahnt, 1965), as suggested by the higher ratio of 17OH-pregnenolone to 17OH-progesterone than of DHEA to androstenedione.

In summary, in four genetic errors of steroidogenesis, i.e. deficiencies of 21-hydroxylase, 11β-hydroxylase, 17β-hydroxylase and 3β-hydroxysteroid dehydrogenase, there is evidence that the fasciculata and glomerulosa can demonstrate different degrees of enzymatic defect. This raises the interesting possibility of separate genetic loci for the regulation of these enzymes in the fasciculata and the glomerulosa.

TREATMENT OF CONGENITAL ADRENAL HYPERPLASIA

Endocrine treatment

The fundamental aim of endocrine therapy in CAH is to replace the deficient hormones. Glucocorticoid administration both replaces deficient cortisol and suppresses ACTH overproduction, and thus ameliorates the noxious effects of inappropriate adrenal steroids. Excessive glucocorticoid administration should be avoided since this produces cushingoid facies, growth retardation and inhibition of epiphyseal maturation. In the enzyme deficiencies which impair mineralocorticoid synthesis, treatment with salt-retaining steroids is required to maintain adequate sodium balance.

Increasing attention has been focused on the role of the renin–angiotensin system in the treatment of CAH. Although circulating levels of aldosterone may not always be subnormal, it has long been recognized that PRA is increased in the simple virilizing as well as the salt-wasting form (Simopoulos *et al.*, 1971; Edwin *et al.*, 1979). However, it was not customary to supplement glucocorticoid replacement with salt-retaining steroids in cases of simple virilizing 21-hydroxylase deficiency. In 1977 Rösler *et al.* demonstrated that addition of salt-retaining hormone to glucocorticoid therapy in patients with simple virilizing CAH and elevated PRA does in fact improve control of the disease.

In patients with 21-hydroxylase deficiency there is a close correlation between PRA and ACTH levels (*Figure 6.9*). When PRA is normalized by administration of

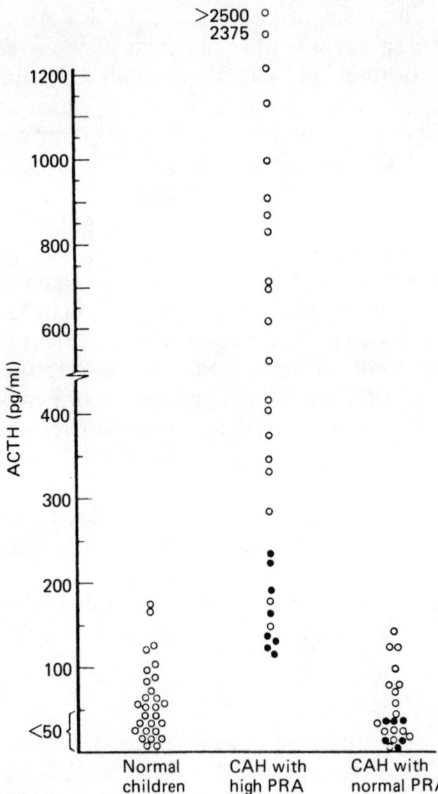

Figure 6.9 Correlation between ACTH and plasma renin activity (PRA) in patients with CAH treated with constant replacement doses of glucocorticoids equivalent to 25 mg/m^2 per day of hydrocortisone. Patients were studied during different states of sodium balance. (○) Salt losers; (●) non-salt losers. (From Rösler *et al.*, 1977, by permission of the *Endocrine Society*)

9α-fludrocortisone acetate, a steroid with salt-retaining activity, ACTH levels fall and androgen secretion decreases. The addition of salt-retaining steroids to the therapeutic regimen often allows a reduction in the glucocorticoid dose and also improves statural growth (Winter, 1980; Kuhnle *et al.*, 1983; *Figure 6.10*). Measurement of PRA can also be used to monitor efficacy of treatment in other salt-losing disorders, such as cholesterol desmolase and 3β-HSD deficiencies, and is useful as a therapeutic index in those forms of CAH with mineralocorticoid excess, such as 11β-hydroxylase and 17α-hydroxylase deficiencies (*Figure 6.11*).

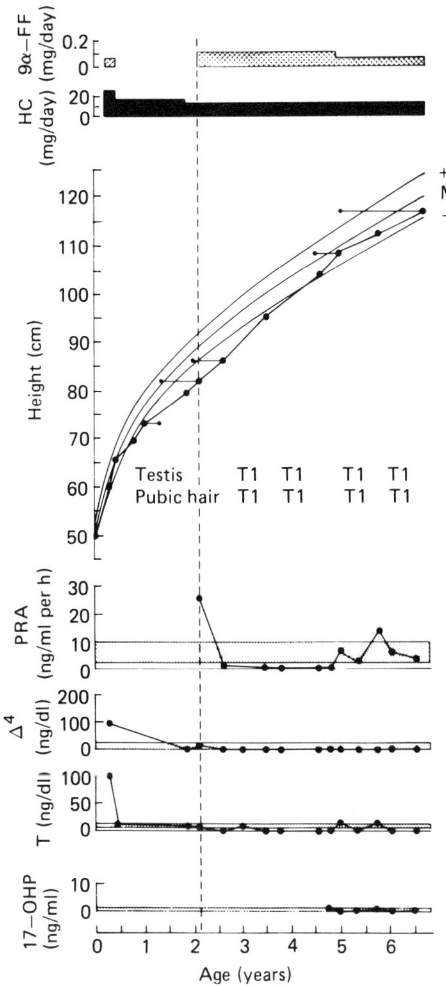

Figure 6.10 Growth curve of a male patient with salt-wasting congenital adrenal hyperplasia (CAH) before and after therapeutic control of plasma renin activity (PRA). Note that with renin suppression, hydrocortisone (HC) dose could be lowered, growth improved and androgen suppression was maintained despite the decrease in hydrocortisone dose. FF = fludrocortisone; 17-OHP = 17-hydroxyprogesterone; T = testosterone; Δ^4 = Δ^4-androstenedione; T1 = Tanner stage 1. (From New *et al.*, 1981, courtesy of the Editor and Publishers, *Recent Progress in Hormone Research*)

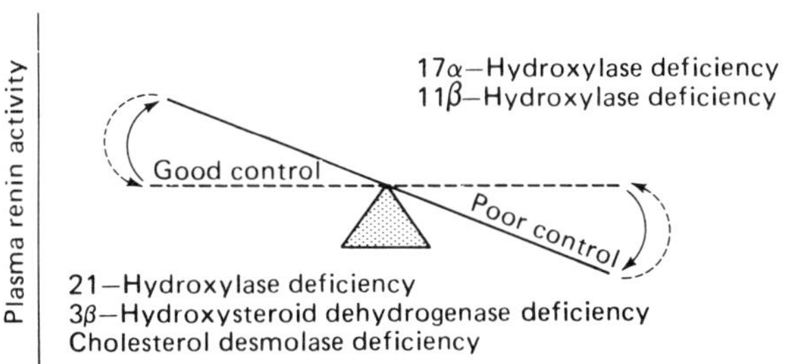

Figure 6.11 The pivotal role of monitoring plasma renin activity in various forms of congenital adrenal hyperplasia. In poor control, renin is elevated in 21-hydroxylase deficiency, 3β-hydroxysteroid dehydrogenase deficiency and cholesterol desmolase deficiency, and decreases with proper mineralocorticoid treatment. In contrast, in 17α-hydroxylase deficiency and 11β-hydroxylase deficiency, renin is suppressed in poor control and rises with proper treatment. (From New *et al.*, 1981, courtesy of the Editor and Publishers, *Recent Progress in Hormone Research*)

GENETICS OF CONGENITAL ADRENAL HYPERPLASIA

Several surveys have established that 21-hydroxylase deficiency is transmitted as an autosomal recessive trait with males and females equally at risk. Although there are exceptions (Rosenbloom and Smith, 1966), the clinical presentation – whether simple virilizing or salt-wasting – often appears to be consistent within one family. Deficiencies of 11β-hydroxylase, 3β-HSD and 17α-hydroxylase are also transmitted as autosomal recessive traits but are much rarer than 21-hydroxylase deficiency.

Screening for congenital adrenal hyperplasia

In Europe and the USA the estimated incidence of 21-hydroxylase deficiency ranges from 1 in 5000 to 1 in 15 000 live births. Screening for CAH in newborns is now possible, using a microfilter paper method developed for measuring 17-OHP (Pang *et al.,* 1977). We and others have proposed that all newborn infants should be screened for 21-hydroxylase deficiency, since both classical and non-classical 21-hydroxylase deficiencies have gone unrecognized (Lorenzen *et al.,* 1979; Levine *et al.,* 1980b). The reliability and feasibility of this method have been tested in a pilot study of all infants born in Alaska during a 30-month period (Pang *et al.,* 1982). This revealed that four of 1131 newborn Yupik Eskimos and one of 13 733 newborn Caucasians screened had salt-wasting 21-hydroxylase deficiency. Based on 95% confidence limits, the incidence of salt-wasting CAH appears to be 1 in 110 to 1 in 1038 live births in Yupiks and 1 in 2465 to 1 in 524 000 live births in Caucasians. This incidence in Yupiks agrees with that predicted by case survey (1 in 292 to 1 in 896) in this population (Hirschfeld and Fleshman, 1969). In Alaskan Caucasians, the highest incidence of salt-wasting CAH based on the 95% confidence level of the screening data (1 in 2465) is far greater than the highest incidence predicted by case surveys (1 in 18 454 for salt-wasting CAH, and 1 in 14 798 for salt-wasting and simple virilizing CAH combined). A larger screening sample is needed to permit more definite conclusions about the frequency of this disorder. The false-positive and recall rates for the pilot screening program were 0.05% and 0.1% respectively, figures that compare favorably with the rates in screening programs for other disorders. In Italy, Natoli *et al.* (1983) found four cases of CAH in 22 400 births, corresponding to an incidence of 1 in 5580, while Cacciari *et al.* (1982) found five cases in 42 930 births, an incidence of 1 in 8586. In Japan, Igarashi *et al.* (1982) detected one patient in 7550 newborns screened.

HLA LINKAGE

The genes for HLA (human leukocyte antigens), cell surface antigens important in transplantation, are located on chromosome 6. The HLA complex consists of at least four genetic loci: HLA-A, HLA-B, HLA-C and HLA-D/DR, with multiple alleles at each locus. Several other loci have been mapped on chromosome 6 in close linkage with HLA (Farid and Bear, 1981). The HLA genes are co-dominantly expressed, and the gene products of the HLA-A, B, C and D/DR loci from both parents are expressed on the surface of all nucleated cells. An example of an HLA genotype is written as follows:

$$\frac{\text{A3, Bw47(w4), Cw6, DR7}}{\text{A28, Bw35(w6), Cw4, DR5}}$$

in which one haplotype, the set of A3, Bw47(w4), Cw6, DR7, is inherited from one parent, while the other haplotype, i.e. A28, Bw35(w6), Cw4, DR5, is inherited from the other parent.

Close genetic linkage between HLA and 21-hydroxylase deficiency was first described by Dupont *et al.* in 1977. In five of the six families studied, all affected offspring were HLA-identical and different from their unaffected siblings. In the sixth family, the two affected children were HLA-B identical. Subsequently, a further 34 unrelated families with a total of 48 patients were reported from New York and Zurich (Levine *et al.*, 1978). The findings of this study are illustrated in the two typical pedigrees shown in *Figure 6.12.* In one family, the three affected siblings are HLA-identical. In the other, the unaffected siblings are all HLA-different from their affected sister. The HLA genotype is thus a marker for the CAH genotype.

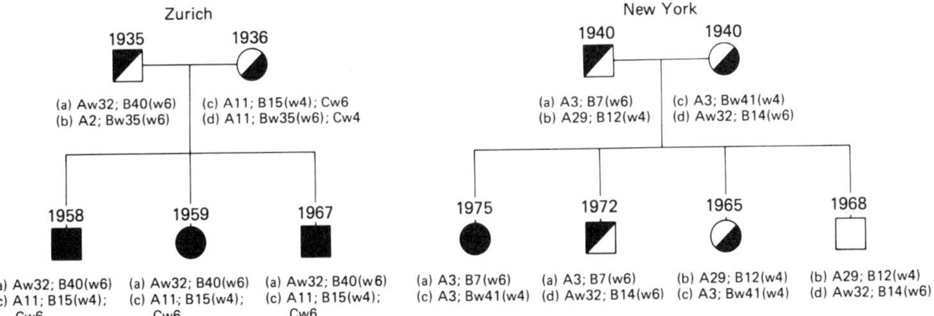

Figure 6.12 Pedigrees of two families with classical 21-hydroxylase deficiency. The HLA haplotypes for the HLA-A, HLA-B and HLA-C alleles are given in each family. The paternal haplotypes are labeled (a) and (b), and the maternal haplotypes (c) and (d). The parents are obligate heterozygous carriers for the 21-hydroxylase deficiency gene (denoted by the half-black symbols). In one family (*left*), three affected siblings (solid black symbols) are HLA genotypically identical. In the other (*right*), the one affected child (●) is HLA genotypically different from the three unaffected siblings. The sibling who carries the parental (a) and (d) haplotypes is presumed to be a heterozygous carrier for the 21-hydroxylase deficiency because he shares the (a) haplotype with the patient. The sibling with the parental (b) and (c) haplotypes shares the (c) haplotype with the patient, and should be a carrier of the classical 21-hydroxylase deficiency gene. The child with the (b) and (d) haplotypes should be normal for the gene. (From Levine *et al.*, 1978, courtesy of the Editor and Publishers, *New England Journal of Medicine*)

Findings from subsequent international studies (Dupont *et al.*, 1980) provided more detailed genetic mapping of the specific location of the 21-hydroxylase deficiency gene relative to the loci that constitute the HLA complex (*Figure 6.13*). Examination of cases of genetic recombination between HLA and the 21-hydroxylase deficiency gene established that the 21-hydroxylase deficiency gene is located between the HLA-A locus and the centromere, but further away from the centromere than the glyoxalase I locus (Dupont *et al.*, 1980). Studies of genetic recombination also suggest that there may be more than one HLA-linked genetic locus for 21-hydroxylase deficiency (Dupont *et al.*, 1980).

Statistical methods of genetic analysis also demonstrate close linkage between the 21-hydroxylase deficiency gene and HLA. Dupont *et al.* (1980) found a peak

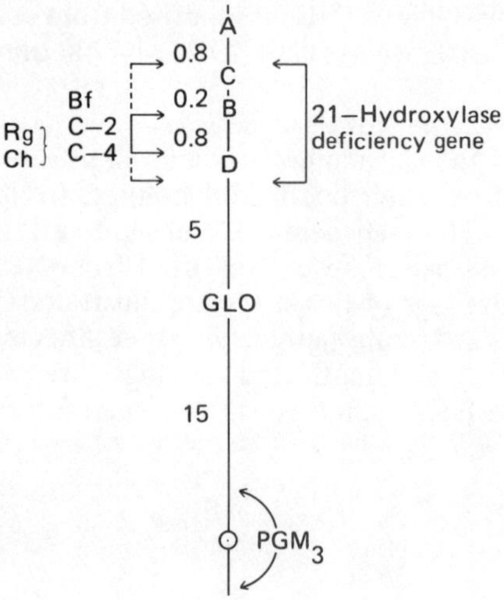

Figure 6.13 HLA linkage group on chromosome 6. The recombinant fractions for the known linkages between A:C, C:B, B:D, B:GLO and GLO:PGM_3 are shown. The position of the genes for factor B (Bf), complement C-2, complement C-4, and Rodgers (Rg) and Chido (Ch) blood groups is also indicated. The 21-hydroxylase deficiency gene can be mapped between HLA-A and glyoxalase I (GLO). The most likely position of the 21-hydroxylase deficiency gene is very close to HLA-B. (From Levine *et al.*, 1978, courtesy of the Editor and Publishers, *New England Journal of Medicine*)

LOD score of 15.65 at a recombination frequency of θ=0.00 for linkage between HLA and 21-hydroxylase deficiency. The LOD score is a statistical index of the certainty with which one can arrive at a conclusion of genetic linkage: a score of 15.65 means that the odds are $10^{15.65}$ to 1 that linkage exists. In humans, genetic linkage is considered established if the LOD score exceeds 3.00 at θ=0.00.

Heterozygote detection

Before the discovery that 21-hydroxylase deficiency is genetically linked to HLA, attempts were made to test hormonally for heterozygosity in the obligate heterozygote parents of children with CAH. Several investigators demonstrated a mild deficiency of 21-hydroxylase in parents, but similar hormonal studies in the siblings of patients with CAH were more difficult to interpret because of the inability to ascertain which were carriers of the gene and which were unaffected (Krensky *et al.,* 1977; Gutai *et al.,* 1979). With the demonstration of linkage between the genes for HLA and 21-hydroxylase deficiency, HLA genotyping makes it possible to predict which siblings are carriers and which are unaffected (*see Figure 6.12*). We have found that family members predicted by HLA genotyping to be heterozygotes demonstrate a higher 17-OHP response to ACTH stimulation than family members predicted to be genetically unaffected (Lorenzen *et al.*, 1979, 1980). Although baseline 17OH-progesterone concentrations in heterozygotes are indistinguishable from those in genetically unaffected subjects, ACTH stimulation can unmask the presence of mild 21-hydroxylase deficiency (Grosse-Wilde *et al.*, 1979; Mauseth *et al.*, 1980; Sobel *et al.*, 1980). Distinction of CAH heterozygotes from unaffected subjects in the adult population can be improved by pretreating with dexamethasone before ACTH stimulation (Grosse-Wilde *et al.*, 1979; Child, Bu'Lock and Anderson, 1979; Weil *et al.*, 1979; Lejeune-Lenain *et al.*, 1980).

Genetic linkage disequilibrium

It is important to distinguish between genetic linkage, e.g. that demonstrated between HLA and 21-hydroxylase deficiency, and genetic linkage *disequilibrium*, which is the non-random association of particular alleles at different genetic loci. In patients with 21-hydroxylase deficiency, certain HLA antigens appear with either a significantly increased or decreased frequency relative to their frequency in the general population. Thus, not only are 21-hydroxylase deficiency and HLA genetically linked, but there is genetic linkage disequilibrium between the 21-hydroxylase deficiency alleles and certain HLA alleles.

The most significant association with classical 21-hydroxylase deficiency has been found for HLA-Bw47, where the combined relative risk for Caucasians from different geographical areas is 15.4 (Dupont *et al.*, 1980). Slight increases have also been reported for Bw51, Bw53, Bw60 and DR7. A review of the Bw47 positive haplotypes in patients with 21-hydroxylase deficiency reveals that this antigen frequently occurs on one particular haplotype, namely A3, Cw6, Bw47, DR7. Although the gene frequency for Bw47 in different Caucasian populations is always very low (less than 0.005), international studies have shown that Bw47 appears with remarkable frequency among classical 21-hydroxylase CAH patients. In one particular region of England the Bw47 frequency among CAH patients is nearly 50%. Several studies have also demonstrated that A1, B8 and DR3 are consistently decreased in patients with 21-hydroxylase deficiency (Levine *et al.*, 1978; Grosse-Wilde *et al.*, 1979; Klouda, Harris and Price, 1980; Pollack *et al.*, 1979a).

Non-classical 21-hydroxylase deficiency

A milder form of 21-hydroxylase deficiency may also occur. Genetic females with this non-classical variant (also referred to as 'late-onset' or 'acquired' adrenal hyperplasia) do not have prenatal virilization. However, affected males and females have varying degrees of postnatal androgen excess, giving rise to virilization, menstrual disturbances, acne and endocrinological features consistent with 21-hydroxylase deficiency in later childhood or early adulthood (Rosenwaks *et al.*, 1979; Kohn *et al.*, 1982). A totally asymptomatic form of non-classical 21-hydroxylase deficiency (termed 'cryptic' 21-hydroxylase deficiency) has been described in family members of patients with classical 21-hydroxylase deficiency. Although the clinical hallmarks of classical CAH, i.e. virilization, abnormal puberty and growth, and infertility, are absent, the biochemical profile is characteristic of 21-hydroxylase deficiency (Levine *et al.*, 1980b, 1981; Zachmann and Prader, 1978, 1979).

The hormonal profiles of patients with symptomatic and asymptomatic non-classical 21-hydroxylase deficiency are similar, suggesting that both have an enzymatic defect which is less severe than that in classical 21-hydroxylase deficiency. Patients with the symptomatic form present with variable clinical signs of androgen excess; fewer than 50% of children studied demonstrate advanced bone age (Kohn *et al.*, 1982). Not only is there marked phenotypic variability among these patients, but symptoms related to androgen excess may regress or even completely disappear despite persistence of the underlying biochemical defect (*Figure 6.14*). One patient remained asymptomatic after hydrocortisone treatment was discontinued. In males, signs of androgen excess may be subtle, such as advanced stature and bone age in childhood. Kohn *et al.* (1982) reported five HLA-identical siblings of index cases who all had the biochemical abnormalities of 21-hydroxylase deficiency, yet three of the five showed no clinical evidence of androgen excess, although the biochemical abnormalities were similar to those in the symptomatic patient population with late-onset adrenal hyperplasia.

The late presentation of a biochemical defect raises the question whether this is the same inherited disorder as CAH, but with delayed presentation, or an 'acquired' disorder distinct from CAH. Although initial studies suggested that the late-onset disorder was not genetically linked to HLA (Morillo and Gardner, 1979; New *et al.*, 1979), more recent reports have provided evidence that this is indeed so (Rosenwaks *et al.*, 1979; Laron *et al.*, 1980; Lobo and Goebelsmann, 1980; Migeon *et al.*, 1980). It has thus been proposed that classical and non-classical 21-hydroxylase deficiency are allelic variants (Blankstein *et al.*, 1980; Migeon *et al.*, 1980; New *et al.*, 1981; Pollack *et al.*, 1981). Pedigrees and HLA genotypes of families of patients with symptomatic 21-hydroxylase deficiency are shown in

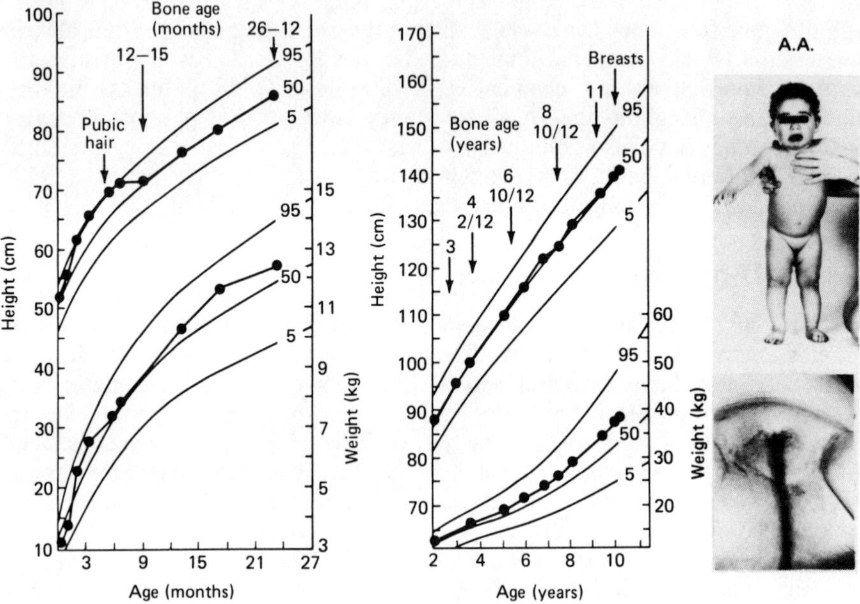

Figure 6.14 The changing clinical profile from symptomatic to asymptomatic non-classical 21-hydroxylase deficiency. In this female, postnatal onset (at 5 months) of virilization and growth acceleration was followed by a remission of symptoms during childhood despite persistence of hormonal abnormalities. (From New *et al.*, 1982, courtesy of the Publishers, *The Metabolic Basis of Inherited Disease*)

Figure 6.15. In our initial family study of symptomatic 21-hydroxylase deficiency (New *et al.*, 1979), the father's serum 17OH-progesterone response to ACTH was in the range of the general population and did not fall in the heterozygote range. In addition, the patient had an HLA-identical, though glyoxalase (GLO)-different, sibling whose stimulated serum 17OH-progesterone was in the heterozygote range for 21-hydroxylase deficiency, clearly below that of her affected sibling. These findings had suggested that the symptomatic disorder was not an HLA-linked autosomal recessive trait. Since the original report, ACTH-stimulation studies have been performed on many additional family members of classical CAH patients. Re-interpretation of the father's stimulated serum 17OH-progesterone response based on more current data places the father's response in the overlapping range between the general population and obligate heterozygote family members of CAH patients. Furthermore, the HLA–GLO recombination found in the HLA-identical sibling had been considered not significant, since most of the recombinations involving the HLA-B-D/DR-GLO region placed the 21-hydroxylase deficiency allele between HLA-B and HLA-D/DR (Dupont *et al.*, 1980). However, in view of recent reports which map the 21-hydroxylase deficiency allele outside the HLA-D/DR region on the GLO side (Grosse-Wilde *et al.*, 1979; Dupont *et al.*, 1980; Klouda, Harris and Price, 1980), the HLA-B–GLO recombination in the family of our original report may be significant and may explain the hormonal response in the HLA-identical but GLO-different sister.

The asymptomatic form of non-classical 21-hydroxylase deficiency had been considered a distinct entity (Zachmann and Prader, 1978, 1979; Levine *et al.*,

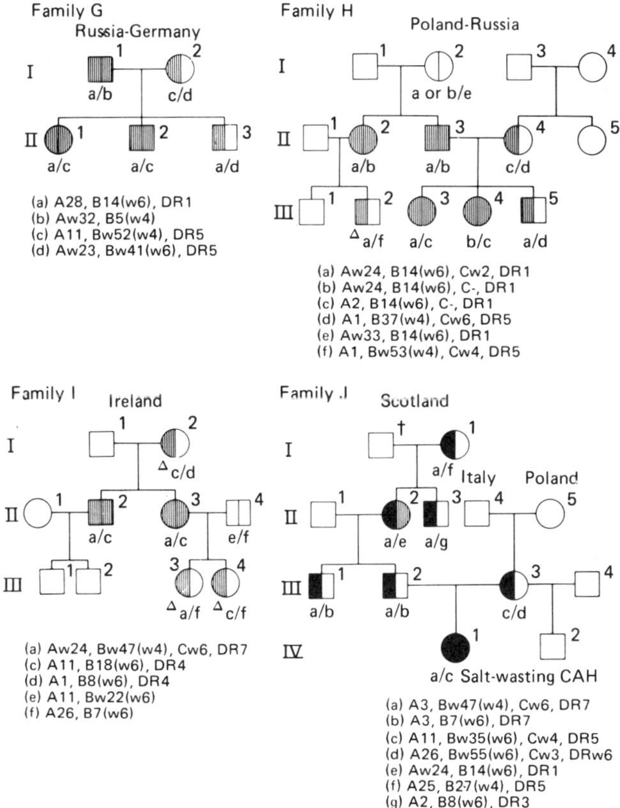

Figure 6.15 Pedigrees of four families with symptomatic non-classical adrenal hyperplasia. (■) and (●) Family members with non-classical 21-hydroxylase deficiency, which is less severe than the classical deficiency; (◧) and (◐) heterozygous carriers of the mild deficiency gene (the assignment is based on HLA genotyping and 17-hydroxyprogesterone response to ACTH stimulation); (●) infant with classical 21-hydroxylase deficiency; (◧) and (◐) heterozygous carriers of the severe deficiency gene; (●) patient with symptomatic non-classical 21-hydroxylase deficiency having both a severe classical 21-hydroxylase deficiency and a mild non-classical deficiency gene. The small open triangle (△) indicates assignment of heterozygosity on the basis of HLA genotyping alone. (From Kuhnle *et al.*, 1982, courtesy of the Editor and Publishers, *Journal of Clinical Endocrinology and Metabolism*)

1980b, 1981) until it was confirmed that the symptomatic and asymptomatic forms of this variant are different clinical manifestations of the same disorder. Pedigrees and HLA genotypes of families of patients with asymptomatic 21-hydroxylase deficiency are given in *Figure 6.16*. We concluded that these asymptomatic relatives of patients with the classical deficiency were genetic compounds, or mixed heterozygotes, who had 21-hydroxylase deficiency as a result of inheriting two recessive genetic defects: (1) a severe 21-hydroxylase defect gene shared with the index case with classical CAH (21-OH^{severe}); and (2) a mild, non-classical variant of the 21-hydroxylase defect gene (21-OH^{mild}). A 21-OH^{severe}/21-OH^{mild} genotype was thus assigned to family members with asymptomatic 21-hydroxylase deficiency.

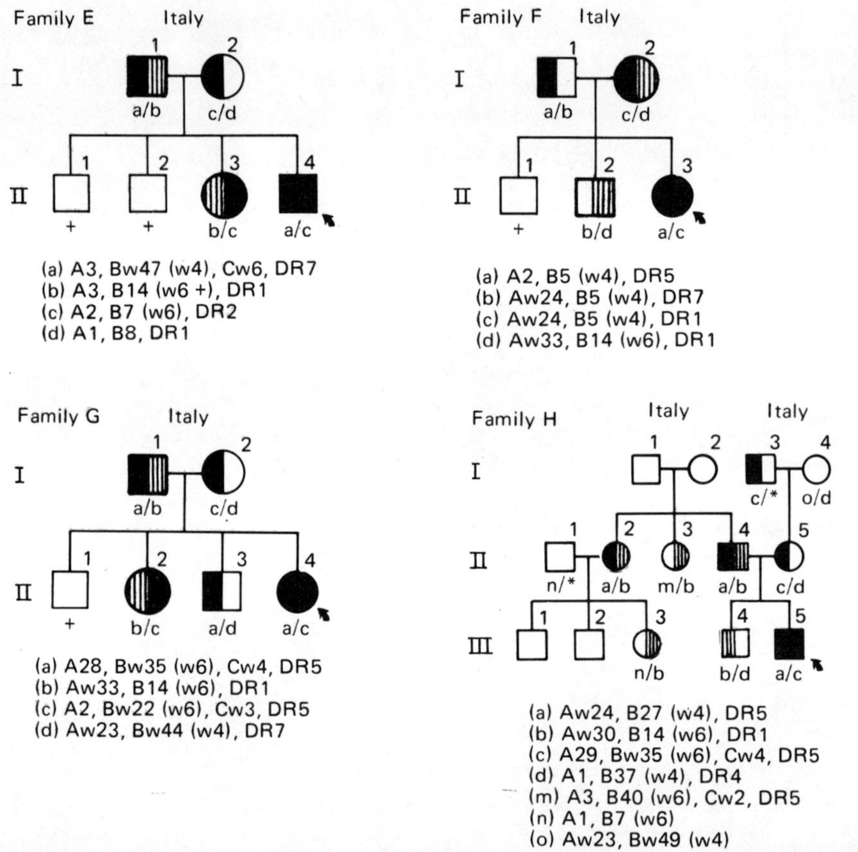

Figure 6.16 Pedigrees of four families with classical and asymptomatic non-classical 21-hydroxylase deficiency. The HLA genotypes for HLA-A, HLA-B, HLA-C and HLA-DR are indicated for each family member tested. The index case with congenital adrenal hyperplasia (arrows) is assigned the haplotypes a/c. The asterisks (*) indicate haplotypes deduced from offspring. ( ) and ( ) Heterozygous carriers of the severe classical deficiency gene for CAH; ( ) and ( ) Family members with asymptomatic non-classical 21-hydroxylase deficiency having both a severe and a mild deficiency gene; ( ) and ( ) heterozygous carriers of the mild 21-hydroxylase deficiency gene; ( ) indicates offspring, number if known. (From New *et al.*, 1981, courtesy of the Editor and Publishers, *Recent Progress in Hormone Research*)

Zachmann and Prader (1978, 1979) reported relatives of patients with classical 21-hydroxylase deficiency whose hormonal responses were consistent with asymptomatic 21-hydroxylase deficiency. They proposed that these members were heterozygotes for the classical gene (21-OH^{severe}/21-OH^{normal}) with unusual manifestation, rather than genetic compounds (21-OH^{severe}/21-OH^{mild}), as we have proposed. However, our demonstration that family members predicted by HLA genotyping to be heterozygous for the cryptic gene (21-OH^{mild}/21-OH^{normal}) have a 17-OHP response to ACTH identical to that of family members predicted to be heterozygous for the classical gene (21-OH^{severe}/21-OH^{normal}) argues against this suggestion. In family H in *Figure 6.16*, for example, members II-3, III-3 and III-4 all demonstrated hormonal responses characteristic of heterozygous carriers for a 21-hydroxylase deficiency gene. HLA genotyping revealed that they shared one HLA haplotype (b) with the asymptomatic family members (II-2 and II-4) but no HLA haplotype with the index case with classical CAH (III-5). If, as proposed by Zachmann and Prader, the asymptomatic members II-2 and II-4 were simply unusual heterozygotes for the classical gene, members II-3, III-3 and III-4 would be predicted to be hormonally unaffected (21-OH^{normal}/21-OH^{normal}). This prediction conflicts with the observed hormonal responses in these members, which clearly fall in the range for heterozygotes. Our proposal that these members have inherited a non-classical 21-OH^{mild} gene from a family member with non-classical 21-hydroxylase deficiency is consistent with the hormonal findings. Subsequent studies of additional asymptomatic 21-hydroxylase deficient patients demonstrated that this phenotype may also be an expression of a homozygous genotype (21-OH^{mild}/21-OH^{mild}) (Kohn *et al.*, 1982).

Genetic linkage disequilibrium has also been reported for the non-classical forms of 21-hydroxylase deficiency (Dupont *et al.*, 1980; Laron *et al.*, 1980). Our recent study has shown a significant increase in the frequency of HLA-B14, DR1 and complement factor BfS in both symptomatic and asymptomatic non-classical 21-hydroxylase deficiency (Pollack *et al.*, 1981). The fact that these alleles tend to appear together on the same haplotype in patients with non-classical 21-hydroxylase deficiency suggests that the HLA haplotype segment HLA-B14, DR1, BfS – rather than the individual HLA alleles – is highly associated with both these non-classical forms of 21-hydroxylase deficiency.

Our analysis of genetic linkage disequilibrium provides provocative evidence that some patients with symptomatic 21-hydroxylase deficiency, similar to those with asymptomatic 21-hydroxylase deficiency, may be genetic compounds possessing both classical and non-classical 21-hydroxylase deficiency genes. A few symptomatic patients have the HLA type HLA-Bw47, an allele occurring with strikingly high frequency in association with the gene for classical 21-hydroxylase deficiency. In the family shown in *Figure 6.17*, the appearance of Bw47 on haplotype c makes the presence of a classical deficiency gene highly likely. Because heterozygotes for the classical gene and those for the non-classical gene are biochemically indistinguishable, hormonal tests provide no indication whether the mother (I-2) carries a classical or non-classical gene on haplotype c. Conclusive proof for the presence of a classical 21-hydroxylase deficiency gene rests on the eventual appearance of a patient with classical 21-hydroxylase deficiency in the family pedigree. Indeed, in studying the family of a patient recently diagnosed as having congenital salt-wasting 21-hydroxylase deficiency (*see Figure 6.15*, family J), we found that the paternal grandmother had a hormonal response and medical history consistent with symptomatic non-classical 21-hydroxylase deficiency. Particularly noteworthy is that the HLA haplotype (a), which the grandmother shares with the patient with classical CAH, contains Bw47, while her other HLA haplotype contains B14, the HLA antigen most often associated with non-classical 21-hydroxylase deficiency. It has also been demonstrated that both symptomatic and asymptomatic patients may be homozygous for the mild allelic mutation which is in genetic linkage disequilibrium with HLA-B14 (Kohn *et al.*, 1982; *see Figure 6.15*, family H).

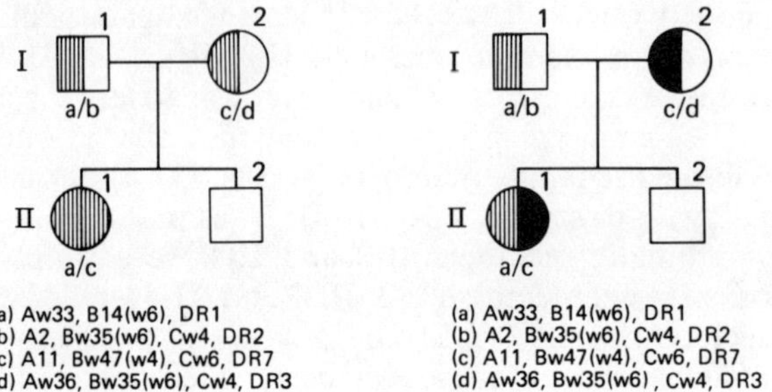

Figure 6.17 Two possible genotypes for symptomatic non-classical 21-hydroxylase deficiency in a family from Greece. Possibility A (*left*): both of the patient's HLA haplotypes are linked to mild 21-hydroxylase deficiency alleles (striped symbols). Possibility B (*right*): the patient is a genetic compound, possessing a Bw47-linked classical 21-hydroxylase deficiency allele (solid symbol) in addition to a mild deficiency allele. Note that hormonal testing cannot distinguish between the two different heterozygote genotypes proposed for the mother in A and B. (Modified by New *et al.*, 1982, courtesy of the Publishers, *The Metabolic Basis of Inherited Disease*)

An interesting biological paradox which remains to be elucidated is the absence of clinical signs in patients with asymptomatic 21-hydroxylase deficiency, in contrast to the virilization noted in patients with the symptomatic form, despite the fact that both groups demonstrate similar biochemical abnormalities. Patients with the symptomatic disorder are detected because of their clinical symptoms, whereas those with the asymptomatic variant are detected only as a result of genetic and hormonal studies of families with classical and symptomatic non-classical 21-hydroxylase deficiency.

Hormonal standards for genotyping 21-hydroxylase deficiency

We have developed nomograms relating the baseline and ACTH-stimulated levels of 17-OHP, Δ^4-androstenedione (Δ^4), DHEA, the ratio DHEA/Δ^4 and testosterone (T) (New *et al.*, 1983). The 17-OHP and Δ^4-androstenedione nomograms provide hormonal standards for assignment of the 21-hydroxylase deficiency genotype, i.e. patients whose hormonal values fall on the regression line within a defined group are assigned to that group. *Figures 6.18* and *6.19* give hormonal reference data using a 60-min ACTH test. There is a strong correlation for the prediction of CAH genotype between 60-min and 360-min ACTH stimulation tests, so that the shorter, less cumbersome 60-min test may be used with the same confidence as the longer test. Clinical symptoms and signs must distinguish between the symptomatic and asymptomatic forms of non-classical

21-hydroxylase deficiency, as these are indistinguishable biochemically. The nomograms show that the degree of the biochemical defect is the same in both groups, providing further evidence that both the symptomatic and the asymptomatic non-classical forms of 21-hydroxylase deficiency result from the same genetic mutation. Heterozygotes for classical and non-classical 21-hydroxylase deficiencies all demonstrate a similar hormonal response and are thus phenotypically indistinguishable from one another.

The distribution of responses along a regression line suggests that there is a spectrum of enzymatic deficiency in these groups. Patients with classical CAH have the most severe deficiency, those with non-classical forms a less severe deficiency,

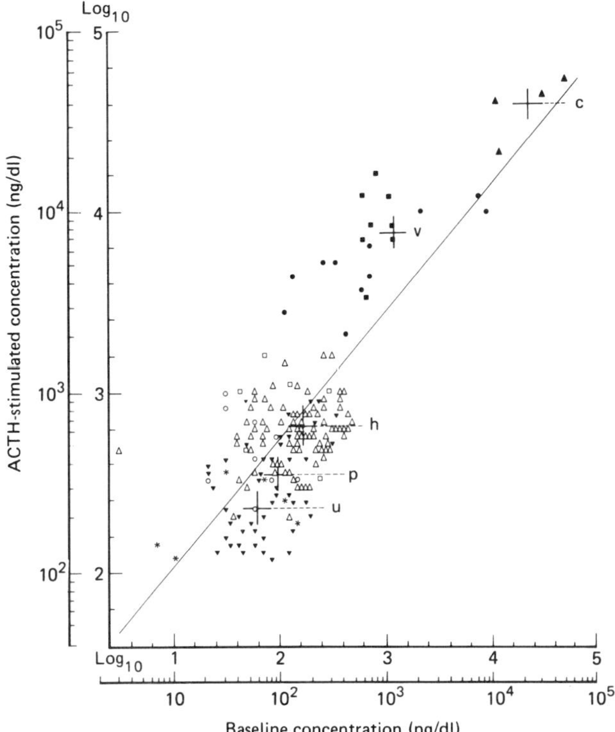

Figure 6.18 Nomogram relating baseline to 60-minute ACTH-stimulated 17-hydroxyprogesterone (17-OHP) concentration. Mean values for each group are indicated as follows: c = congenital adrenal hyperplasia (CAH); v = patients with symptomatic or asymptomatic non-classical (cryptic, acquired or late-onset) 21-hydroxylase deficiency; h = heterozygotes for classical CAH, non-classical symptomatic CAH (acquired or late-onset adrenal hyperplasia) and non-classical asymptomatic CAH (cryptic 21-hydroxylase deficiency); u = family members predicted by HLA genotyping to be unaffected; p = general population (not HLA genotyped). (▼) General population; (*) genetically unaffected. Patients with CAH (▲), acquired adrenal hyperplasia (●) and cryptic 21-hydroxylase deficiency (■). Open symbols indicate heterozygotes for CAH (△), acquired adrenal hyperplasia (○) and cryptic 21-hydroxylase deficiency (□). (From New, 1982)

while heterozygotes for both forms have an even milder deficiency, which is unmasked only by ACTH stimulation. Hormonal values in family members predicted by HLA genotyping to be unaffected by 21-hydroxylase deficiency fall at the lowest point of the regression line and serve as the best control population for normal 21-hydroxylase activity. Members of the general population whose responses are in the heterozygote range may actually be carriers of a gene for 21-hydroxylase deficiency. The incidence of classical 21-hydroxylase deficiency in a homogeneous Caucasian population has been estimated by screening to be between 1 in 5000 and 1 in 10 000 (Cacciari *et al.*, 1982; Natoli *et al.*, 1983). The

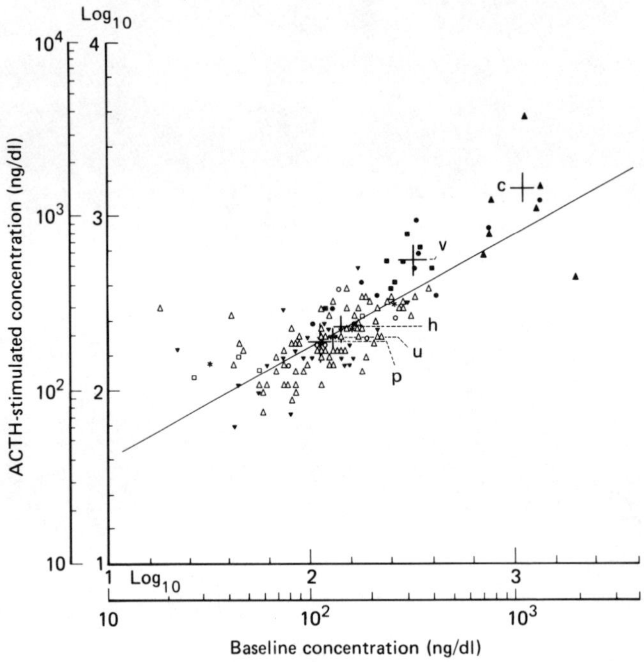

Figure 6.19 Nomogram relating baseline to 60-min ACTH-stimulated concentration of serum Δ^4-androstenedione. For key *see Figure 6.18.* (From New, 1982)

corresponding carrier frequency is calculated to be 1 in 35. However, this estimate of heterozygote frequency applies only to classical 21-hydroxylase deficiency. The gene frequency for non-classical 21-hydroxylase deficiency remains to be determined, but our experience suggests that the non-classical defect may be more frequent than the classical defect.

Allelic variants

We propose that there are allelic variants at the 21-hydroxylase locus which produce different degrees of 21-hydroxylase deficiency, resulting in the phenotypic diversity of this disorder (*Table 6.2*). According to the model described earlier in which the adrenal fasciculata and glomerulosa are assumed to function separately, the clinical spectrum of virilization may be the result of allelism at the

Table 6.2 Glossary of terms for disorders of steroid 21-hydroxylase (21-OH) deficiency. (From Kohn *et al.*, 1982, courtesy of the Editor and Publishers, *Journal of Clinical Endocrinology and Metabolism*)

Form of 21-OH deficiency	*Clinical phenotype*	*Hormonal phenotype*	*21-OH deficiency genotype**
Classical	Virilized prenatally and symptomatic	Markedly elevated serum 17-OHP and Δ^4-androstenedione concentrations	$\frac{\text{21-OH deficiency}^{\text{severe}}}{\text{21-OH deficiency}^{\text{severe}}}$
Non-classical	Symptomatic: virilized postnatally; symptomatic Asymptomatic: not virilized prenatally or postnatally; asymptomatic	Modestly elevated serum 17-OHP and Δ^4-androstenedione concentrations	$\frac{\text{21-OH deficiency}^{\text{severe}}}{\text{21-OH deficiency}^{\text{mild}}}$ or $\frac{\text{21-OH deficiency}^{\text{mild}}}{\text{21-OH deficiency}^{\text{mild}}}$

* The allelic variant which transmits severe 21-OH deficiency is in genetic linkage disequilibrium with HLA-Bw47. The allelic variant which transmits mild 21-OH deficiency is in genetic linkage disequilibrium with HLA-B14.
17-OHP = 17-hydroxyprogesterone.

21-hydroxylase locus in the fasciculata, while salt-wasting may be the result of a separate genetic defect. We further suggest that the non-classical forms result from either a genetic compound of the classical (or more severe) 21-hydroxylase deficiency allele and the mild non-classical 21-hydroxylase deficiency allele, or homozygosity for the mild non-classical 21-hydroxylase deficiency allele. Ultimate proof may be obtained by sequencing the gene by recombinant DNA techniques and demonstrating the specific nature of the genetic mutation for each allelic variant.

It is to be anticipated that a similar degree of genetic heterogeneity will be found in other inherited enzyme defects of steroidogenesis. The phenotypic variability already recognized in 11β-hydroxylase and 3β-HSD deficiencies may be other examples of allelic variants.

HLA linkage to other enzyme defects of steroidogenesis

Studies of families of patients with CAH due to 11β-hydroxylase deficiency (Dupont *et al.,* 1980), 17α-hydroxylase deficiency (Mantero *et al.,* 1980) and 3β-HSD deficiency (Pang *et al.,* 1983) have demonstrated that these enzyme defects are not genetically linked to HLA. In addition, hormonal studies have not provided a means for detecting heterozygote carriers for 11β-hydroxylase deficiency (Pang *et al.,* 1980a).

PRENATAL DIAGNOSIS OF CONGENITAL ADRENAL HYPERPLASIA

21-Hydroxylase deficiency

Elevated levels of 17OH-progesterone and Δ^4-steroids have been reported in the amniotic fluid of fetuses with CAH due to 21-hydroxylase deficiency (Pang *et al.,* 1980b). HLA genotyping of amniotic cells is an additional method for prenatal diagnosis of 21-hydroxylase deficiency in a pregnancy at risk and allows the prediction of a heterozygous fetus (Pollack *et al.,* 1979b). A fetus shown to be HLA-identical to the affected sibling is predicted to be affected. The HLA prediction of CAH genotype should always be corroborated by hormonal measurement of 17OH-progesterone and Δ^4-androstenedione in amniotic fluid. Caution must be exercised in interpreting results if multiple births are expected or if there is antigen sharing in the parents (Pollack *et al.,* 1979b).

11β-Hydroxylase deficiency

Increased levels of and tetrahydro-11-deoxycortisol (THS) in amniotic fluid and in maternal urine have been found in pregnancies with fetuses affected with 11β-hydroxylase deficiency (Rösler *et al.,* 1979). This suggests that prenatal diagnosis of this disorder by hormonal measurement is feasible. Since this disorder is not genetically linked to HLA, HLA typing of amniotic cells is not helpful.

SUMMARY

In summary, the enzyme defects of steroidogenesis appear to be monogenic disorders. The clinical heterogeneity of these disorders suggests allelic variations at the loci for these disorders, as has been reported for other genetic inborn errors. When the genes for these enzymes are cloned and sequenced, the final proof of allelism will be obtained. Prenatal diagnosis of one of the enzyme defects is possible by biochemical and HLA studies of the amniotic fluid. In the others, DNA restriction fragment polymorphism may provide a future tool for prenatal diagnosis. Since all these disorders are compatible with normal intelligence and a productive life, the more common ones at least are worthy of screening for early diagnosis and treatment.

Acknowledgements

This work was supported by the following NIH grants: HD 00072, HD 15084, HL 17749; and by a grant (RR 47) from the General Clinical Research Centers Program of the Division of Research Resources, NIH.

References

ANTONIPILLAI, I., MOGHISSI, E., HAWKS, D., SCHNEIDER, T. and HORTON, R. (1983) The origin of plasma deoxycorticosterone in men and in women during the menstrual cycle. *Journal of Clinical Endocrinology and Metabolism,* **56,** 93–98

BARTTER, F. C., HENKIN, R. I. and BRYAN, G. T. (1968) Aldosterone hypersecretion in 'non-salt-losing' congenital adrenal hyperplasia. *Journal of Clinical Investigation,* **47,** 1742–1752

BIGLIERI, E. G. (1979) Mechanisms establishing the mineralocorticoid hormone patterns in the 17α-hydroxylase deficiency syndrome. *Journal of Steroid Biochemistry,* **11,** 653–657

BIGLIERI, E. G., CHANG, B., HIRAI, J., BRUST, N., ROST, C. R. and SCHAMBELAN, M. (1979) Adrenocorticotropin inhibition of mineralocorticoid hormone production. *Clinical Science,* **57,** 307s–311s

BINOUX, M., PHAM-HUU-TRUNG, M. T., GOURMELEN, M., GIRARD, F. and CANLORBE, P. (1972) Plasma ACTH in adrenogenital syndrome. *Acta Paediatrica Scandinavica,* **61,** 269–270

BLANKSTEIN, J., FAIMAN, C., REYES, F. I., SCHROEDER, M. L. and WINTER, J. S. D. (1980) Adult-onset familial 21-hydroxylase deficiency. *American Journal of Medicine,* **68,** 441–448

BLUNCK, W. and BIERICH, J. R. (1968) Congenital adrenal hyperplasia with 11-hydroxylase deficiency. A case report and contribution to diagnosis. *Acta Paediatrica Scandinavica,* **57,** 157–161

BONGIOVANNI, A. M. (1981) Acquired adrenal hyperplasia: with special reference to 3β-hydroxysteroid dehydrogenase. *Fertility and Sterility,* **35,** 599–608

CACCIARI, E., BALSAMO, A., CASSIO, A. *et al.* (1982) Neonatal screening for congenital adrenal hyperplasia. *Lancet,* **1,** 1069

CATHELINEAU, G., BRERAULT, J. L., FIET, J., JULIEN, R., DREUX, C. and CANIVET, J. (1980) Adrenocortical 11β-hydroxylation defect in adult women with postmenarchial onset of symptoms. *Journal of Clinical Endocrinology and Metabolism,* **51,** 287–291

CHILD, D. F., BU'LOCK, D. E. and ANDERSON, D. C. (1979) Adrenal steroidogenesis in heterozygotes for 21-hydroxylase deficiency. *Clinical Endocrinology,* **11,** 391–398

DEGENHART, H. G., VISSER, H. K. A., WILMINK, R. and CROUGHS, W. (1965) Aldosterone and cortisol secretion rates in infants and children with congenital adrenal hyperplasia suggesting different 21-hydroxylation defects in 'salt-losers' and 'non-salt-losers'. *Acta Endocrinologica,* **48,** 587–601

DE PERETTI, E., FOREST, M. G., FEIT, J. P. and DAVID, M. (1980) Endocrine studies in two children with male pseudohermaphroditism due to 3β-hydroxysteroid dehydrogenase defect. In *Adrenal Androgens,* edited by A. R. Genazzani, J. H. H. Thijssen and P. K. Siiteri, pp. 141–145. New York: Raven Press

DUPONT, B., OBERFIELD, S. E., SMITHWICK, E. M., LEE, T. D. and LEVINE, L. S. (1977) Close genetic linkage between HLA and congenital adrenal hyperplasia (21-hydroxylase deficiency). *Lancet,* **2,** 1309–1311

DUPONT, B., POLLACK, M. S., LEVINE, L. S., O'NEILL, G. J., HAWKINS, B. R. and NEW, M. I. (1980) Congenital adrenal hyperplasia. In *Histocompatibility Testing 1980,* edited by P. I. Terasaki, pp. 693–706. Los Angeles: UCLA Tissue Typing Laboratory

EDWIN, C., LANES, R., MIGEON, C. J., LEE, P. A., PLOTNICK, L. P. and KOWARSKI, A. A. (1979) Persistence of the enzymatic block in adolescent patients with salt-losing congenital adrenal hyperplasia. *Journal of Pediatrics,* **95,** 534–537

FARID, N. R. and BEAR, J. C. (1981) The human major histocompatibility complex and endocrine disease. *Endocrine Reviews,* **2,** 50–86

FEDERMAN, D. D. (1968) *Abnormal Sexual Development,* pp. 121–133. Philadelphia: W. B. Saunders

FINKELSTEIN, M. and SCHAEFER, J. M. (1979) Inborn errors of steroid biosynthesis. *Physiology Reviews,* **59,** 353–406

FUKUSHIMA, D. K., FINKELSTEIN, J. W., YOSHIDA, K., BOYAR, R. M. and HELLMAN, L. (1975) Pituitary–adrenal activity in untreated congenital adrenal hyperplasia. *Journal of Clinical Endocrinology and Metabolism,* **40,** 1–12

GLENTHØJ, A., DAMKJAER NIELSEN, M. and STARUP, J. (1980) Congenital adrenal hyperplasia due to 11β-hydroxylase deficiency: final diagnosis in adult age in three patients. *Acta Endocrinologica,* **93,** 94–99

GREGORY, T. and GARDNER, L. I. (1976) Hypertensive virilizing adrenal hyperplasia with minimal impairment of synthetic route to cortisol. *Journal of Clinical Endocrinology and Metabolism,* **43,** 769–774

GROSSE-WILDE, H., WEIL, J., ALBERT, E. *et al.* (1979) Genetic linkage studies between congenital adrenal hyperplasia and the HLA blood group system. *Immunogenetics,* **8,** 41–49

GRUMBACH, M. M. and CONTE, F. A. (1981) Disorders of sex differentiation. In *Textbook of Endocrinology,* 6th edn., edited by R. H. Williams, pp. 422–514. Philadelphia: W. B. Saunders

GUTAI, J. P., LEE, P. A., JOHNSONBAUGH, R. E., GAREIS, F., URBAN, M. D. and MIGEON, C. J. (1979) Detection of the heterozygous state in siblings of patients with congenital adrenal hyperplasia due to 21-hydroxylase deficiency. *Journal of Pediatrics,* **94,** 770–772

HIRSCHFELD, A. J. and FLESHMAN, J. K. (1969) An unusually high incidence of salt-losing congenital adrenal hyperplasia in the Alaskan Eskimo. *Journal of Pediatrics,* **75,** 492–494

IGARASHI, Y., HIKITA, Y., TAKEHIRO, A. *et al.* (1982) A pilot neonatal mass-screening study for congenital adrenal hyperplasia (CAH) due to 21-hydroxylase deficiency in Japan. Clinical evaluation of a detected patient and false positive cases. In *Program and Abstracts of the 7th Asia and Oceania Congress of Endocrinology,* (Tokyo, 1982), p. 164 (abstract)

JÄNNE, O., PERHEENTUPA, J. and VIHKO, R. (1970) Plasma and urinary steroids in an eight-year-old boy with 3β-hydroxysteroid-dehydrogenase deficiency. *Journal of Clinical Endocrinology and Metabolism,* **31,** 162–165

JANOSKI, A. H. (1977) Naturally occurring adrenal steroids with salt-losing properties: relationship to congenital adrenal hyperplasia. In *Congenital Adrenal Hyperplasia,* edited by P. A. Lec, L. P. Plotnick., A. A. Kowarski and C. J. Migeon, pp. 99–112. Baltimore: University Park Press

JOSSO, N., PICARD, J. Y. and TRAN, D. (1977) The anti Müllerian hormone. *Recent Progress Hormone Research,* **33,** 117–168

KATER, C. E., BIGLIERI, E. G., BRUST, N., CHANG, B. and HIRAI, J. (1982) The unique patterns of plasma aldosterone and 18-hydroxycorticosterone concentrations in the 17α-hydroxylase deficiency syndrome. *Journal of Clinical Endocrinology and Metabolism,* **55,** 295–302

KLEIN, R. (1960) Evidence for and against the existence of a salt-losing hormone. *Journal of Pediatrics,* **57,** 452–460

KLOUDA, P. T., HARRIS, R. and PRICE, D. A. (1980) Linkage and association between HLA and 21-hydroxylase deficiency. *Journal of Medical Genetics,* **17,** 337–341

KOHN, B., LEVINE, L. S., POLLACK, M. S. *et al.* (1982) Late-onset steroid 21-hydroxylase deficiency: a variant of classical congenital adrenal hyperplasia. *Journal of Clinical Endocrinology and Metabolism,* **55,** 817–827

KOTCHEN, T. A. and GUTHRIE, G. P. (1980) Renin–angiotensin–aldosterone and hypertension. *Endocrine Reviews,* **1,** 78–99

KOWARSKI, A., FINKELSTEIN, J. W., SPAULDING, J. S., HOLMAN, G. H. and MIGEON, C. J. (1965) Aldosterone secretion rate in congenital adrenal hyperplasia. A discussion of the theories on the pathogenesis of the salt-losing form of the syndrome. *Journal of Clinical Investigation,* **44,** 1505–1513

KRENSKY, A. M., BONGIOVANNI, A. M., MARINO, J., PARKS, J. and TENORE, A. (1977) Identification of heterozygote carriers of congenital adrenal hyperplasia by radioimmunoassay of serum 17-OH progesterone. *Journal of Pediatrics,* **90,** 930–933

KUHNLE, U., CHOW, D., RAPAPORT, R., PANG, S., LEVINE, L. S. and NEW, M. I. (1981a) The 21-hydroxylase activity in the glomerulosa and fasciculata of the adrenal cortex in congenital adrenal hyperplasia. *Journal of Clinical Endocrinology and Metabolism,* **52,** 534–544

KUHNLE, U., LAND, M., MARVER, D., NEW, M. I. and ULICK, S. (1981b) Evidence for the secretion of mineralocorticoid antagonist in congenital adrenal hyperplasia. In *Program and Abstracts of the Annual Meeting of the Endocrine Society* (Cincinnati, 1981) (abstract)

KUHNLE, U., ROSLER, A., PAREIRA, J. A., GUNCZLER, P., LEVINE, L. S. and NEW, M. I. (1983) The effects of long-term normalization of sodium balance on linear growth in disorders with aldosterone deficiency. *Acta Endocrinologica,* **102,** 577–582

LAFRANCHI, S. (1980) Plasma adrenocorticotrophic hormone in congenital adrenal hyperplasia: importance in long-term management. *American Journal of Diseases in Childhood,* **134,** 1068–1072

LARON, Z., POLLACK, M. S., ZAMIR, R. *et al.* (1980) Late onset 21-hydroxylase deficiency and HLA in the Ashkenazi population: new allele at the 21-hydroxylase locus. *Human Immunology,* **1,** 55–66

LAWRENCE, R. D. (1943) Gynaecomastia produced by deoxycorticosterone acetate (DOCA). *British Medical Journal,* **1,** 12

LEJEUNE-LENAIN, C., CANTRAINE, F., DUFRASNES, M., PREVOT, F., WOLTER, R. and FRANCKSON, J. R. M. (1980) An improved method for the detection of heterozygosity of congenital virilizing adrenal hyperplasia. *Clinical Endocrinology,* **12,** 525–535

LEVINE, L. S., ZACHMANN, M., NEW, M. I. *et al.* (1978) Genetic mapping of the 21-hydroxylase-deficiency gene within the HLA linkage group. *New England Journal of Medicine,* **299,** 911–915

LEVINE, L. S., RAUH, W., GOTTESDIENER, K. *et al.* (1980a) New studies of the 11β-hydroxylase and 18-hydroxylase enzymes in the hypertensive form of congenital adrenal hyperplasia. *Journal of Clinical Endocrinology and Metabolism,* **50,** 258–263

LEVINE, L. S., DUPONT, B., LORENZEN, F. *et al.* (1980b) Cryptic 21-hydroxylase deficiency in families of patients with classical congenital adrenal hyperplasia. *Journal of Clinical Endocrinology and Metabolism,* **51,** 1316–1324

LEVINE, L. S., DUPONT, B., LORENZEN, F. *et al.* (1981) Genetic and hormonal characterization of cryptic 21-hydroxylase deficiency. *Journal of Clinical Endocrinology and Metabolism,* **53,** 1193–1198

LIM, N. Y., MIMICA, N. and DINGMAN, J. F. (1969) Hyperaldosteronism in hypertensive congenital adrenal hyperplasia. *Journal of Clinical Endocrinology and Metabolism,* **29,** 1564–1568

LOBO, R. A. and GOEBELSMANN, U. (1980) Adult manifestation of congenital adrenal hyperplasia due to incomplete 21-hydroxylase deficiency mimicking polycystic ovarian disease. *American Journal of Obstetrics and Gynecology,* **138,** 720–726

LORENZEN, F., PANG, S., NEW, M. I. *et al.* (1979) Hormonal phenotype and HLA-genotype in families of patients with congenital adrenal hyperplasia (21-hydroxylase deficiency). *Pediatric Research,* **13,** 1356–1360

LORENZEN, F., PANG, S., NEW, M. I. *et al.* (1980) Studies of the C-21 and C-19 steroids and HLA genotyping in siblings and parents of patients with congenital adrenal hyperplasia due to 21-hydroxylase deficiency. *Journal of Clinical Endocrinology and Metabolism,* **50,** 572–577

MANTERO, F., BUSNARDO, B., RIONDEL, A., VEYRAT, R and AUSTONI, M. (1971) Hypertension artérielle, alcalose hypokaliémique et pseudohermaphrodisme male par déficit en 17α-hydroxylase. *Schweizerische Medizinische Wochenschrift,* **101,** 38–43

MANTERO, F., SCARONI, C., PASINI, C. V. and FAGIOLO, U. (1980) No linkage between HLA and congenital adrenal hyperplasia due to 17α-hydroxylase deficiency. *New England Journal of Medicine,* **303,** 530

MAUSETH, R. S., HANSEN, J. A., SMITH, E. K., GIBLETT, E. R. and KELLEY, V. C. (1980) Detection of heterozygotes for congenital adrenal hyperplasia: 21-hydroxylase deficiency – a comparison of HLA typing and 17-OH progesterone response to ACTH infusion. *Journal of Pediatrics,* **97,** 749–753

MCCALL, A. L., STERN, J., DALE, S. I. and MELBY, J. C. (1978) Adrenal steroidogenesis in methylandrostenediol-induced hypertension. *Endocrinology,* **103,** 1–5

MIGEON, C. J., ROSENWAKS, Z., LEE, P. A., URBAN, M. D. and BIAS, W. B. (1980) The attenuated form of congenital adrenal hyperplasia as an allelic form of 21-hydroxylase deficiency. *Journal of Clinical Endocrinology and Metabolism,* **51,** 647–649

MOLTENI, A., SKELTON, F. R. and BROWNIE, A. C. (1970) Effect of adrenalectomy on the development of androgen-induced hypertension. *Laboratory Investigation,* **23,** 429–435

MORILLO, E. and GARDNER, L. I. (1979) Genetics of acquired and congenital adrenal hyperplasia. *Lancet,* **2,** 202–203

NATOLI, G., MOSCHINI, L., ACCONCIA, P., ALBINO, G., COSTA, P. and PANSA, G. (1983) Neonatal screening by microassay of 17-α-hydroxyprogesterone in congenital adrenal hyperplasia. In *Recent Progress in Pediatric Endocrinology,* edited by G. Chiumello and M. Sperling. New York: Raven Press, pp. 285–290

NEHER, R. and KAHNT, F. W. (1965) On the biosynthesis of testicular steroids *in vitro* and its inhibition. *Experientia,* **21,** 310–312

NEW, M. I. (1970) Male pseudohermaphroditism due to 17α-hydroxylase deficiency. *Journal of Clinical Investigation,* **49,** 1930–1941

NEW, M. I. and LEVINE, L. S. (1973) Congenital adrenal hyperplasia. In *Advances in Human Genetics,* Vol. 4, edited by H. Harris and K. Hirschhorn, pp. 251–326. London: Plenum Press

NEW, M. I. and PETERSON, R. E. (1966) Disorders of aldosterone secretion in childhood. *Pediatric Clinics of North America,* **13,** 43–58

NEW, M. I. and SEAMAN, M. P. (1970) Secretion rates of cortisol and aldosterone precursors in various forms of congenital adrenal hyperplasia. *Journal of Clinical Endocrinology and Metabolism,* **30,** 361–371

NEW, M. I., LORENZEN, F., PANG, S., GUNCZLER, P., DUPONT, B. and LEVINE, L. S. (1979) 'Acquired' adrenal hyperplasia with 21-hydroxylase deficiency is not the same genetic disorder as congenital adrenal hyperplasia. *Journal of Clinical Endocrinology and Metabolism,* **48,** 356–359

NEW, M. I., DUPONT, B., PANG, S., POLLACK, M. S. and LEVINE, L. S. (1981) An update of congenital adrenal hyperplasia. *Recent Progress in Hormone Research,* **37,** 105–181

NEW, M. I., DUPONT, B., GRUMBACH, K. and LEVINE, L. S. (1982) Congenital adrenal hyperplasia and related conditions. In *The Metabolic Basis of Inherited Disease,* 5th edn., edited by J. B. Stanbury, J. B. Wyngaarden, D. S. Fredrickson, J. F. Goldstein and M. S. Brown, pp. 973–1000. New York: McGraw-Hill

NEW, M. I., LORENZEN, F., LERNER, A. J. *et al.* (1983) Genotyping steroid 21-hydroxylase deficiency: hormonal reference data. *Journal of Clinical Endocrinology and Metabolism,* **57,** 320–326

PANG, S., HOTCHKISS, J., DRASH, A. L., LEVINE, L. S. and NEW, M. I. (1977) Microfilter paper method for 17α-hydroxyprogesterone radioimmunoassay: its application for rapid screening for congenital adrenal hyperplasia. *Journal of Clinical Endocrinology and Metabolism,* **45,** 1003–1008

PANG, S., LEVINE, L. S., CHOW, D. M., FAIMAN, C. and NEW, M. I. (1979) Serum androgen concentrations in neonates and young infants with congenital adrenal hyperplasia due to 21-hydroxylase deficiency. *Clinical Endocrinology,* **11,** 575–584

PANG, S., LEVINE, L. S., LORENZEN, F. *et al.* (1980a) Hormonal studies with obligate heterozygotes and siblings of patients with 11β-hydroxylase deficiency congenital adrenal hyperplasia. *Journal of Clinical Endocrinology and Metabolism,* **50,** 586–589

PANG, S., LEVINE, L. S., CEDERQVIST, L., FUENTES, M. *et al.* (1980b) Amniotic fluid concentrations of Δ^5 and Δ^4 steroids in fetuses with congenital adrenal hyperplasia due to 21-hydroxylase deficiency and in anencephalic fetuses. *Journal of Clinical Endocrinology and Metabolism,* **51,** 223–229

PANG, S., MURPHEY, W., LEVINE, L. S. *et al.* (1982) A pilot newborn screening for congenital adrenal hyperplasia in Alaska. *Journal of Clinical Endocrinology and Metabolism,* **55,** 413–420

PANG, S., LEVINE, L. S., STONER, E., OPITZ, J. M. and NEW, M. I. (1983) Non-salt-losing congenital adrenal hyperplasia due to 3β-hydroxysteroid dehydrogenase deficiency with normal glomerulosa function. *Journal of Clinical Endocrinology and Metabolism,* **56,** 808–818

PARKS, G. A., BERMUDEZ, A. J., ANAST, C. S., BONGIOVANNI, A. M. and NEW, M. I. (1971) Pubertal boy with the 3β-hydroxysteroid dehydrogenase defect. *Journal of Clinical Endocrinology and Metabolism,* **33,** 269–278

PHAM-HUU-TRUNG, M. T., GOURMELEN, M., RAUX-EURIN, M. C. and GIRARD, F. (1978) Pituitary-adrenal axis activity in treated congenital adrenal hyperplasia: static and dynamic studies. *Journal of Clinical Endocrinology and Metabolism,* **47,** 422–427

POLLACK, M. S., LEVINE, L. S., ZACHMANN, M. *et al.* (1979a) Possible genetic linkage disequilibrium between HLA and the 21-hydroxylase deficiency gene (congenital adrenal hyperplasia). *Transplant Proceedings,* **11,** 1315–1316

POLLACK, M. S., LEVINE, L. S., PANG, S. *et al.* (1979b) Prenatal diagnosis of congenital adrenal hyperplasia (21-hydroxylase deficiency) by HLA typing. *Lancet,* **1,** 1107–1108

POLLACK, M. S., LEVINE, L. S., O'NEILL, G. J. *et al.* (1981) HLA linkage and B14,DR1, BfS haplotype association with the genes for late onset and cryptic 21-hydroxylase deficiency. *American Journal of Human Genetics,* **33,** 540–550

ROSENBLOOM, A. L. and SMITH, D. W. (1966) Varying expression for salt losing in related patients with congenital adrenal hyperplasia. *Pediatrics,* **38,** 215–219

ROSENFIELD, R. L., RICH, B. H., WOLFSDORF, J. I. *et al.* (1980) Pubertal presentation of congenital Δ^5-3β-hydroxysteroid dehydrogenase deficiency. *Journal of Clinical Endocrinology and Metabolism,* **51,** 345–353

ROSENWAKS, Z., LEE, P. A., JONES, G. S., MIGEON, C. J. and WENTZ, A. C. (1979) An attenuated form of congenital virilizing adrenal hyperplasia. *Journal of Clinical Endocrinology and Metabolism,* **49,** 335–339

RÖSLER, A. and LEIBERMAN, E. (1980) The clinical heterogeneity of 11β-hydroxylase deficiency congenital adrenal hyperplasia. In *Program and Abstracts of the Annual Meeting of the Endocrine Society* (Washington, DC, 1980)

RÖSLER, A. and LEIBERMAN, E. (1984) 11β-Hydroxylase deficiency congenital adrenal hyperplasia. In *Adrenal Diseases in Children,* edited by M. I. New and L. S. Levine, pp. 47–71. Basel: S. Karger

RÖSLER, A., LEVINE, L. S., SCHNEIDER, B., NOVOGRODER, M. and NEW, M. I. (1977) The interrelationship of sodium balance, plasma renin activity and ACTH in congenital adrenal hyperplasia. *Journal of Clinical Endocrinology and Metabolism,* **45,** 500–512

RÖSLER, A., LEIBERMAN, E., ROSENMANN, A., BEN-UZILIO, R. and WEIDENFELD, J. (1979) Prenatal diagnosis of 11β-hydroxylase deficiency congenital adrenal hyperplasia. *Journal of Clinical Endocrinology and Metabolism,* **49,** 546–551

SCHAISON, G., COUZINET, B., GOURMELEN, M., ELKIK, F. and BOUGNERES, P. (1980) Angiotensin and adrenal steroidogenesis: study of 21-hydroxylase-deficient congenital adrenal hyperplasia. *Journal of Clinical Endocrinology and Metabolism,* **51,** 1390–1394

SCHAMBELAN, M., ROST, G. R., SEBASTIAN, A. and BIGLIERI, E. G. (1980) Influence of the renin–angiotensin system (RAS), potassium (K) and ACTH on plasma levels of 18-hydroxycorticosterone (18-OHB) and aldosterone (A) in man. In *Program and Abstracts of the 6th International Congress of Endocrinology* (Melbourne, 1980), p. 241

SHARMA, D. C., FORCHIELLI, E. and DORFMAN, R. I. (1963) Inhibition of enzymatic steroid 11β-hydroxylation by androgens. *Journal of Biological Chemistry,* **238,** 572–575

SIMOPOULOS, A. P., MARSHALL, J. R., DELEA, C. S. and BARTTER, F. C. (1971) Studies on the deficiency of 21-hydroxylation in patients with congenital adrenal hyperplasia. *Journal of Clinical Endocrinology and Metabolism,* **32,** 438–443

SIZONENKO, P-C., RIONDEL, A. M., KOHLBERG, I. J. and PAUNIER, L. (1972) 11β-Hydroxylase deficiency: steroid response to sodium restriction and ACTH stimulation. *Journal of Clinical Endocrinology and Metabolism,* **35,** 281–287

SOBEL, D. O., GUTAI, J. P., JONES, J. C., WAGENER, D. K. and SMITH, W. (1980) Detection of heterozygote of 21-hydroxylase deficiency. *Lancet,* **1,** 47

SYDNOR, K. L., KELLEY, V. C., RAILE, R. B., ELY, R. S. and SAYERS, G. (1953) Blood adrenocorticotrophin in children with congenital adrenal hyperplasia. *Proceedings of the Society for Experimental Biology and Medicine,* **82,** 695–697

ULICK, S. (1976) Adrenocortical factors in hypertension, I. Significance of 18-hydroxy-11-deoxycorticosterone. *American Journal of Cardiology,* **38,** 814–824

VILLEE, D. B. (1973) Changes in fetal steroid metabolism with age. *Clinical Pharmacology and Therapeutics,* **14,** 705–713

WEIL, J., BIDLINGMAIER, F., SIPPELL, W. G., BUTENANDT, O. and KNORR, D. (1979) Comparison of two tests for heterozygosity in congenital adrenal hyperplasia (CAH). *Acta Endocrinologica,* **91,** 109–121

WEST, C. D., ATCHESON, J. B., STANCHFIELD, J. B., RALLISON, M. L., CHAVRE, V. J. and TYLER, F. H. (1979) Multiple or single 21-hydroxylases in congenital adrenal hyperplasia. *Journal of Steroid Biochemistry,* **11,** 1413–1419

WILSON, J. D. and WALSH, P. C. (1979) Disorders of sexual differentiation. In *Cambell's Urology,* 4th edn, Vol. 2, edited by J. H. Harrison, R. F. Gittes, A. D. Perlmutter, T. A. Stamey and P. C. Walsh, pp. 1484–1532. Philadelphia: W. B. Saunders

WINKEL, C. A., PARKER, C. R. JR, SIMPSON, E. R. and MACDONALD, P. C. (1980) Production rate of deoxycorticosterone in women during the follicular and luteal phases of the ovarian cycle: the role of extra-adrenal 21-hydroxylation of circulating progesterone in deoxycorticosterone production. *Journal of Clinical Endocrinology and Metabolism,* **51,** 1354–1358

WINTER, J. S. D. (1980) Current approaches to the treatment of congenital adrenal hyperplasia. *Journal of Pediatrics,* **97,** 81–82

WOLFF, P. B., WILBOIS, R. P., WELDON, V. V. and HAYMOND, M. W. (1977) Posterior fusion without clitoromegaly in a female with partial 21-hydroxylase deficiency. *Journal of Pediatrics,* **91,** 951–952

ZACHMANN, M. and PRADER, A. (1975) Gynecomastia with congenital virilizing adrenal hyperplasia (11β-hydroxylase deficiency). *Journal of Pediatrics,* **87,** 839–840

ZACHMANN, M. and PRADER, A. (1978) Unusual heterozygotes of congenital adrenal hyperplasia due to 21-hydroxylase deficiency. *Acta Endocrinologica,* **87,** 557–565

ZACHMANN, M. and PRADER, A. (1979) Unusual heterozygotes of congenital adrenal hyperplasia due to 21-hydroxylase deficiency confirmed by HLA tissue typing. *Acta Endocrinologica,* **92,** 542–546

7
The etiology and management of Cushing's syndrome

Charles Faiman

INTRODUCTION

The dismal mortality statistics for untreated Cushing's syndrome can only be improved by early recognition and treatment of this disorder (Ross and Linch, 1982). In one series, 17 of 32 patients died after an average of 4.5 years of known disease (Plotz, Knowlton and Ragan, 1952). The striking clinical features of patients with long-standing hypercortisolism were first described at the beginning of this century by Osler (*see* Altschule, 1980) and Cushing (1912). Cushing (1932) later provided the definitive description of the clinicopathological features of the syndrome now bearing his name. Of the eight patients studied by him *post mortem*, six had pituitary adenomas, five of which were shown to contain basophilic elements.

Even though non-iatrogenic glucocorticoid excess is not a common condition, it has long been a source of interest and controversy. One major issue, still the subject of lively debate, is whether the hypothalamus or the pituitary plays the primary role in the pathophysiology of Cushing's disease or pituitary ACTH-induced adrenocortical hyperplasia. This argument has continued ever since Cushing's first description of the association with pituitary adenomas. A strong case can be made that the disease, at least initially, is due to faulty secretion of hypothalamic corticotropin-releasing factor (CRF) (Besser and Edwards, 1972; Krieger, Amorosa and Linick, 1975); on the other hand, the majority of patients do have ACTH-secreting pituitary tumors by the time of diagnosis (Salassa *et al.*, 1978; Hardy, 1982; Boggan, Tyrrell and Wilson, 1983).

Dramatic advances over the last two decades in hormone assay methodology, neuropharmacology, neuroradiological technology and pituitary microsurgical techniques have had a major impact on the investigation and management of patients with Cushing's syndrome. It is the purpose of this chapter to review some of these major advances, to discuss their practical implications in terms of current patient management and to provide some ideas as to future directions in this field.

CLINICAL CLASSIFICATION

A classification of Cushing's syndrome is shown in *Table 7.1*. Only the endogenous causes will be discussed here. Excessive alcohol intake is now well-recognized as a cause of ACTH-mediated glucocorticoid excess, and therefore of true (rather than pseudo-) Cushing's syndrome. It is reversible on stopping alcohol, and its pathogenesis and diagnosis are discussed further in Chapter 9. *Cushing's disease*, the inappropriate or excessive production of ACTH by the pituitary gland, with resultant bilateral adrenocortical hyperplasia, makes up approximately 70–80% of cases in adults (Besser and Edwards, 1972; Gold, 1979; Ross and Linch, 1982). Of these, about 10–20% show *nodular adrenal hyperplasia*, with nodules generally less than 2.5 cm in diameter (Aron *et al.*, 1981; Urbanic and George, 1981). A recent series suggests that this variant may be even more prevalent and that individual nodules may be larger (Smals *et al.*, 1984). About 10–15% of cases result from *adrenal tumors*, of which half are malignant (Bertagna and Orth, 1981); another

Table 7.1 Classification of Cushing's syndrome

Exogenous	*Endogenous*
(1) Iatrogenic	(1) Cushing's disease
(a) ACTH	(2) Adrenal tumors
(b) Glucocorticoids	(a) benign
	(b) malignant
(2) Alcohol-induced	(3) Ectopic ACTH syndrome
	(4) Ectopic CRF syndrome
	(5) Uncertain or unknown etiology
	(a) nodular adrenal hyperplasia
	(b) micronodular adrenal dysplasia

10–15% result from *ectopic ACTH* production by non-pituitary tumors (Besser and Edwards, 1972). An apparently rare cause is *ectopic corticotropin-releasing factor* production. In the main, cases to date described have been of tumors containing both ACTH and CRF-like activity (Imura *et al.*, 1975; Upton and Amatruda, 1971). A recent report, however, described a case of prostatic carcinoma in which there was isolated production of ectopic CRF (Carey *et al.*, 1984). A rare cause in young adults is *micronodular adrenocortical dysplasia* (Ruder, Loriaux and Lipsett, 1974).

Both Cushing's disease and Cushing's syndrome resulting from adrenal neoplasms show a 3:1 to 5:1 female preponderance. This contrasts with a male preponderance in ectopic ACTH cases, most commonly due to bronchogenic carcinoma.

PATHOPHYSIOLOGY OF CUSHING'S SYNDROME

Cushing's disease

Krieger (1983) recently reviewed the controversy regarding the etiology and pathogenesis of this disorder. The evidence in favor of a primary hypothalamic abnormality, with hypersecretion of CRF or some other secretagogue leading to corticotroph hyperplasia and/or neoplasia and ACTH overproduction, includes the following:

(1) The presence of relative but not absolute resistance to feedback suppression by glucocorticoids (Liddle, 1960);

(2) The presence of associated neuroendocrine abnormalities, including abnormal circadian periodicity of plasma growth hormone (Krieger and Glick, 1974) and prolactin concentrations (Krieger, Howanitz and Frantz, 1976), as well as suppression of basal luteinizing hormone (LH) secretion (Fitzgerald *et al.*, 1982);
(3) The therapeutic response to neuropharmacological agents with a presumed hypothalamic locus of action (Krieger, Howanitz and Frantz, 1976; Lamberts *et al.*, 1977; Dornhorst *et al.*, 1983);
(4) Persistence of a subtle abnormality during acute inhibition of ACTH secretion by cortisol despite cure of the hypercorticoid state by adrenalectomy or selective pituitary microadenomectomy (Fehm *et al.*, 1977), and correction of this defect by the neuropharmacological agents cyproheptadine (Lankford, Tucker and Blackard, 1981) or reserpine (Fehm *et al.*, 1983);
(5) Recurrence of disease after pituitary microadenomectomy (Pont and Gutierrez-Hartman, 1979; Bigos *et al.*, 1980);
(6) The finding in some cases of corticotroph hyperplasia, with or without a tumor (Lamberts *et al.*, 1980b; Schnall *et al.*, 1980).

Arguments conflicting with a primary hypothalamic defect have accumulated over the past decade, and the available evidence now strongly suggests that pituitary tumors are the cause of the majority of cases of Cushing's disease:

(1) It has been shown that glucocorticoid feedback can occur directly at the level of the pituitary corticotroph (Lagerquist *et al.*, 1974; Jones, Hillhouse and Burden, 1977; Yasuda, Greer and Aizawa, 1982), indicating that relative glucocorticoid feedback resistance does not necessarily imply faulty secretion of hypothalamic CRF.
(2) Correction of ACTH–cortisol overproduction, for example by pituitary microadenomectomy, normalizes the circadian rhythms of growth hormone and prolactin and restores LH secretory dynamics (Schnall *et al.*, 1978; Fitzgerald *et al.*, 1982), suggesting that these associated abnormalities result from cortisol overproduction *per se* rather than hypothalamic dysfunction.
(3) Neuropharmacological agents, including bromocriptine, cyproheptadine and reserpine, originally thought to act solely at a hypothalamic locus, also appear to have a *direct* action on the ACTH-secreting cell (Lamberts *et al.*, 1980a; Suda *et al.*, 1983); however, the demonstration that sodium valproate, which thus far has not been shown to have a direct pituitary action, also lowers ACTH levels in some patients with Nelson's syndrome, does maintain the possibility that these individuals have disordered hypothalamic regulation (Dornhorst *et al.*, 1983).
(4) Since some neuropharmacological agents may not be nervous system specific, the assumption that their ability to normalize the acute ACTH response to cortisol infusion indicates hypothalamic dysfunction (Lankford, Tucker and Blackard, 1981; Fehm *et al.*, 1983) may be fallacious.
(5) There is some evidence that tumor recurrence usually results from incomplete tumor removal (Pont and Gutierrez-Hartman, 1979).
(6) Approximately 80–90% of patients with Cushing's disease are found to have pituitary tumors and in the vast majority resection is followed by apparent cure (Hardy, 1982; Boggan, Tyrrell and Wilson, 1983); indeed, not a single case of hyperplasia was observed in the latter series of 100 pituitary explorations.

The fact that successful pituitary surgery is followed by subnormal function of the hypothalamic–pituitary–adrenal axis for up to one year or longer (Kuwayama *et al.*, 1981; Fitzgerald *et al.*, 1982) also argues against a primary hypothalamic abnormality. If the latter were true, one would have expected more rapid restoration of function of the residual non-adenomatous corticotrophs under the influence of a supranormal hypothalamic drive. Long-term ACTH–cortisol deficiency following tumor removal does not reflect merely surgically-induced trauma, since similarly treated patients with acromegaly do not show such postoperative impairment of hypothalamic–pituitary–adrenal function (Fitzgerald *et al.*, 1982).

Correlation of pituitary tumor histology with preoperative responsiveness to dexamethasone and bromocriptine has provided some evidence for two different tumor types in patients with Cushing's disease (Lamberts, de Lange and Stefanko, 1982). In one group, the adenomas appeared to be of anterior pituitary cell origin and contained no argyrophilic (presumably neural) elements; patients in this group responded to dexamethasone but not to bromocriptine. The second group had multiple adenomatous and/or hyperplastic cell nests containing nervous tissue, which in three of six cases appeared to arise from a vestigial neuro-intermediate lobe; these patients tended to be relatively resistant to dexamethasone but responsive to bromocriptine. These observations raise the intriguing possibility that some tumors may arise by true neoplasia of anterior pituitary ACTH-secreting cells whereas others may be of neural, i.e. hypothalamic, origin. These findings must be confirmed by other workers and the origin and putative neurotransmitter function of the argyrophilic fibers identified before one can properly interpret their significance.

In the final analysis it may prove impossible to determine the original etiology of Cushing's disease when the patient first presents for study. It is possible, however, that measurements of CRF levels and secretory dynamics, either in the peripheral circulation or in the hypothalamopituitary portal vasculature at the time of surgery, will provide a clearer indication of the relative importance of hypothalamic influences in the pathophysiology of ACTH hypersecretion. Regrettably, the present reliance on inferences drawn from *in vivo* neuropharmacological manipulations throws more heat than light on this important question. The frequency and latency period of disease recurrence might also have some bearing on our views as to the likelihood of extrapituitary influences leading to the development of pituitary tumors, but such data must await longer-term clinical studies than are currently available.

Nodular adrenal hyperplasia

The pathophysiology of this variant is unclear. It is characterized by the presence of one or more nodules in both adrenal glands and diffuse bilateral cortical hyperplasia (Neville and Symington, 1967). Individual nodules are usually less than 2.5 cm in diameter (Urbanic and George, 1981) but may be larger (Smals *et al.*, 1984) and mimic adrenocortical tumors on computerized tomography (Aron *et al.*, 1981). Indeed, adrenal carcinoma may develop within a macronodular adrenal cortex (Anderson *et al.*, 1978). ACTH levels may be undetectable, normal or elevated (Crapo, 1979; Aron *et al.*, 1981). Relative or absolute insensitivity to suppression by high-dose dexamethasone has been reported (Crapo, 1979; Smals *et al.*, 1984) but is certainly not the experience of all investigators (Urbanic and George, 1981).

Evidence that this disorder is fundamentally of pituitary origin, with secondary nodular hyperplasia, includes reports of non-suppressed ACTH levels which increase even further after adrenalectomy, the more frequent occurrence of nodules in long-standing disease, and the finding of pituitary tumors in some patients either before or after adrenalectomy. Others argue that although initial hypersecretion of ACTH may have existed such nodules ultimately become autonomous (tertiary nodular hyperplasia).

Not surprisingly, these conflicting views have led to opposing recommendations for therapy, being directed primarily at pituitary exploration by some (Aron *et al.*, 1981) and bilateral adrenalectomy by others (Smals *et al.*, 1984). What is clear is that the present data base does not allow a decision as to the best primary treatment for this condition.

Micronodular adrenal dysplasia

This rare disorder of unknown etiology, also termed *primary adrenocortical nodular dysplasia* (Meador *et al.*, 1967) or *micronodular adrenal disease* (Ruder, Loriaux and Lipsett, 1974), appears to be an entirely different syndrome. Adrenal weight is usually normal or reduced. The striking pathological features are the small nodules, usually less than 3 mm in diameter, containing brown, black, blue or green pigment. Microscopic examination shows non-encapsulated sheets of hyperplastic cells, atrophy of the intervening adrenal remnant and lymphocytic infiltration. It is typically a disease of young adults, with onset before or during puberty. It may be familial (Schweizer-Cagianut, Froesch and Hedinger, 1980). Osteopenia is a striking characteristic.

Ruder, Loriaux and Lipsett (1974) reviewed the laboratory findings in 12 patients with this disorder and found that seven had shown no cortisol response to ACTH, two had shown suppression, and three stimulation. ACTH levels were usually low. None of 10 patients tested showed suppression by high-dose dexamethasone, attesting to the autonomous nature of the disorder; indeed, *paradoxical stimulation* by high-dose *dexamethasone* was observed in four of these patients. Such paradoxical responses are occasionally also seen in association with pituitary tumors, ectopic ACTH-secreting bronchial adenomas or adrenal adenomas (Gold, 1979) and, accordingly, are not diagnostic of micronodular dysplasia. A similar response may occur as an artifact because of periodic hormonogenesis (Brown *et al.*, 1973; Green and Van't Hoff, 1975; Liberman *et al.*, 1976). We have recently observed a reproducible, paradoxical response to dexamethasone in a patient with a pituitary microadenoma (later cured by surgery) and no evidence of periodic hormonogenesis. In this context it is of interest that glucocorticoid-dependent ACTH secretion has been suggested as the mechanism of prolonged remission after metyrapone treatment in two patients with Cushing's syndrome apparently due to ectopic ACTH secretion (source unknown) (Beardwell, Adamson and Shalet, 1981).

The treatment of choice for micronodular adrenal dysplasia is bilateral adrenalectomy.

TREATMENT OF CUSHING'S SYNDROME

Objectives

The objectives of therapy today are no different from those espoused by Orth and Liddle in 1971.

(1) Correction of the syndrome by lowering cortisol secretion to normal.
(2) Eradication of any tumor threatening the health of the patient.
(3) Avoidance of permanent endocrine deficiency.
(4) Avoidance of permanent dependency on medication.

Since the various therapeutic modalities have been reviewed recently in a companion volume (Beardwell and Robertson, 1981), as well as by Krieger (1983), only recent contributions to the field will be emphasized. These mainly concern accumulating experience with the trans-sphenoidal microsurgical approach to the pituitary as primary treatment for Cushing's disease.

Definition of cure

Orth and Liddle (1971) also suggested criteria for the definition of 'cure' in adults. Modified in the light of current views, these include:

(1) Normalization of urinary excretion of cortisol and its metabolites, as shown by
 (a) urinary 17-hydroxycorticosteroid excretion less than 7 mg (19 μmol)/g creatinine;
 (b) urinary free cortisol excretion less than 110 μg (315 nmol)/day (Crapo, 1979);
(2) Normalization of plasma cortisol levels, with restoration of diurnal rhythm and integrated values of less than 10 μg/dl (285 nmol/l);
(3) Restoration of normal dexamethasone suppressibility;
(4) Maintenance or restoration of normal anterior pituitary reserve, as gauged by appropriate testing, for example with a combined stimulus of insulin-induced hypoglycemia, gonadotropin-releasing factor and thyrotropin-releasing factor.

Without guidelines such as these, which can be followed in most patients, comparisons of results from different centers or of different therapeutic modalities will not be meaningful, and the differentiation between recurrence and incomplete cure (Pont and Gutierrez-Hartman, 1979) will be impossible. Sufficiently long follow-up, perhaps for a decade or more, must be a component of any therapeutic regimen. For example, we are only now beginning to observe the gradual onset of hypopituitarism in some patients with prolactinomas treated by conventional cobalt-60 radiotherapy over a decade ago. Similarly, recurrence after micro-adenomectomy for prolactinoma may be delayed for several years (Serri *et al.*, 1983).

Treatment of Cushing's syndrome (excluding Cushing's disease)

Optimal treatment depends on an accurate diagnosis (*see Table 7.1*). *Adrenal adenomas* and *micronodular adrenal dysplasia* are invariably cured by adrenal surgery (Ruder, Loriaux and Lipsett, 1974; Bertagna and Orth, 1981). Surgical results with histologically unequivocal *adrenal carcinoma* are poor. Long-term survival after incomplete tumor resection is rare; most patients die of tumor recurrence or metastatic disease within seven years, often within one or two years of diagnosis (Bertagna and Orth, 1981). Tumors are radio-resistant and usually aggressive. Occasionally, treatment with the adrenolytic agent *o,p'*DDD (mitotane) may be followed by an objective but temporary remission. The value of other agents such as *cis*-platinum, either alone or in combination with *o,p'*DDD, remains unclear. Hypercortisolism can be controlled with inhibitors of steroid biosynthesis such as metyrapone or aminoglutethimide (Beardwell, 1981). The majority of patients with the *ectopic ACTH syndrome* have non-resectable malignant neoplasms. Again, hypercortisolism can be controlled by metyrapone or aminoglutethimide; hypokalemia and hypertension are ameliorated by the mineralocorticoid antagonist spironolactone (aldactone). In some patients with more indolent but inoperable tumors, bilateral adrenalectomy or chemical adrenalectomy with *o,p'*DDD may be considered.

A therapeutic approach to Cushing's disease

It is generally agreed that the preferred treatment of this disorder in *adults* is tumor resection by trans-sphenoidal microsurgery. However, *children and adolescents* are often treated by conventional radiotherapy. In the series of Jennings, Liddle and Orth (1977), for example, 12 of 15 patients were cured by radiotherapy, with a median response time of five months (range 2–18 months). Follow-up (median three years; range 1–18 years) revealed adult stature (156 to 166 cm), no apparent neurological sequelae and, in most cases, normal anterior pituitary hormone reserve. More recently, children and adolescents have begun to be treated by pituitary microadenomectomy, with initial cure rates comparable to those reported for radiotherapy (Styne *et al.*, 1984). Longer-term follow up, with emphasis on final stature, recurrence and future development of hypopituitarism, is needed for a more definitive comparison of these two modalities (Bergstrand and Nilsson, 1982).

A decision to explore the pituitary gland must rest on endocrinological criteria. Less than 10% of cases show gross enlargement of the pituitary fossa by standard radiography at the time of diagnosis (Salassa *et al.*, 1959). Even with high-resolution computerized tomography scanning techniques, no tumor is seen in about one-third of cases, and in about 20% the results are false positives (either no tumor found or tumor in another location at exploration).

Two large series of patients treated by pituitary microsurgery have recently been reported (Hardy, 1982; Boggan, Tyrrell and Wilson, 1983). Although not strictly defined, a cure rate of approximately 85% was reported if the tumor was confined to the sella. This figure included cases in whom a central wedge resection (Hardy, 1982) or total hypophysectomy (Boggan, Tyrrell and Wilson, 1983) was performed after initial exploration had failed to reveal an obvious tumor. Hypophysectomy should not be considered in children, in whom conventional radiotherapy appears to be so effective. Results from other centers are variable (Burch, 1983), but cure rates for microadenomas of 70% or better are usual for most experienced neurosurgeons. Not surprisingly, the cure rate for macroadenomas is much lower – about 50% (Hardy, 1982; Boggan, Tyrrell and Wilson, 1983). A recent survey by Zervas (1984) of the results in over 500 patients world-wide is summarized in *Table 7.2*.

Kuwayama *et al.* (1981) and Fitzgerald *et al.* (1982) have stressed that postoperative hypoadrenalism occurs in patients with a favorable response, although occasional exceptions do occur (Bigos *et al.*, 1980). Therefore, cortisol replacement therapy may be required for up to 1–2 years. Fitzgerald *et al.* (1982) suggest that adrenal function studies be performed two months after surgery and thereafter at 4- to 6-monthly intervals until adrenal function normalizes; ACTH stimulation testing appears to be an adequate technique for such a purpose.

Those who are not cured by trans-sphenoidal surgery or develop recurrences may be treated by radiotherapy or bilateral adrenalectomy. Anti-adrenal agents such as metyrapone or aminoglutethimide, or neuropharmacological agents such as cyproheptadine or bromocriptine, may be used to ameliorate hypercortisolism during the 6 to 12 months or longer required to obtain the effects of irradiation.

Table 7.2 Survey of transphenoidal surgery in Cushing's disease*

Radiological features	*Findings at surgery*	*Number of cases*	*Cures***		*Follow-up at five years*	*Recurrence at five years*	
			(*n*)	(%)	(*n*)	(*n*)	(%)
Small sella	Microadenoma	350	320	91	51	4	8
	No microadenoma	76	45†	59	10	0	0
	Diffuse hyperplasia	5	2†	40	1	0	0
	Overall	431	367	85	62	4	6
Large sella	Macroadenoma	70	43	61	11	4	36

* See Zervas (1984).
** Urinary free cortisol less than 100 μg (285 nmol)/day *or* urinary hydroxycorticosteroid less than 8 mg (22 μmol)/day *or* normal diurnal plasma cortisol.
† Cured by wedge resection or total ablation.

An outline of the various treatment modalities for Cushing's disease is given in *Table 7.3*. Readers who are interested in a more detailed account of the results of radiotherapy for Cushing's disease, and for the prevention and treatment of *Nelson's syndrome*, i.e. hyperpigmentation associated with a pituitary tumor following bilateral adrenalectomy for Cushing's disease (Nelson, Meakin and Thorn, 1960), are referred to the recent review by Sheline (1981). Neuropharmacological agents may also play a role in the treatment of this latter condition (Dornhorst *et al.*, 1983; Krieger, 1983). In one patient, sustained remission of Nelson's syndrome was reported after long-term therapy with cyproheptadine (Aronin and Krieger, 1980). In another, there was radiographic evidence of tumor shrinkage after three months of sodium valproate treatment (Dornhorst *et al.*, 1983).

Notwithstanding the fact that the pathogenesis of *nodular adrenal hyperplasia* is obscure, therapy directed towards a pituitary tumor in patients with this disorder may achieve results similar to those obtained in patients with Cushing's disease but without adrenal nodules (Aron *et al.*, 1981; Urbanic and George, 1981). However, Smals *et al.* (1984) have suggested that initial therapy should be bilateral adrenalectomy; their evidence against treatment directed at the pituitary consists of three patients who underwent either hypophysectomy or pituitary irradiation and were identified as treatment failures. They also allude to four patients who were not cured following 'selective pituitary surgery'.

Table 7.3 Summary of current treatment modalities in Cushing's disease (modified from Krieger, 1983)

Mode of therapy	*Major advantages*	*Major disadvantages*	*Specific indications*
Pituitary surgery	Rapid cure in 85%	Operative morbidity (small); hypopituitarism if hypophysectomy performed; recurrence rate (?)	Severe disease, for rapid cure; Nelson's syndrome with obvious tumor
Pituitary irradiation			
Radiotherapy (cobalt or linear accelerator)	Minimal side effects Efficacy 80% in children	Slow – up to 18 months; efficacy 20–50% in adults	Moderate disease in children
Proton beam or α-particles	Efficacy 90%	Hypopituitarism in ⩾10%; risk of parasellar injury; limited availability	
Neuropharmacological agents	No major side effects; no adverse effect on other pituitary functions	Slow onset; not permanent cure; efficacy up to 30–50%; weight gain, especially in children (cyproheptadine)	Adjunct to pituitary irradiation or pretreatment for surgery; mild–moderate disease; (?) prevention of progression of Nelson's syndrome
Adrenalectomy			
Surgical	Rapid	Need for replacement therapy; associated morbidity and mortality; development of Nelson's syndrome (10–15%); failure or recurrence in 5% (residual adrenal tissue)	Rapid amelioration where pituitary approach unsuccessful
Medical		Frequent escape (except for *o,p'*DDD); need for replacement therapy; side-effects	Adjunct to pituitary irradiation or pretreatment for surgery

FUTURE DIRECTIONS IN CUSHING'S SYNDROME

An animal model

Spontaneous Cushing's disease has been described in two non-human species, the dog and the horse (Krieger, 1983; McNicol, Thomson and Stewart, 1983). In the dog, tumors have been described in the anterior lobe and the pars intermedia, and are sometimes associated with corticotroph hyperplasia. Corticotroph hyperplasia, alone, has also been described. Whether the two pituitary sites of origin in canine disease correspond to those described recently in humans (Lamberts, de Lange and Stefanko, 1982) remains to be seen. Cushing's disease in the horse is almost invariably associated with adenomas or adenomatous hyperplasia of the pars intermedia.

The fact that, unlike in man, there is no female preponderance in Cushing's disease of dogs makes one wonder if this species is an ideal human analogue. Furthermore, a major stumbling block is the present inability to induce the disease at will and to identify spontaneously affected animals readily. Perhaps long-term CRF administration may prove to be an effective means of inducing pituitary hyperplasia or adenomas, and also lead to clarification of the pathogenesis of Cushing's disease. In the meantime, spontaneously affected animals can be used to test the safety and efficacy of newer chemotherapeutic agents.

Refinements in diagnosis

ACTH and other pro-opiomelanocortin (POMC)-related assays

One of the most pressing problems for the clinician is the difficulty of obtaining reliable and interpretable plasma ACTH values. Even though appropriately sensitive, reliable and accurate techniques for ACTH determination are now available, they are not provided by many endocrine laboratories. Additional problems stem from the biological nature of ACTH, namely its pulsatile release, diurnal rhythmicity and short half-life in the circulation. Accordingly, investigators have developed techniques to look at other cleavage products of the parent POMC molecule, which generally are secreted in equimolar amounts with ACTH, in the hope that these will turn out to be more helpful in differentiating ACTH-dependent (pituitary or ectopic) from adrenal causes of hypercortisolism. Theoretical advantages of measuring β-endorphin (Wilkes *et al.*, 1980) or the *N*-terminal fragment (1–76) of pro-opiomelanocortin (Chan, Seidah and Chrétien, 1983) include the ease of assay and the more prolonged serum half-life of these peptides (*see also* Chapter 9).

Selective venous sampling

Findling *et al.* (1981) suggested that inferior petrosal sinus sampling (using a transfemoral vein approach through the internal jugular vein and jugular bulb) is helpful to differentiate Cushing's disease from the ectopic ACTH syndrome. In five of the six cases with Cushing's disease in whom the inferior petrosal sinus could be catheterized, ACTH levels were more than twice as high at this site than simultaneously obtained peripheral levels, whereas in three patients with the ectopic ACTH syndrome this ratio was less than 1.5 to 1. In one patient, a central venous–systemic arterial gradient of less than 1 correctly predicted the presence of an ACTH-secreting bronchial carcinoid two years before it became clinically apparent. These authors stress the futility of venous sampling at sites more distal from the pituitary than the inferior petrosal sinus. More recently, Manni *et al.* (1983) have confirmed the usefulness of selective venous sampling, but indicated that in two of three of their patients the correct diagnosis might still have been missed had *bilateral* inferior petrosal samples not been taken. Obviously, a larger experience is needed to put this technique into proper perspective. Perhaps a combination of selective venous sampling and assay of other POMC peptides in addition to ACTH may circumvent the problems inherent in pulsatile hormonal release. Lastly, this 'jugular act' seems to require the presence of a particularly skilled and experienced vascular radiologist, which may not always be possible.

Corticotropin-releasing factor testing

The 41-amino-acid peptide CRF has recently been evaluated as an aid in the differential diagnosis of Cushing's syndrome (Nakahara *et al.*, 1983; Chrousos *et al.*, 1984; Lytras *et al.*, 1984). It can clearly separate patients with adrenal neoplasms, who have low baseline ACTH concentrations and an absent or blunted response to CRF, analogous to that of thyroid stimulating hormone (TSH) to thyrotropin-releasing factor (TRF) in patients with hyperthyroidism. More importantly, it may differentiate patients with Cushing's disease from those with ectopic ACTH production. In one study, all 13 patients with Cushing's disease responded with a further increment in their already elevated levels of ACTH and cortisol; none of six patients with the ectopic ACTH syndrome responded (Chrousos *et al.*, 1984). However, responses may be variable in Cushing's disease both before and after pituitary microadenomectomy (Orth *et al.*, 1982; Chrousos *et al.*, 1984; Orth, 1984), and at least one patient with the ectopic ACTH syndrome has had a response to CRF which was indistinguishable from that observed in patients with Cushing's disease. Thus, the initial results of Chrousos *et al.* (1984) require further confirmation. Advantage might also be taken of the synergistic effect of vasopressin and CRF (Lamberts *et al.*, 1984) in order to magnify subtle differences of ACTH release among diagnostic categories. (For a more detailed discussion of CRF *see* Chapter 9.)

Newer drug therapy

Although steroid antagonists such as antiandrogens (cyproterone acetate), anti-estrogens (clomiphene citrate, tamoxifen) and an anti-mineralocorticoid (spironolactone) have been available for some time, no safe and effective anti-glucocorticoid has yet been developed for the palliation of otherwise untreatable hypercortisolism. Such a drug might be useful as an adjunct in the preparation of patients for surgery, in the interval while a slow treatment such as pituitary irradiation is taking effect, or in primary therapeutic failures. Two types of agent have been studied recently. The first, effective both *in vitro* and *in vivo*, is a steroid derivative, RU38486, which is potent, rapid in onset and effective by mouth; it reversibly binds to the glucocorticoid receptor in a competitive fashion and has no apparent agonistic properties (Côté *et al.*, 1982; Chobert *et al.*, 1983). Preliminary results indicate its efficacy in man (Bertagna *et al.*, 1984) and its successful use in the treatment of one patient with the ectopic ACTH syndrome (Nieman *et al.*, 1984). The second group, imidazole antimycotic agents, bind to glucocorticoid receptors in target tissues and inhibit dexamethasone-induced enzyme activity; they, too, have no agonistic properties (Loose, Stover and Feldman, 1983). In addition, the latter have been shown both *in vitro* and *in vivo* to inhibit steroid biosynthesis. Obviously, further work must be done in this fruitful area.

Meanwhile, as with any new therapy, we should be cautious of the 'latest' additions to our therapeutic armamentarium and remember the words of Galen: 'All who drink of this remedy recover in a short time except those whom it does not help, who all die. Therefore, it is obvious that it fails only in incurable cases'.

Acknowledgement

The author wishes to thank Miss Emma Gulbis for her excellent secretarial assistance.

References

ALTSCHULE, M. D. (1980) A near miss – Osler's early description of Cushing's syndrome with, regrettably, no *post-mortem* examination. *New England Journal of Medicine,* **302,** 1153 1155

ANDERSON, D. C., CHILD, D. F., SUTCLIFFE, C. H., BUCKLEY, C. H., DAVIES, D. and LONGSON, D. (1978) Cushing's syndrome, nodular adrenal hyperplasia and virilizing carcinoma. *Clinical Endocrinology,* **9,** 1–14

ARON, D. C., FINDLING, J. W., FITZGERALD, P. A. *et al.* (1981) Pituitary ACTH dependency of nodular adrenal hyperplasia in Cushing's syndrome: report of two cases and review of the literature. *American Journal of Medicine,* **71,** 302–306

ARONIN, N. and KRIEGER, D. T. (1980) Sustained remission of Nelson's syndrome after stopping cyproheptadine treatment. *New England Journal of Medicine,* **302,** 453–455

BEARDWELL, C. (1981) Medical therapy in the management of pituitary adenomas. In *The Pituitary, Clinical Endocrinology,* Vol. 1, edited by C. Beardwell and G. L. Robertson, pp. 140–174. London: Butterworths

BEARDWELL, C. and ROBERTSON, G. L. (Eds) (1981) *The Pituitary, Clinical Endocrinology* Vol. 1. London: Butterworths

BEARDWELL, C. G., ADAMSON, A. R. and SHALET, S. M. (1981) Prolonged remission in florid Cushing's syndrome following metyrapone treatment. *Clinical Endocrinology,* **14,** 485–492

BERGSTRAND, C. G. and NILSSON, K. O. (1982) Treatment of Cushing's disease in children. *Acta Paediatrica Scandinavica,* **71,** 1–6

BERTAGNA, C. and ORTH, D. N. (1981) Clinical and laboratory findings and results of therapy in 58 patients with adrenocortical tumors admitted to a single medical center (1951 to 1978). *American Journal of Medicine,* **71,** 855–875

BERTAGNA, X., BERTAGNA, C., LUTON, J. P., HUSSON, J. M. and GIRARD, F. (1984) The new steroid analogue RU 486 inhibits glucocorticoid action in man. *Journal of Clinical Endocrinology and Metabolism,* **59,** 25–28

BESSER, G. M. and EDWARDS, C. R. W. (1972) Cushing's syndrome. *Clinics in Endocrinology and Metabolism,* **1,** 451–490

BIGOS, S. T., SOMMA, M., RASIO, E. *et al.* (1980) Cushing's disease: management by transsphenoidal pituitary microsurgery. *Journal of Clinical Endocrinology and Metabolism,* **50,** 348–354

BOGGAN, J. E., TYRRELL, J. B. and WILSON, C. B. (1983) Transsphenoidal microsurgical management of Cushing's disease: report of 100 cases. *Journal of Neurosurgery,* **59,** 195–200

BROWN, R. D., VAN LOON, G. R., ORTH, D. N. and LIDDLE, G. W. (1973) Cushing's disease with periodic hormonogenesis: one explanation for paradoxical response to dexamethasone. *Journal of Clinical Endocrinology and Metabolism,* **36,** 445–451

BURCH, W. (1983) A survey of results with transsphenoidal surgery in Cushing's disease. *New England Journal of Medicine,* **308,** 103–104

CAREY, R. M., VARMA, S. K., DRAKE, C. R. JR, *et al.* (1984) Ectopic secretion of corticotropin-releasing factor as a cause of Cushing's syndrome: a clinical, morphologic, and biochemical study. *New England Journal of Medicine,* **311,** 13–20

CHAN, J. S. D., SEIDAH, N. G. and CHRÉTIEN, M. (1983) Measurement of *N*-terminal (1–76) of human pro-opiomelanocortin in human plasma: correlation with adrenocorticotropin. *Journal of Clinical Endocrinology and Metabolism,* **56,** 791–796

CHOBERT, M. N., BAROUKI, R., FINIDORI, J. *et al.* (1983) Antiglucocorticoid properties of RU 38486 in a differentiated hepatoma cell line. *Biochemical Pharmacology,* **32,** 3481–3483

CHROUSOS, G. P., SCHULTE, H. M., OLDFIELD, E. H., GOLD, P. W., CUTLER, G. B. JR and LORIAUX, D. L. (1984) The corticotropin-releasing factor stimulation test: an aid in the evaluation of patients with Cushing's syndrome. *New England Journal of Medicine,* **310,** 622–626

CÔTÉ, J., PROULX-FERLAND, L., LABRIE, F., PHILIBERT, D. and DERAEDT, R. (1982) Potent antiglucocorticoid activity of RU38486 on ACTH secretion *in vitro* and *in vivo. Clinical and Investigative Medicine,* **5,** 11B

CRAPO, L. (1979) Cushing's syndrome: a review of diagnostic tests. *Metabolism,* **28,** 955–977

CUSHING, H. (1912) *The Pituitary Body and Its Disorders: Clinical Studies Produced by Disorders of the Hypophysis Cerebri,* p. 217. Philadelphia and London: Lippincott

CUSHING, H. (1932) The basophil adenomas of the pituitary body and their clinical manifestations (pituitary basophilism). *Bulletin of the Johns Hopkins Hospital,* **50,** 137–195

DORNHORST, A., JENKINS, J. S., LAMBERTS, S. W. J. *et al.* (1983) The evaluation of sodium valproate in the treatment of Nelson's syndrome. *Journal of Clinical Endocrinology and Metabolism,* **56,** 985–991

FEHM, H. L., STECK, R., HOHNLOSER, J., VOIGT, K. H. and PFEIFFER, E. F. (1983) Influence of neuroactive drugs on corticosteroid feedback regulation of ACTH secretion inman. *Hormone and Metabolic Research,* **15,** 29–32

FEHM, H. L., VOIGT, K. H., LANG, R. E., BEINERT, K. E., KUMMER, G. W. and PFEIFFER, E. F. (1977) Paradoxical ACTH response to glucocorticoids in Cushing's disease. *New England Journal of Medicine,* **297,** 904–907

FINDLING, J. W., ARON, D. C., TYRRELL, J. B. *et al.* (1981) Selective venous sampling for ACTH in Cushing's syndrome. Differentiation between Cushing's disease and ectopic ACTH syndrome. *Annals of Internal Medicine,* **96,** 647–652

FITZGERALD, P. A., ARON, D. C., FINDLING, J. W. *et al.* (1982) Cushing's disease: transient secondary adrenal insufficiency after selective removal of pituitary microadenomas: evidence for a pituitary origin. *Journal of Clinical Endocrinology and Metabolism,* **54,** 413–422

GOLD, E. M. (1979) The Cushing syndromes: changing views of diagnosis and treatment. *Annals of Internal Medicine,* **90,** 829–844

GREEN, J. R. B. and VAN'T HOFF, W. (1975) Cushing's syndrome with fluctuation due to adrenal adenoma. *Journal of Clinical Endocrinology and Metabolism,* **41,** 235–240

HARDY, J. (1982) Cushing's disease: 50 years later. *Canadian Journal of Neurological Sciences,* **9,** 375–379

IMURA, H., MATSUKURA, S., YAMAMOTO, H. *et al.* (1975) Studies on ectopic ACTH-producing tumors. II. Clinical and biochemical features of 30 cases. *Cancer,* **35,** 1430–1437

JENNINGS, A. S., LIDDLE, G. W. and ORTH, D. N. (1977) Results of treating childhood Cushing's disease with pituitary irradiation. *New England Journal of Medicine,* **297,** 957–962

JONES, M. T., HILLHOUSE, E. W. and BURDEN, H. L. (1977) Dynamics and mechanics of corticosteroid feedback at the hypothalamus and anterior pituitary gland. *Journal of Endocrinology,* **73,** 405–417

KRIEGER, D. T. (1983) Physiopathology of Cushing's disease. *Endocrine Reviews,* **4,** 22–43

KRIEGER, D. T., AMOROSA, L. and LINICK, F. (1975) Cyproheptadine-induced remission of Cushing's disease. *New England Journal of Medicine,* **293,** 893–896

KRIEGER, D. T. and GLICK, S. M. (1974) Sleep EEG stages and plasma growth hormone concentration in states of endogenous and exogenous hypercortisolemia or ACTH elevation. *Journal of Clinical Endocrinology and Metabolism,* **39,** 986–1000

KRIEGER, D. T., HOWANITZ, P. J. and FRANTZ, A. G. (1976) Absence of nocturnal elevation of plasma prolactin concentration in Cushing's disease. *Journal of Clinical Endocrinology and Metabolism,* **42,** 260–272

KUWAYAMA, A., KAGEYAMA, N., NAKANE, T., WATANABE, M. and KANIE, N. (1981) Anterior pituitary function after transsphenoidal selective adenomectomy in patients with Cushing's disease. *Journal of Clinical Endocrinology and Metabolism,* **53,** 165–173

LAGERQUIST, L. G., MEIKLE, A. W., WEST, C. D. and TYLER, F. H. (1974) Cushing's disease with cure by resection of a pituitary adenoma: evidence against a primary hypothalamic defect. *American Journal of Medicine,* **57,** 826–830

LAMBERTS, S. W. J., DE LANGE, S. A. and STEFANKO, S. Z. (1982) Adrenocorticotropin-secreting pituitary adenomas originate from the anterior or the intermediate lobe in Cushing's disease: differences in the regulation of hormone secretion. *Journal of Clinical Endocrinology and Metabolism,* **54,** 286–291

LAMBERTS, S. W. J., KLIJN, J. G. M., DE QUIJADA, M. *et al.* (1980a) The mechanism of the suppressive action of bromocriptine on adrenocorticotropin secretion in patients with Cushing's disease and Nelson's syndrome. *Journal of Clinical Endocrinology and Metabolism,* **51,** 307–311

LAMBERTS, S. W. J., STEFANKO, S. Z., DE LANGE, S. A. *et al.* (1980b) Failure of clinical remission after transsphenoidal removal of a microadenoma in a patient with Cushing's disease: multiple hyperplastic and adenomatous cell nests in surrounding pituitary tissue. *Journal of Clinical Endocrinology and Metabolism,* **50,** 793–795

LAMBERTS, S. W. J., TIMMERMANS, H. A. T., DE JONG, F. H. and BIRKENHÄGER, J. C. (1977) The role of dopaminergic depletion in the pathogenesis of Cushing's disease and the possible consequences for medical therapy. *Clinical Endocrinology,* **7,** 185–193

LAMBERTS, S. W. J., VERLEUN, T., OOSTEROM, R., DE JONG, F. and HACKENG, W. H. L. (1984) Corticotropin-releasing factor (ovine) and vasopressin exert a synergistic effect on adrenocorticotropin release in man. *Journal of Clinical Endocrinology and Metabolism,* **58,** 298–303

LANKFORD, H. V., TUCKER, H. ST. G. and BLACKARD, W. G. (1981) A cyproheptadine-reversible defect in ACTH control persisting after removal of the pituitary tumor in Cushing's disease. *New England Journal of Medicine,* **305,** 1244–1248

LIBERMAN, B., WAJCHENBERG, B. L., TAMBASCIA, M. A. and MESQUITA, C. H. (1976) Periodic remission in Cushing's disease with paradoxical dexamethasone response: an expression of periodic hormonogenesis. *Journal of Clinical Endocrinology and Metabolism,* **43,** 913–918

LIDDLE, G. W. (1960) Tests of pituitary–adrenal suppressibility in the diagnosis of Cushing's syndrome. *Journal of Clinical Endocrinology and Metabolism,* **20,** 1539–1560

LOOSE, D. S., STOVER, E. P. and FELDMAN, D. (1983) Ketoconazole binds to glucocorticoid receptors and exhibits glucocorticoid antagonist activity in cultured cells. *Journal of Clinical Investigation,* **72,** 404–408

LYTRAS, N., GROSSMAN, A., PERRY, L. *et al.* (1984) Corticotrophin releasing factor: responses in normal subjects and patients with disorders of the hypothalamus and pituitary. *Clinical Endocrinology,* **20,** 71–84

MANNI, A., LATSHAW, R. F., PAGE, R. and SANTEN, R. J. (1983) Simultaneous bilateral venous sampling for adrenocorticotropin in pituitary-dependent Cushing's disease: evidence for lateralization of pituitary venous drainage. *Journal of Clinical Endocrinology and Metabolism,* **57,** 1070–1073

MC NICOL, A. M., THOMSON, H. and STEWART, C. J. R. (1983) The corticotrophic cells of the canine pituitary gland in pituitary-dependent hyperadrenocorticism. *Journal of Endocrinology,* **96,** 303–309

MEADOR, C. K., BOWDOIN, B., OWEN, W. C. JR and FARMER, T. A. JR (1967) Primary adrenocortical nodular dysplasia: a rare cause of Cushing's syndrome. *Journal of Clinical Endocrinology and Metabolism,* **27,** 1255–1263

NAKAHARA, M., SHIBASAKI, T., SHIZUME, K. *et al.* (1983) Corticotropin-releasing factor test in normal subjects and patients with hypothalamic–pituitary–adrenal disorders. *Journal of Clinical Endocrinology and Metabolism,* **57,** 963–968

NELSON, D. H., MEAKIN, J. W. and THORN, G. W. (1960) ACTH-producing pituitary tumors following adrenalectomy for Cushing's syndrome. *Annals of Internal Medicine,* **52,** 560–569

NEVILLE, A. M. and SYMINGTON, T. (1967) The pathology of the adrenal gland in Cushing's syndrome. *Journal of Pathology and Bacteriology,* **93,** 19–35

NIEMAN, L. K., CHROUSOS, G. P., SPITZ, I. *et al.* (1984) Successful treatment of Cushing's syndrome with the glucocorticoid antagonist RU 38486. *Clinical Research,* **32,** 271A

ORTH, D. N. (1984) The old and the new in Cushing's syndrome. *New England Journal of Medicine,* **302,** 649–651

ORTH, D. N., DEBOLD, C. R., DECHERNEY, G. S. *et al.* (1982) Pituitary microadenomas causing Cushing's disease respond to corticotropin-releasing factor. *Journal of Clinical Endocrinology and Metabolism,* **55,** 1017–1019

ORTH, D. N. and LIDDLE, G. W. (1971) Results of treatment in 108 patients with Cushing's syndrome. *New England Journal of Medicine,* **285,** 243–247

PLOTZ, C. M., KNOWLTON, A. I. and RAGAN, C. (1952) The natural history of Cushing's syndrome. *American Journal of Medicine,* **13,** 597–614

PONT, A. and GUTIERREZ-HARTMAN, A. (1979) Cushing's disease: recurrence after surgically induced remission. *Archives of Internal Medicine,* **139,** 938–940

ROSS, E. J. and LINCH, D. C. (1982) Cushing's syndrome – killing disease: discriminatory value of signs and symptoms aiding early diagnosis. *Lancet,* **2,** 646–649

RUDER, H. J., LORIAUX, D. L. and LIPSETT, M. B. (1974) Severe osteopenia in young adults associated with Cushing's syndrome due to micronodular adrenal disease. *Journal of Clinical Endocrinology and Metabolism,* **39,** 1138–1147

SALASSA, R. M., KEARNS, T. P., KERNOHAN, J. W., SPRAGUE, R. G. and MAC CARTY, C. S. (1959) Pituitary tumors in patients with Cushing's syndrome. *Journal of Clinical Endocrinology and Metabolism,* **19,** 1523–1539

SALASSA, R. M., LAWS, E. R. JR, CARPENTER, P. C. and NORTHCUTT, R. C. (1978) Transsphenoidal removal of pituitary microadenoma in Cushing's disease. *Mayo Clinic Proceedings,* **53,** 24–28

SCHNALL, A. M., BRODKEY, J. S., KAUFMAN, B. and PEARSON, O. H. (1978) Pituitary function after removal of pituitary microadenomas in Cushing's disease. *Journal of Clinical Endocrinology and Metabolism,* **47,** 410–417

SCHNALL, A. M., KOVACS, K., BRODKEY, J. S. and PEARSON, O. H. (1980) Pituitary Cushing's disease without adenoma. *Acta Endocrinologica,* **94,** 297–303

SCHWEIZER-CAGIANUT, M., FROESCH, E. R. and HEDINGER, C. (1980) Familial Cushing's syndrome with primary adrenocortical microadenomatosis (primary adrenocortical nodular dysplasia). *Acta Endocrinologica,* **94,** 529–535

SERRI, O., RASIO, E., BEAUREGARD, H., HARDY, J. and SOMMA, M. (1983) Recurrence of hyperprolactinemia after selective transsphenoidal adenomectomy in women with prolactinoma. *New England Journal of Medicine,* **309,** 280–283

SHELINE, G. E. (1981) Pituitary tumors: radiation therapy. In *The Pituitary, Clinical Endocrinology,* Vol. 1, edited by C. Beardwell and G. L. Robertson, pp. 106–139. London: Butterworths

SMALS, A. G. H., PIETERS, G. F. F. M., VAN HAELST, V. J. G. and KLOPPENBORG, P. W. C. (1984) Macronodular adrenocortical hyperplasia in long-standing Cushing's disease. *Journal of Clinical Endocrinology and Metabolism,* **58,** 25–31

STYNE, D. M., GRUMBACH, M. M., KAPLAN, S. L., WILSON, C. B. and CONTE, F. A. (1984) Treatment of Cushing's disease in childhood and adolescence by transsphenoidal microadenomectomy. *New England Journal of Medicine,* **310,** 889–893

SUDA, T., TOZAWA, F., MOURI, T. *et al.* (1983) Effects of cyproheptadine, reserpine and synthetic corticotropin-releasing factor on pituitary glands from patients with Cushing's disease. *Journal of Clinical Endocrinology and Metabolism,* **56,** 1094–1099

UPTON, G. V. and AMATRUDA, T. T. JR (1971) Evidence for the presence of tumor peptides with corticotropin-releasing-factor-like activity in the ectopic ACTH syndrome. *New England Journal of Medicine,* **285,** 419–424

URBANIC, R. C. and GEORGE, J. M. (1981) Cushing's disease – 18 years' experience. *Medicine,* **60,** 14–24

WILKES, M. M., STEWART, R. D., BRUNI, J. F. *et al.* (1980) A specific homologous radioimmunoassay for human β-endorphin: direct measurement in biological fluids. *Journal of Clinical Endocrinology and Metabolism,* **50,** 309–315

YASUDA, N., GREER, M. A. and AIZAWA, T. (1982) Corticotropin releasing factor. *Endocrine Reviews,* **3,** 123–140

ZERVAS, N. T. (1984) Surgical results for pituitary adenomas: results of an international survey. In *Secretory Tumors of the Pituitary Gland, Progress in Endocrine Research and Therapy,* Vol. 1, edited by P. McL. Black, N. T. Zervas, E. C. Ridgeway and J. B. Martin, pp. 377–385. New York: Raven Pess

8
The adrenal cortex and hypertension

M. B. Vallotton and L. Favre

INTRODUCTION

The role of steroids in the genesis of hypertension has long been the focus of intense research and is the subject of several excellent reviews (Genest *et al.*, 1977; Stockigt, 1979; Fraser *et al.*, 1983; Melby, Dale and Griffing, 1983). As a result of both fundamental research activity and astute clinical observations a number of conditions associated with hypertension have been identified in which steroids, particularly mineralocorticoids, play the major role in generating and maintaining an elevated blood pressure (*Table 8.1*). In addition to these well-defined conditions, there are others in which mineralocorticoids are believed to play a role – albeit a poorly understood one. This review will deal with both, concentrating in the main on hypertension in man, and alluding only when necessary to animal models and experiments, on which so much of our current knowledge is based.

Poiseuille's law of fluid dynamics tells us that pressure results from the effects of both flow and resistance. Translated into physiological terms and applied to the cardiovascular system, this means that blood pressure is proportional to cardiac output and peripheral resistance, both of which depend on several physiological parameters (Guyton *et al.*, 1974). Cardiac output depends on heart rate, systolic ejection volume and blood volume, while vascular resistance depends on arteriolar sensitivity and reactivity to vasoactive humoral agents and neurogenic factors, and on wall thickness and compliance. Since mineralocorticoids by definition retain sodium chloride followed by water, and thereby increase blood volume, it was believed at first that they acted on the cardiovascular system via changes in blood volume. Yet recent experimental studies and clinical observations suggest that their action is more complex. This will be discussed later, in connection with the different forms of human hypertension in which steroids are believed to play a role.

HYPERTENSION AND KNOWN MINERALOCORTICOIDS

Hyperaldosteronism

The best example of hypertension caused by a known steroid seems to be primary aldosteronism (Weinberger, 1983). Yet many unanswered questions raise doubt about aldosterone excess as the primary cause of this disorder. First, it has been impossible to induce reproducible hypertension in healthy subjects by administration of mineralocorticoids. In some studies, no changes of blood pressure were achieved by administration of aldosterone (Rosemberg, Demany and Budnitz, 1962; Rovner *et al.*, 1965), deoxycorticosterone (DOC) (Shade and Grim, 1975) or fludrocortisone (Adetuyibi and Mills, 1972). In others, even when blood pressure did rise with aldosterone (Kassirer, London and Goldman, 1970), DOC (Perera, Knowlton and Lowell, 1944) or fludrocortisone administration (August, Nelson and Thorn, 1958; Chobanian, Burrows and Hollander, 1961), the increase was minimal, amounting to only about 10 mmHg systolic, with no significant change in diastolic pressure (Nicholls *et al.*, 1979). This corroborates the findings in rats, in which hypertension could be induced only in those sensitized by unilateral nephrectomy, by concomitant administration of isotonic saline by mouth, or by using pharmacological doses of aldosterone or DOC (Fregly, Kim and Hood, 1969). This difficulty in inducing hypertension is most likely explained by the ability

Table 8.1 Hypertension associated with known mineralocorticoids

Disease	*Lesion*	*Steroid involved*
Primary aldosteronism		
Conn's syndrome	Adrenal adenoma	Aldosterone (+ other steroids?)
Idiopathic aldosteronism	Bilateral hyperplasia of the zona glomerulosa	Aldosterone (+ other steroids?)
Dexamethasone suppressible aldosteronism	Bilateral hyperplasia of the zona glomerulosa	Aldosterone (+ other steroids?)
Cushing's syndrome	Hypothalamic and/or pituitary dysfunction Adrenal adenoma Adrenal carcinoma Ectopic production of ACTH	(DOC, cortisol, corticosterone)
Congenital adrenal hyperplasia	Enzymic defect	
Biglieri's syndrome	17α-Hydroxylase	DOC, 18-OH DOC
New's syndrome	11β-Hydroxylase	DOC, deoxycortisol

DOC = deoxycorticosterone.

of the kidney to escape the mineralocorticoid action, implying that patients with mineralocorticoid-induced hypertension may have another defect, either an inability to escape or another hypertensinogenic factor (Wenting *et al.*, 1977).

Suppressed plasma renin activity, unresponsive to a variety of stimuli, is the hallmark of primary aldosteronism (Conn, 1964). It is therefore tempting to conclude that the rise in blood pressure is secondary to sodium retention and consequent volume expansion, yet in many studies no correlation has been found between the rise in blood pressure and body weight. The simple view of mineralocorticoid hypertension as a volume-dependent form of hypertension therefore has to be revised (Laragh, 1973; Kaplan, 1977; Laragh, Letcher and Pickering, 1979) and alternative mechanisms must be considered. Pressor activity of aldosterone in predisposed subjects, either directly through activation of the sympathetic nervous system or potentiation of norepinephrine or angiotensin II, or indirectly through a shift of sodium into the intracellular compartment (reviewed in Nicholls *et al.,* 1979; Weber, Purdy and Drayer, 1983), has been suggested but not proven. Whatever the mechanism may be, one is forced to admit that the pathophysiology of mineralocorticoid-induced hypertension is unclear. Another indication that aldosterone is not the only responsible factor is that removal of an aldosterone-secreting adenoma is not always followed by cure. Conn (1964) reported a cure rate of hypertension of 70%, but figures as low as 50–60% (Biglieri *et al.,* 1970; George *et al.,* 1970; Ferriss *et al.,* 1975; Weinberger *et al.,* 1979), and in one case as high as 84% (Lund, Nielson and Gammelgaard, 1977) have been reported after long-term follow-up (reviewed by Ganguly and Donohue, 1983).

The origin of these adenomas and the control and pattern of their steroid secretion further complicate the picture. Histological examination by conventional light and electron microscopy reveals that they do not, as expected, consist of typical zona glomerulosa cells but rather of cells resembling those of the zona fasciculata or zona reticularis (Priestley *et al.,* 1968; Reidbord and Fisher, 1969; Sommers and Terzakis, 1970; Kovacs *et al.,* 1974; Mazzocchi *et al.,* 1982). Hybrid cells or intermediate cells with features of both zona glomerulosa and fasciculata have been described. In the majority of cases the adenomas consists of more than one cell type (Favre *et al.,* 1980; Sigg, Schweizer-Cagianut and Hedinger, 1983); lipofuchsin-rich cells of the reticularis type which may form black adenomata have also been reported (Caplan and Virata, 1974).

Chu and Ulick (1982) have recently isolated the 20, 18-hemiketal form of 11β, 17α, 18, 21-tetrahydroxy-4-pregnene-3,20-dione (18-hydroxycortisol) from the urine of patients with primary aldosteronism. Since 17α-hydroxylation is a distinctive feature of the zona fasciculata and reticularis this finding lends biochemical support to the view that aldosterone-producing adrenocortical adenomas have the unique ability to combine hydroxylations which are normally exclusive features of different zones, and thus to the concept of the hybrid nature of the cells forming such adenomas. The role of 18-hydroxycortisol, perhaps combined with that of aldosterone, in producing hypertension remains to be proven. The same considerations also apply to 18-hydroxycorticosterone, considered to be a good marker of primary aldosteronism (Biglieri and Schambelan, 1979; Vescei *et al.,* 1982; Bravo *et al.,* 1983).

Studies of the regulation of aldosterone secretion provide further evidence for the hybrid nature of these cells. Instead of continuous autonomous aldosterone secretion, patients with aldosterone-producing adenomas display a circadian rhythm of secretion, mediated by ACTH (Kem *et al.,* 1973, 1978). Furthermore, a subgroup of aldosterone-producing adenomas has been described, characterized by dexamethasone suppressibility (Schambelan *et al.,* 1976; Ganguly *et al.,* 1977) or enhanced responsiveness to ACTH (Kem *et al.,* 1978). Such forms are encountered in children and are often of a familial nature (Giebink *et al.,* 1973; Sonino, Levine and New, 1981; Lee *et al.,* 1982); familial hyperaldosteronism not suppressed by dexamethasone has also been described (Greco *et al.,* 1982).

The discovery that some patients presenting with all the criteria of primary aldosteronism do not harbour an adenoma, but rather have bilateral cortical hyperplasia, posed a puzzling problem, as they did not benefit from adrenalectomy. This condition has been termed non-adenomatous primary aldosteronism (Ferriss *et al.,* 1978), pseudo-primary aldosteronism (Baer *et al.,* 1970) or idiopathic aldosteronism (Biglieri *et al.,* 1970). The regulation of aldosterone in these patients differs from that in cases of tumorous aldosteronism. Instead of decreasing during the day, as in patients with adenoma, the already increased plasma aldosterone concentration rises further with standing (Ganguly *et al.,* 1973a, 1973b). This postural response suggests that, despite suppression of plasma renin activity, the cells secreting aldosterone in excess still respond to angiotensin II. Aldosterone-secreting adenomas may also respond to stimulation by angiotensin II (Spark *et al.,* 1969), but increased sensitivity to angiotensin II appears to be characteristic of idiopathic aldosteronism (Wisgerhof, Carpenter and Brown, 1978; Marks *et al.,* 1979). The postural response and the presence of a nycthemeral rhythm occasionally fail to distinguish between tumorous and non-tumorous forms of aldosteronism but this manoeuvre still has considerable clinical application (Kem *et al.,* 1976; Vetter *et al.,* 1978; Herf *et al.,* 1979; Streeten, Tomycz and Anderson, 1979). A few cases of idiopathic aldosteronism due to hyperplasia of the zona glomerulosa but not responsive to stimulation by angiotensin have been reported (Lufkin, Katz and Herman, 1972).

To complicate diagnostic evaluation further, a subtype of idiopathic aldosteronism, relieved by glucocorticoid, has been described (Sutherland, Ruse and Laidlaw, 1966; Salti *et al.,* 1969a), apparently different from that not relieved by glucocorticoid (Salti *et al.,* 1969b). Patients with this dexamethasone-suppressible form of aldosteronism have an anomalous postural aldosterone response (Ganguly, Grim and Weinberger, 1981): before glucocorticoid treatment, plasma aldosterone decreases with standing (as in the tumorous form of aldosteronism) while after glucocorticoid treatment it rises with standing. All these observations raise serious doubts about the accepted classification and, indeed, the existence of idiopathic aldosteronism as a separate pathological entity (Editorial, 1979). Depending on the criteria used (degree of renin suppression, responsiveness of aldosterone to angiotensin II, or correlation between plasma angiotensin II and plasma aldosterone), the classification of the various forms of hypertension will change (Davies *et al.,* 1979; Vallotton, 1980); yet a clear distinction between low-renin essential hypertension and idiopathic aldosteronism will always be difficult.

Liddle's syndrome

Rare and still poorly understood, this disorder presents with the hallmarks of primary aldosteronism (hypertension, hypokalaemia and suppressed plasma renin activity), except for hypoaldosteronism and a failure of spironolactone to correct these abnormalities (Gardner *et al.,* 1971). Renal tubular cells (but not salivary or sweat glands) function as if exposed to excess of mineralocorticoid. Only triamterene administration is followed by normalization of kalaemia and lowering of blood pressure. Abnormal membrane sodium transport has been described, reflecting possibly a more generalized membrane defect.

Other adrenal diseases associated with hypertension

Cushing's syndrome

Hypertension is encountered in 50 to 90% of patients with Cushing's syndrome (Gold, 1979; Greminger *et al.,* 1982), but is much less common during ACTH or glucocorticoid treatment (Christy, 1971). Only 30% of patients with ectopic production of ACTH are hypertensive, despite higher levels of ACTH and cortisol. Differences in blood pressure are difficult to relate to either ACTH, cortisol, corticosterone or deoxycorticosterone excess, and are observed despite similar hypokalaemic alkalosis (Schambelan, Slaton and Biglieri, 1971). Marked differences in urinary excretion of DOC and aldosterone have been reported in all forms of Cushing's syndrome, but no attempt at correlation with blood pressure has been made (Cassar *et al.,* 1980). Therefore, the pathogenesis of hypertension characterized by an excess of glucocorticoid remains enigmatic. A unique case of a corticosterone-secreting adrenocortical tumour inducing hypertension, hypokalaemic alkalosis, muscle weakness and polyuria has been described by Fraser *et al.* (1968). In patients with adrenocortical carcinoma various patterns of steroid secretion have been reported (Powell-Jackson *et al.,* 1974; Tan *et al.,* 1977).

Congenital adrenal hyperplasia (CAH)

Two forms of CAH are characterized by an elevation of blood pressure: that due to a deficiency of 11β-hydroxylase and that due to a deficiency of 17α-hydroxylase (Mantero *et al.,* 1982). In these inborn errors of metabolism one would expect to find a relation between steroid levels and hypertension, with a distinctive pattern for each enzymatic deficiency. Thus, in 17α-hydroxylase deficiency, concentrations of DOC, 18-hydroxycorticosterone and 18-hydroxy DOC are increased and those of aldosterone decreased, while in 11β-hydroxylase deficiency concentrations of DOC are elevated while the three other steroids are reduced. In 21-hydroxylase deficiency, which is not accompanied by hypertension, levels of aldosterone, 18-hydroxycorticosterone and DOC are elevated, while those of 18-hydroxy DOC are normal (Kater and Biglieri, 1983). In these three CAH forms, ACTH levels are elevated while those of cortisol tend to be low. It is still uncertain which of these steroids is responsible for the increase in blood pressure. DOC, which is markedly elevated in both 17α-hydroxylase and 11β-hydroxylase deficiency, is the most likely candidate, but this remains to be proven. The target cells of these steroids must also be considered; it is conceivable that weak mineralocorticoids become more effective when mineralo- and glucocorticoid receptors are not occupied by their natural agonists (*see below*).

New *et al.* (1982) have carried out extensive studies in search of an unidentified steroid that might be responsible for the syndrome of apparent mineralocorticoid excess responsive to spironolactone. They suggested that cortisol itself, despite its low levels, might be the culprit, by acting *in vivo* as a mineralocorticoid and binding to altered mineralocorticoid receptors. This disorder represents one of the forms of ACTH-induced hypertension described in animals (Scoggins *et al.,* 1982) but rare in man.

HYPERTENSION INDUCED BY EXOGENOUS MINERALOCORTICOIDS OR MINERALOCORTICOID-LIKE SUBSTANCES

A number of hypertensive conditions are caused by ingestion or administration of mineralocorticoid or mineralocorticoid-like substances (Mantero, 1981). These cases are of great interest, as they reproduce experimental models of mineralocorticoid-induced hypertension and allow a better understanding of the

Table 8.2 Hypertension induced by exogenous mineralocorticoid or mineralocorticoid-like substances

Substance	*Source*	*Effective dose*	*Reference*
Glycyrrhizinic acid	(1) Liquorice (extract from *Glycyrrhiza glabra*)	0.7–4 g/day	More than 60 cases reported (Molhuysen *et al.*, 1950; Conn *et al.*, 1968)
	(2) Chewing tobacco	1 g/day	One case report (Blachley and Knochel, 1980)
	(3) Non-alcoholic 'pastis'	More than 1 g/week	Several cases in France and Switzerland (Trono, Cereda and Favre, 1983)
18β-Glycyrrhizinic acid	Carbenoxolone (drug used for the treatment of gastroduodenal ulcer)	More than 300 mg/day	Turpie and Thomson, 1965)
9α-Fluoroprednisolone	Nasal spray (1 mg/ml) Topical cream (1 g%)	2 mg/day	Several cases in Europe (Ambruster *et al.*, 1975; Mantero *et al.*, 1981)
Fludrocortisone	Treatment of orthostatic hypotension	0.3–1 mg/day	(Chobanian *et al.*, 1979)

factors required for the occurrence of hypertension in man. These substances (true mineralocorticoids or synthetic derivatives) and the drugs or foods which contain them, are listed in *Table 8.2.* Although the clinical manifestations of so-called 'pseudo-aldosteronism' can mimic primary aldosteronism, the main feature is hypokalaemia and potassium depletion; hypertension is much less common. The reason for this discrepancy may be the brief duration of exposure to mineralocorticoid, since hypertension develops only after prolonged administration combined with high salt intake in predisposed subjects.

The effective dose of each agent and the individual response can vary considerably. The effects of mineralocorticoid intoxication are usually reversible if exposure is discontinued. However, it can take a few days to two weeks for plasma potassium concentrations and blood pressure to return to normal, suggesting that other factors (baroreceptor resetting and vascular sensitivity) are involved. These substances act by competition for renal binding sites, i.e. mineralo- and glucocorticoid receptors (Armanini *et al.,* 1983), and also by displacement of aldosterone from non-specific binding proteins (Humphrey *et al.,* 1979). As with true mineralocorticoids, an escape phenomenon occurs within a few days, preventing the formation of oedema. Sodium retention provokes renin and aldosterone suppression, which helps to distinguish pseudo-hyperaldosteronism from primary aldosteronism (Epstein *et al.,* 1977).

MINERALOCORTICOIDS IN ESSENTIAL HYPERTENSION

As essential hypertension is a broad concept comprising a variety of causes and pathogenic mechanisms, it is not surprising that so many divergent patterns of steroid secretion have been reported. These have been reviewed by Ulick (1976) and Melby and Dale (1979). The contribution of aldosterone, the most potent of the mineralocorticoids, and other known or as yet unidentified mineralocorticoids, has been examined in several studies. Essential hypertension differs from hypermineralocorticism or primary aldosteronism in that there is no overt excess of sodium-retaining hormones. However, subtle abnormalities in the proportions of some mineralocorticoids have been observed, suggesting a mild primary or secondary disorder, possibly related to the increase in blood pressure (Genest *et al.,* 1977; Williams and Dluhy, 1977; Melby, Dale and Griffing, 1983).

Role of aldosterone

With the advent of techniques to measure aldosterone secretion or excretion came several attempts to define aldosterone abnormalities in essential hypertension – albeit with different conclusions (George, Gillespie and Bartter, 1968; Channick, Adlin and Marks, 1969; Collins *et al.,* 1970; Kotchen *et al.,* 1971). Some of the discrepancies resulted from differences in methodology or clinical conditions, but most were certainly related to the heterogeneity of the hypertensive population studied.

These new techniques also allowed repeated determinations for acute dynamic studies. Analysis of the relationships between aldosterone, renin and volume or sodium factors identified subsets of hypertensive subjects with characteristic responses of the renin–angiotensin–aldosterone system to stimulation or suppression tests (Coghlan *et al.,* 1972; Laragh, Sealey and Brunner, 1972;

Table 8.3 Plasma aldosterone in normal-renin and low-renin essential hypertension

Test	*Normal renin** (*n*)	*Low renin** (*n*)	*Reference*
Supine			
Normal sodium diet	N (21)	N or ↑ (13)	Wallach, Nyarai and Dawson (1975)
	↑ (55)	↑ (26)	Genest *et al.* (1977)
	ND	↓ (12) N or ↑ A/PRA	Re *et al.* (1978)
	N (91)	N (26)	Davies *et al.* (1979)
	ND	↑ (17) N (3) ↓ (11)	Melby and Dale (1979)
	N (16)	N (12)	Witzgall, Thayil and Weber (1982)
	N (44)	ND	Safar *et al.* (1982)
Upright			
Normal sodium diet	N (51)	N (34)	Crane (1973)
	ND	↓ (12) ↑ A/PRA	Re *et al.* (1978)
Low sodium diet	N (52) ↓ (12)	ND	Moore *et al.* (1977)
	ND	↓ (12) ↑ A/PRA	Re *et al.* (1978)
Response to frusemide	N (40)	ND	Williams and Dluhy (1977)
	N (10)	N (6)	Wisgerhof and Brown (1978)
	↓ (16)	↓ (12)	Witzgall, Thayil and Weber (1982)
Response to sodium	N or ↓ (62)	ND	Tuck *et al.* (1976)
loading	ND	↑ (12)	Re *et al.* (1978)
Sensitivity to angiotensin II			
Normal sodium diet	↑ (12)	ND	Kisch, Dluhy and Williams (1976)
	N (9)	↑ (7)	Mitchell (1977)
	N (13)	↑ (8)	Marks *et al.* (1979)
	↑ (56)	↑ (16)	Davies *et al.* (1979)
High sodium diet plus dexamethasone	N (10)	↑ (6)	Wisgerhof and Brown (1978)
Low sodium diet	N (14) ↓ (5)	ND	Moore *et al.* (1977)
	N (7) ↓ (8)	ND	Dawson-Hughes *et al.* (1981)
Sensitivity to ACTH			
Normal sodium diet	↑ (5)	↑ (4)	Honda *et al.* (1977)
	N (6)	N (12)	Kem *et al.* (1978)
Low sodium diet	N (7) ↓ (8)	ND	Dawson-Hughes *et al.* (1981)

* Values in hypertensive subjects compared with those in normotensive subjects in the same study.
A/PRA = ratio of plasma aldosterone to plasma renin activity.
N = normal; ND = no data available.
Number of subjects (*n*) are given in parentheses.

Wallach, Nyarai and Dawson, 1975; Genest *et al.*, 1977; Williams and Dluhy, 1977; Re *et al.*, 1978). Of the several abnormalities described in patients with essential hypertension, the commonest is the syndrome of low-renin essential hypertension (LREH) (Woods *et al.*, 1969; Jose, Crout and Kaplan, 1970; Crane, Harris and Varner, 1972; Sealey and Laragh, 1973; Dunn and Tannen, 1977; Ganguly and Weinberger, 1979). This condition is characterized by a blunted response to acute stimulation of renin by sodium depletion or upright posture, and is found in about 30% of patients with essential hypertension, particularly those who are elderly or black. It is generally accepted that the inhibition of renin results from the combined effect of age and duration of hypertension leading to decreased adrenergic renal sensitivity and nephrosclerosis. Interestingly, plasma and urinary aldosterone are not suppressed along with renin, but are paradoxically normal, with an increased aldosterone/renin ratio (*Table 8.3*). Moreover, the aldosterone response to sodium restriction may be enhanced in LREH, reflecting increased adrenal sensitivity to angiotensin II (Wisgerhof and Brown, 1978). By contrast, about one-third of patients with normal-renin essential hypertension show reduced aldosterone responsiveness after sodium depletion, and about one-half exhibit delayed suppression of the renin–angiotensin–aldosterone axis after acute or chronic sodium loading (Williams and Dluhy, 1977). Other abnormalities of adrenal steroids have been reported in response to angiotensin II, ACTH or frusemide (furosemide). The vascular pressor response to angiotensin II also differs between normotensive and hypertensive subjects. Responsiveness is increased in hypertensives when similar doses are compared but is unchanged at threshold doses (Williams and Dluhy, 1977). Comparative values of plasma aldosterone in normal-renin and low-renin essential hypertension are summarized in *Table 8.3*.

Role of other mineralocorticoids

Whereas there is little evidence for steroid involvement in normal-renin essential hypertension, it is tempting to postulate that excess of mineralocorticoid plays an important role in low-renin essential hypertension. As no known mineralocorticoid is produced in amounts sufficient to maintain the elevation of blood pressure, investigation has focused on the detection of inappropriate secretion of unknown mineralocorticoids (Spark, 1972).

The hypothesis of excess mineralocorticoid secretion is based on the following evidence:

(1) Renin secretion is suppressed, as in primary aldosteronism.
(2) Some authors have observed increased blood volume or exchangeable sodium space (Woods *et al.*, 1969; Jose *et al.*, 1970).
(3) The hypertension responds to mineralocorticoid antagonists such as spironolactone (Adlin, Marks and Channick, 1972; Carey *et al.*, 1972) or inhibitors of steroidogenesis such as metyrapone and aminoglutethimide (Woods *et al.*, 1969; Taylor *et al.*, 1978).

However, other findings are difficult to reconcile with mineralocorticoid excess:

(1) Levels of non-aldosterone mineralocorticoids are usually within the normal range or only mildly elevated (*Table 8.4*). Moreover, the sodium-retaining effect of these steroids is too small to account for the presumed mineralocorticoid excess.

(2) To date no hitherto unknown mineralocorticoid has been identified in low-renin essential hypertension despite extensive search. The few steroids found were either not specific for the syndrome or not confirmed in subsequent studies (Sennett *et al.*, 1977; Melby, Dale and Griffing, 1983). The mineralocorticoids detected in experimental low-renin hypertension, e.g. adrenal regeneration hypertension (Gomez-Sanchez *et al.*, 1983), do not contribute to hypertension in man. However, it has been argued that if no single substance could reproduce the syndrome, a synergistic action of multiple weak steroids might do so (Coghlan *et al.*, 1976; Sennett *et al.*, 1977; Adam *et al.*, 1979). The syndrome may in any case cover a heterogeneous group of disorders (Sekihara *et al.*, 1979).

Table 8.4 Major non-aldosterone mineralocorticoids in normal-renin and low-renin essential hypertension

Steroid and test	*Normal renin** (*n*)	*Low renin** (*n*)	*Reference*
11-DOC			
Plasma levels	N (10)	N or ↑ (21)	Brown *et al.* (1972)
	N (5)	N (4)	Honda *et al.* (1977)
	N (25)	N (23)	Tan and Mulrow (1979)
	ND	N (11)	Griffing *et al.* (1983c)
Response to ACTH	N (5)	↑ (4)	Honda *et al.* (1977)
Excretion rate	N (25)	N (23)	Tan and Mulrow (1979)
	ND	↑ (11)	Griffing *et al.* (1983c)
Secretion rate	N (11)	N (6)	Messerli *et al.* (1976)
18-OH-11-DOC			
Plasma levels	N (16)	N (12)	Witzgall, Thayil and Weber (1982)
	ND	N (11)	Griffing *et al.* (1983c)
Excretion rate	N (70)	N (18)	Agrin *et al.* (1978)
	ND	N (11)	Griffing *et al.* (1983c)
Response to low salt	N (37)	N (12)	Agrin *et al.* (1978)
Secretion rate	N (33)	↑ (11)	Messerli *et al.* (1976)
	ND	N (11)	Griffing *et al.* (1983c)
19-nor-DOC			
Excretion rate	ND	↑ (11)	Griffing *et al.* (1983c)
16β-OH DHEA			
Plasma levels	ND	N (12)	Sekihara *et al.* (1976)
Excretion rate	N (14)	↑ (15)	Liddle and Sennett (1975)
	ND	N (12)	Sennett *et al.* (1977)

* Values in hypertensive subjects compared to those in normotensive subjects in the same study.
DOC = deoxycorticosterone; 16β-OH DHEA = 16β-hydroxydehydroepiandrosterone.
ND = no data available.
Number of subjects (*n*) in each study are given in parentheses.

(3) The 'renin suppression' reported by some investigators may represent no more than a deficiency in renin stimulation or response due to adrenergic insufficiency or sclerosis of the juxtaglomerular apparatus, as commonly seen in older patients (Weidman *et al.*, 1975).

(4) The therapeutic effect of diuretics in low-renin essential hypertension is not restricted to spironolactone, but also occurs with other diuretics (Adlin, Marks

and Channick, 1972; Vaughan *et al.*, 1973). Indeed, this favourable response is generally seen in elderly patients. Whether it is related to low renin, excess mineralocorticoid or independent factors is still a matter of conjecture.
(5) If a mineralocorticoid were secreted to provoke prolonged elevation of blood pressure, concomitant manifestations of hypermineralocorticism would be expected. However, sodium retention is hardly ever detectable and an increase in blood volume has not been demonstrated (Dunn and Tannen, 1977). More surprisingly, hypokalaemia is never observed and potassium balance appears to be normal.

Accordingly, the evidence for mineralocorticoid excess as the primary abnormality in LREH is far from convincing, although the hypothesis deserves further investigation.

In conclusion, many functional disorders have been described in essential hypertension; their heterogeneous distribution within the hypertensive population suggests that several aetiologies account for the different types of essential hypertension and their characteristic response to different forms of treatment.

Possible steroids responsible for hypertension

Because of all these contradictions and the lack of evidence for a clear role of specific steroids, a whole array of steroids have been examined for their potential hypertensive properties: 18-hydroxy DOC (Rapp and Dahl, 1972; Feldman and Funder, 1973; Oliver, *et al.*, 1973; Williams, Braley and Underwood, 1976; Fraser and Lantos, 1978; Huston *et al.*, 1981); 19-nor-DOC and 19-nor-hydroxy-DOC (Sennett *et al.*, 1977; Perrone *et al.*, 1981, 1982; Griffing *et al.*, 1983a, 1983b); 18-hydroxycorticosterone (Feldman and Funder, 1973); 16β-hydroxy-dehydroepiandrosterone (Higgins *et al.*, 1977), 19-hydroxyandrostenedione (Sekihara, 1983) and 19-oxo-DOC (Thomas *et al.*, 1983). There is also some evidence for amplification of the action of aldosterone by 5-dihydrocortisol (Adam *et al.*, 1979) and the synergistic effect of mixtures of steroids such as 17α,20α-dihydroxyprogesterone and 17α-hydroxyprogesterone (Coghlan *et al.*, 1976). The discovery that cells not specialized in the transmembranous transport of sodium have mineralocorticoid receptors led to speculation that mineralocorticoids could exert a hypertensinogenic effect through entirely different mechanisms (Coghlan *et al.*, 1976; De Nicola *et al.*, 1981; Krozowski and Funder, 1981). So far, none of these studies have produced results directly applicable to human forms of hypertension.

For studies of the role of steroid receptors, the availability of a specific antagonist would seem to be of utmost importance. Spironolactone is one such competitive antagonist, and its antimineralocorticoid effects have been reviewed by Corvol *et al.* (1981). At one time, the response to spironolactone was believed to be useful in the detection of aldosteronism and other mineralocorticoid-dependent forms of hypertension (Spark and Melby, 1968), but subsequent studies have shown that it is non-specific as a test and as a form of treatment (Bravo, Dustan and Tarazi, 1973; Ferguson, Turek and Rovner, 1977; Sundsfjord and Oedegaard, 1977; Ogilvie, Piafsky and Ruedy, 1978). Combinations of triamterene and hydrochlorothiazide were found to be as effective as spironolactone (Ganguly and Weinberger, 1981). It has been proposed that spironolactone might also act directly on the adrenal cortex, blocking the synthesis of aldosterone, but such an action could not account for the *in vivo* effect of spironolactone (Gaillard *et al.,* 1980).

Possible causes of abnormal steroid production and/or responsiveness to stimuli of adrenocortical cells

Both biochemical and morphological features point to a common origin of zona glomerulosa and fasciculata cells. We have seen that aldosterone-producing adrenocortical tumours may contain both types of cells, display a circadian rhythm of secretion and show anomalous responses to changes in posture and to angiotensin II and ACTH administration. In many species, both types of cells respond to ACTH and to angiotensin II, and possess receptors with the same characteristics (Hepp *et al.,* 1977; Vallotton *et al.,* 1981). Differentiation of these cells from a common precursor may be determined *in vivo* by ACTH and angiotensin II (McDougall *et al.,* 1980; Mazzocchi *et al.,* 1982; Mazzocchi, Meneghelli and Nussdorfer, 1983). While adrenal cells specialize in producing either aldosterone or cortisol, they still maintain receptors for and responsiveness to the two peptide hormones, one of which becomes the primary regulator. It is therefore conceivable that hypertension could develop as a result of proliferation of abnormal cells with an exquisite sensitivity to an otherwise secondary regulator (ACTH in the case of glomerulosa cells and angiotensin II in the case of fasciculata cells). One could even imagine that these cells may lose the ability to escape the action of ACTH, which is characteristic of normal glomerulosa but not of fasciculata cells (Gaillard *et al.,* 1980, 1983).

CONCLUSION

Our understanding of the relationship between steroids and hypertension, despite enormous scientific efforts, leaves much to be desired. Too many incomplete studies with too few patients and incompletely defined parameters have probably confused the issue. Many promising results, published prematurely, have only served to mislead subsequent investigators. However, current research in the field of peptide receptors on adrenocortical cells and steroid receptors outside the kidney, including the study of their regulation and post-receptor activation events, gives new hope for a better understanding of the role played by steroids in the pathogenesis of hypertension. Some forms of hypertension may turn out to be diseases of receptors or post-receptor activation events. Liddle's syndrome (*see earlier*) may be but one such example.

Acknowledgement

Dedicated to Professor René S. Mach for his 80th birthday.

References

ADAM, W. R., FUNDER, J. W., MERCER, J. and ULICK, S. (1979) Amplification of the action of aldosterone by 5α-dihydrocortisol. *Endocrinology,* **103,** 465–471

ADETUYIBI, A. and MILLS, I. H. (1972) Relation between urinary kallikrein and renal function, hypertension, and excretion of sodium and water in man. *Lancet,* **2,** 203–207

ADLIN, E. V., MARKS, A. D. and CHANNICK, B. J. (1972) Spironolactone and hydrochlorothiazide in essential hypertension. Blood pressure response and plasma renin activity. *Archives of Internal Medicine,* **130,** 855–858

AGRIN, R. J., DALE, S. L., HOLBROOK, M., LAROSSA, J. T. and MELBY, J. C. (1978) Urinary free 18-hydroxy-11-deoxycorticosterone excretion in normal and hypertensive patients. *Journal of Clinical Endocrinology and Metabolism,* **47,** 877–884

AMBRUSTER, H., VETTER, W., RECH, G., BECKERHOFF, R. and SIEGENTHALER, W. (1975) Severe arterial hypertension caused by chronical abuse of a topical mineralocorticoid. *International Journal of Pharmacology,* **12,** 170–173

ARMANINI, D., KARBOWIAK, I. and FUNDER, J. W. (1983) Affinity of liquorice derivatives for mineralocorticoid and glucocorticoid receptors. *Clinical Endocrinology,* **19,** 609–612

AUGUST, J. T., NELSON, D. H. and THORN, G. W. (1958) Response of normal subjects to large amounts of aldosterone. *Journal of Clinical Investigation,* **37,** 1549–1555

BAER, L., SOMMERS, S. C., KRAKOFF, L. R., NEWTON, M. A. and LARAGH, J. H. (1970) Pseudo-primary aldosteronism: an entity distinct from true primary aldosteronism. *Circulation Research,* Suppl. 1, **26/27,** 203–216

BIGLIERI, E. G. and SCHAMBELAN, M. (1979) The significance of elevated levels of plasma 18-hydroxycorticosterone in patients with primary aldosteronism. *Journal of Clinical Endocrinology and Metabolism,* **49,** 87–91

BIGLIERI, E. G., SCHAMBELAN, M., SLATON, P. E. and STOCKIGT, J. R. (1970) The intercurrent hypertension of primary aldosteronism. *Circulation Research,* Suppl. 1, **26/27,** 195–202

BLACHLEY, J. D. and KNOCHEL, J. P. (1980) Tobacco chewer's hypokalemia: licorice revisited. *New England Journal of Medicine,* **302,** 784–785

BRAVO, E. L., DUSTAN, H. P. and TARAZI, R. C. (1973) Spironolactone as a non-specific treatment for primary aldosteronism. *Circulation Research,* **48,** 491–498

BRAVO, E. L., TARAZI, R. C., DUSTAN, H. P. *et al.* (1983) The changing clinical spectrum of primary aldosteronism. *American Journal of Medicine,* **74,** 641–651

BROWN, J. J., FERRISS, J. B., FRASER, R. *et al.* (1972) Apparently isolated excess deoxycorticosterone in hypertension. *Lancet,* **2,** 243–247

CAPLAN, R. H. and VIRATA, R. L. (1974) Functional black adenoma of the adrenal cortex. A rare cause of primary aldosteronism. *American Journal of Clinical Pathology,* **62,** 97–103

CAREY, R. M., DOUGLAS, J. G., SCHWEIKERT, J. R. and LIDDLE, G. W. (1972) The syndrome of essential hypertension and suppressed plasma renin activity. Normalisation of blood pressure with spironolactone. *Archives of Internal Medicine,* **130,** 849–854

CASSAR, J., LOIZOU, S., KELLY, W. F., MASHITER, K. and JOPLIN, G. F. (1980) Deoxycorticosterone and aldosterone excretion in Cushing's syndrome. *Metabolism,* **29,** 115–119

CHANNICK, B. J., ADLIN, E. V. and MARKS, A. D. (1969) Suppressed plasma renin activity in hypertension. *Archives of Internal Medicine,* **123,** 131–140

CHRISTY, N. P. (Ed.) (1971) Iatrogenic Cushing's syndrome. In *The Human Adrenal Cortex,* pp. 395–425. New York: Harper and Row

CHOBANIAN, A. V., BURROWS, B. A. and HOLLANDER, W. (1961) Body fluid and electrolyte composition in arterial hypertension. II. Studies in mineralocorticoid hypertension. *Journal of Clinical Investigation,* **40,** 416–422

CHOBANIAN, A. V., VOLICER, L., TIFFT, C. P., GAVRAS, H., LIANG, C. S. and FAXON, D. (1979) Mineralocorticoid-induced hypertension in patients with orthostatic hypotension. *New England Journal of Medicine,* **301,** 68–73

CHU, M. D. and ULICK, S. (1982) Isolation and identification of 18-hydroxycortisol from the urine of patients with primary aldosteronism. *Journal of Biological Chemistry,* **257,** 2218–2224

COGHLAN, J. P., DENTON, D. A., FAN, J. S. K., MCDOUGALL, J. G. and SCOGGINS, B. A. (1976) Hypertensive effect of 17α,20α-dihydroxyprogesterone and 17α-hydroxyprogesterone in the sheep. *Nature,* **263,** 608–609

COGHLAN, J. P., DOYLE, A. E., JERUMS, G. and SCOGGINS, B. A. (1972) The effects of sodium loading and deprivation on plasma renin and urinary aldosterone in hypertension. *Clinical Science,* **42,** 15–23

COLLINS, R. D., WEINBERGER, M. H., DOWDY, A. J., NOKES, G. W., GONZALES, C. M. and LUETSCHER, J. A. (1970) Abnormally sustained aldosterone secretion during salt loading in patients with various forms of benign hypertension; relation to plasma renin activity. *Journal of Clinical Investigation,* **49,** 1415–1426

CONN, J. W. (1964) Plasma renin activity in primary aldosteronism. *Journal of the American Medical Association,* **190,** 222–225

CONN, J. W., ROVNER, D. R. and COHEN, E. L. (1968) Licorice-induced pseudoaldosteronism. Hypertension, hypokalemia, aldosteronopenia, and suppressed plasma renin activity. *Journal of the American Medical Association,* **205,** 492–496

CORVOL, P., CLAIRE, M., OBLIN, M. E., GEERING, K. and ROSSIER, B. C. (1981) The mechanism of the antimineralocorticoid effects of spirolactones. *Kidney International,* **20,** 1–6

CRANE, M. G. (1973) Hypermineralocorticoidism in hyporeninemic hypertension. In *Essential Hypertension: New Concepts About Mechanisms,* M. P. Sambhi (moderator) (NIH conference). *Annals of Internal Medicine,* **79,** 411–424

CRANE, M. G., HARRIS, J. J. and VARNER, J. J. JR (1972) Hyporeninemic hypertension. *American Journal of Medicine,* **52,** 457–466

DAVIES, D. L., BEEVERS, D. G., BROWN, J. J. *et al.* (1979) Aldosterone and its stimuli in normal and hypertensive man: are essential hypertension and primary hyperaldosteronism without tumour the same condition? *Journal of Endocrinology,* **81,** 79P–91P

DAWSON-HUGHES, B. F., MOORE, T. J., DLUHY, R. G., PODOLSKY, S. and WILLIAMS, G. H. (1981) Alterations in aldosterone biosynthesis in essential hypertensives. *Circulation Research,* **49,** 627–632

DE NICOLA, A. F., TORNELLO, S., WEISENBERG, L., FRIDMAN, O. and BIRMINGHAM, M. K. (1981) Uptake and binding of [^{3}H]-aldosterone by the anterior pituitary and brain regions in adrenalectomized rats. *Hormone and Metabolic Research,* **13,** 103–106

DUNN, M. J. and TANNEN, R. L. (1977) Low renin essential hypertension. In *Hypertension,* 1st Edn, edited by J. Genest, E. Koiw and O. Kuchel, pp. 349–364. New York: McGraw-Hill

EDITORIAL (1979) Idiopathic aldosteronism: a diagnostic artifact. *Lancet,* **2,** 1221–1222

EPSTEIN, M. T., ESPINER, E. A., DONALD, R. A. and HUGHES, H. (1977) Effect of eating liquorice on the renin–angiotensin–aldosterone axis in normal subjects. *British Journal of Medicine,* **1,** 488–490

FAVRE, L., JACOT-DES-COMBES, E., MOREL, PH. *et al.* (1980) Primary aldosteronism with bilateral adrenal adenomas. *Virchows Archiv für Pathologische Anatomie und Histologie,* **388,** 229–236

FELDMAN, D. and FUNDER, J. W. (1973) The binding of 18-hydroxydeoxycorticosterone and 18-hydroxycorticosterone to mineralocorticoid and glucocorticoid receptors in the rat kidney. *Endocrinology,* **92,** 1389–1395

FERGUSON, R. K., TUREK, D. M. and ROVNER, D. R. (1977) Spironolactone and hydrochlorothiazide in normal-renin and low-renin essential hypertension. *Clinical Pharmacology and Therapeutics,* **21,** 62–69

FERRISS, J. B., BROWN, J. J., FRASER, R. *et al.* (1975) Results of adrenal surgery in patients with hypertension, aldosterone excess, and low plasma renin concentration. *British Medical Journal,* **1,** 135–138

FERRIS, J. B., BEEVERS, D. G., BODDY, K. *et al.* (1978) The treatment of low-renin ('primary') hyperaldosteronism. *American Heart Journal,* **96,** 97–109

FRASER, R., JAMES, V. H. T., LANDON, J. *et al.* (1968) Clinical and biochemical studies of a patient with a corticosterone-secreting adrenocortical tumour. *Lancet,* **2,** 1116–1120

FRASER, R. and LANTOS, C. P. (1978) 18-Hydroxycorticosterone: a review. *Journal of Steroid Biochemistry,* **9,** 273–286

FRASER, R., BROWN, J. J., LEVER, A. F. and ROBERTSON, J. I. S. (1983) Control of aldosterone in hypertension. In *Hypertension,* 2nd Edn, edited by J. Genest, O. Kuchel, P. Hamet and M. Cantin, pp. 338–348. New York: McGraw-Hill

FREGLY, M. J., KIM, K. J. and HOOD, C. I. (1969) Development of hypertension in rats treated with aldosterone acetate. *Toxicology and Applied Pharmacology,* **15,** 229–243

GAILLARD, R. C., RIONDEL, A. M., CHABERT, P. and VALLOTTON, M. B. (1980) Effect of spironolactone on aldosterone regulation in man. *Clinical Science,* **58,** 227–233

GAILLARD, R. C., RIONDEL, A. M., FAVROD-COUNE, C. A., VALLOTTON, M. B. and MULLER, A. F. (1983) Aldosterone escape to chronic ACTH administration in man. *Acta Endocrinologica,* **103,** 116–124

GANGULY, A., CHAVARRI, M., LUETSCHER, J. A. and DOWDY, A. J. (1977) Transient fall and subsequent return of high aldosterone secretion by adrenal adenoma during continued dexamethasone administration. *Journal of Clinical Endocrinology and Metabolism,* **44,** 775–779

GANGULY, A. and DONOHUE, J. P. (1983) Primary aldosteronism: pathophysiology, diagnosis and treatment. *Journal of Urology,* **129,** 241–247

GANGULY, A., DOWDY, A. J., LUETSCHER, J. A. and MELADA, G. A. (1973a) Anomalous postural response of plasma aldosterone concentration in patients with aldosterone-producing adrenal adenoma. *Journal of Clinical Endocrinology and Metabolism,* **36,** 401–404

GANGULY, A., MELADA, G. A., LUETSCHER, J. A. and DOWDY, A. J. (1973b) Control of plasma aldosterone in primary aldosteronism: distinction between adenoma and hyperplasia. *Journal of Clinical Endocrinology and Metabolism,* **37,** 765–775

GANGULY, A., GRIM, C. E. and WEINBERGER, M. H. (1981) Anomalous postural aldosterone response in glucocorticoid-suppressible hyperaldosteronism. *New England Journal of Medicine,* **305,** 991–993

GANGULY, A. and WEINBERGER, M. H. (1979) Low renin hypertension: a current review of definitions and controversies. *American Heart Journal,* **98,** 642–652

GANGULY, A. and WEINBERGER, M. H. (1981) Triamterene–thiazide combination: alternative therapy for primary aldosteronism. *Clinical Pharmacology and Therapeutics,* **30,** 246–250

GARDNER, J. D., LAPEY, A., SIMOPOULOS, A. P. and BRAVO, E. L. (1971) Abnormal membrane sodium transport in Liddle's syndrome. *Journal of Clinical Investigation,* **50,** 2253–2258

GENEST, J., NOWACZYNSKI, W., KUCHEL, O., BOUCHER, R. and ROJO-ORTEGA, J. M. (1977) The role of the adrenal cortex in human essential hypertension. *Mayo Clinic Proceedings,* **52,** 291–307

GEORGE, J. M., GILLESPIE, L. and BARTTER, F. C. (1968) Aldosterone secretion in hypertension. *Annals of Internal Medicine,* **69,** 693–701

GEORGE, J. M., WRIGHT, L., BELL, N. H. and BARTTER, F. C. (1970) The syndrome of primary aldosteronism. *American Journal of Medicine,* **48,** 343–356

GIEBINK, G. S., GOTLIN, R. W., BIGLIERI, E. G. and KATZ, F. H. (1973) A kindred with familial glucocorticoid-suppressible aldosteronism. *Journal of Clinical Endocrinology and Metabolism,* **36,** 715–723

GOLD, E. M. (1979) The Cushing syndrome: changing views of diagnosis and treatment. *Annals of Internal Medicine,* **90,** 829–844

GOMEZ-SANCHEZ, C., GOMEZ-SANCHEZ, E., UPCAVAGE, R. J. and HALL, E. B. (1983) Urinary free and serum 19-nor-deoxycorticosterone in adrenal regeneration hypertension. *Hypertension,* Suppl. 1, **5,** 32–34

GRECO, R. G., CARROLL, J. E., MORRIS, D. J., GREKIN, R. J. and MELBY, J. C. (1982) Familial hyperaldosteronism, not suppressed by dexamethasone. *Journal of Clinical Endocrinology and Metabolism,* **55,** 1013–1016

GREMINGER, P., TENSCHERT, W., VETTER, W., LUESCHER, T. and VETTER, H. (1982) Hypertension in Cushing's syndrome. In *Endocrinology of Hypertension,* edited by F. Mantero, E. G. Biglieri and C. R. W. Edwards, pp. 103–110. London and New York: Academic Press

GRIFFING, G. T., DALE, S. L., HOLBROOK, M. M. and MELBY, J. C. (1983a) The regulation of urinary free 19-nordeoxycorticosterone and its relation to systemic arterial blood pressure in normotensive and hypertensive subjects. *Journal of Clinical Endocrinology and Metabolism,* **56,** 99–103

GRIFFING, G. T., DALE, S. L., HOLBROOK, M. M. and MELBY, J. C. (1983b) 19-Nordeoxycorticosterone excretion in primary aldosteronism and low-renin hypertension. *Journal of Clinical Endocrinology and Metabolism,* **56,** 218–221

GRIFFING, G. T., DALE, S. L., HOLBROOK, M. M. and MELBY, J. C. (1983c) Relationship of 19-nor-deoxycorticosterone to other mineralocorticoids in low-renin hypertension. *Hypertension,* **5,** 385–389

GUYTON, A. C., COWLEY, A. W., COLEMAN, T. G., DECLUE, J. W., NORMAN, R. A. and MANNING, R. D. (1974) Hypertension: a disease of abnormal circulatory control. *Chest,* **65,** 328–338

HEPP, R., GRILLET, C., PEYTREMANN, A. and VALLOTTON, M. B. (1977) Stimulation of corticosteroid biosynthesis by angiotensin I, [Des-Asp^1]-angiotensin I, angiotensin II and [Des-Asp^1]-angiotensin II in bovine adrenal fasciculata cells. *Journal of Clinical Endocrinology and Metabolism,* **101,** 717–725

HERF, S. M., TEATES, D. C., TEGTMEYER, C. J., VAUGHAN, E. D., AYERS, C. R. and CAREY, R. M. (1979) Identification and differentiation of surgically correctable hypertension due to primary aldosteronism. *American Journal of Medicine,* **67,** 397–402

HIGGINS, J. R., WAMBACH, G., KEM, D. C., GOMEZ-SANCHEZ, C., HOLLAND, O. B. and KAPLAN, N. M. (1977) Interaction of 16β-hydroxydehydroepiandrosterone with renal mineralocorticoid receptors. *Journal of Laboratory and Clinical Medicine,* **89,** 250–256

HONDA, M., NOWACYNSKI, W., GUTHRIE, G. P. JR *et al.* (1977) Response of several adrenal steroids to ACTH stimulation in essential hypertension. *Journal of Clinical Endocrinology and Metabolism,* **44,** 264–272

HUMPHREY, M. J., LINDUP, W. E., CHAKRABORTY, J. and PARKE, D. V. (1979) Effect of carbenoxolone on the concentration of aldosterone in rat plasma and kidney. *Journal of Endocrinology,* **81,** 143–151

HUSTON, G., MARTIN, V. I., AL-DUJAILI, A. S. and EDWARDS, C. R. W. (1981) Evaluation of the mineralocorticoid activity of 18-hydroxycorticosterone. *Clinical Science,* **61,** 201–206

JOSE, A., CROUT, J. R. and KAPLAN, N. M. (1970) Suppressed plasma renin activity in essential hypertension. *Annals of Internal Medicine,* **72,** 9–16

KAPLAN, N. M. (1977) Renin profiles. The unfulfilled promises. *Journal of the American Medical Association,* **238,** 611–613

KASSIRER, J. P., LONDON, A. M. and GOLDMAN, D. M. (1970) On the pathogenesis of metabolic alkalosis in hyperaldosteronism. *American Journal of Medicine,* **49,** 306–315

KATER, C. E. and BIGLIERI, E. G. (1983) Distinctive plasma aldosterone, 18-hydroxycorticosterone, and 18-hydroxydeoxycorticosterone profile in the 21-, 17α- and 11β-hydroxylase deficiency types of congenital adrenal hyperplasia. *American Journal of Medicine,* **75,** 43–48

KEM, D. C., WEINBERGER, M. H., GOMEZ-SANCHEZ, C. *et al.* (1973) Circadian rhythm of plasma aldosterone concentration in patients with primary aldosteronism. *Journal of Clinical Investigation,* **52,** 2272–2277

KEM, D. C., WEINBERGER, M. H., GOMEZ-SANCHEZ, C., HIGGINS, J. R. and KRAMER, N. J. (1976) The role of ACTH in the episodic release of aldosterone in patients with idiopathic adrenal hyperplasia, hypertension, and hyperaldosteronism. *Journal of Laboratory and Clinical Medicine,* **88,** 261–270

KEM, D. C., WEINBERGER, M. H., HIGGINS, J. R., KRAMER, N. J., GOMEZ-SANCHEZ, C. and HOLLAND, O. B. (1978) Plasma aldosterone response to ACTH in primary aldosteronism and in patients with low renin hypertension. *Journal of Clinical Endocrinology and Metabolism,* **46,** 552–560

KISCH, E. S., DLUHY, R. G. and WILLIAMS, G. H. (1976) Enhanced aldosterone response to angiotensin II in human hypertension. *Circulation Research,* **38,** 502–505

KOTCHEN, T. A., MULROW, P. J., MORROW, L. B., SHUTKIN, P. M. and MARIEB, N. (1971) Renin and aldosterone in essential hypertension. *Clinical Science,* **41,** 321–331

KOVACS, K., HORVATH, E., DELARUE, N. C. and LAIDLAW, J. C. (1974) Ultrastructural features of an aldosterone secreting adrenocortical adenoma. *Hormone Research,* **5,** 47–56

KROZOWSKI, Z. and FUNDER, J. W. (1981) Mineralocorticoid receptors in rat anterior pituitary: toward a redefinition of 'mineralocorticoid hormone'. *Endocrinology,* **109,** 1221–1224

LARAGH, J. H., SEALEY, J. and BRUNNER, H. R. (1972) The control of aldosterone secretion in normal and hypertensive man: abnormal renin–aldosterone patterns in low-renin hypertension. *American Journal of Medicine,* **53,** 649–662

LARAGH, J. H. (1973) Vasoconstriction–volume analysis for understanding and treating hypertension: the use of renin and aldosterone profiles. *American Journal of Medicine,* **55,** 261–274

LARAGH, J. H., LETCHER, R. L. and PICKERING, T. G. (1979) Renin profiling for diagnosis and treatment of hypertension. *Journal of the American Medical Association,* **241,** 151–156

LEE, S. M., LIGHTNER, E., WITTE, M., OBERFIELD, S., LEVINE, L. and NEW, M. (1982) Dexamethasone suppressible hyperaldosteronism in a child with nephrosclerosis. *Acta Endocrinologica,* **99,** 251–255

LIDDLE, G. W. and SENNETT, J. A. (1975) New mineralocorticoids in the syndrome of low renin essential hypertension. *Journal of Steroid Biochemistry,* **6,** 751–753

LUFKIN, E. G., KATZ, F. H. and HERMAN, R. H. (1972) Primary aldosteronism due to hyperplasia of zona glomerulosa: failure of suppression by DOCA or stimulation by angiotensin. *American Journal of Medical Sciences,* **264,** 367–374

LUND, J. O., NIELSON, M. D. and GAMMELGAARD, P. A. (1977) Conn's syndrome. A follow-up of thirteen surgically treated cases. *Acta Medica Scandinavica,* Suppl. **602,** 37–39

MANTERO, F. (1981) Exogenous mineralocorticoid-like disorders. In *Endocrine Hypertension,* edited by E. G. Biglieri and M. Schambelan. *Clinics in Endocrinology and Metabolism,* **10,** 465–478. London: W. B. Saunders

MANTERO, F., ARMANINI, D., OPOCHER, G. *et al.* (1981) Mineralocorticoid hypertension due to a nasal spray containing 9α-fluoroprednisolone. *American Journal of Medicine,* **71,** 352–357

MANTERO, F., SCARONI, C., ARMANINI, D. *et al.* (1982) Hypertension due to adrenal enzymatic defects. In *Endocrinology of Hypertension,* edited by F. Mantero, E. G. Biglieri and C. R. W. Edwards, pp. 111–124. London and New York: Academic Press

MARKS, A. D., MARKS, D. B., KANEFSKY, T. M., ADLIN, V. E. and CHANNICK, B. J. (1979) Enhanced adrenal responsiveness to angiotensin II in patients with low renin essential hypertension. *Journal of Clinical Endocrinology and Metabolism,* **48,** 266–270

MAZZOCCHI, G., MENEGHELLI, V. and NUSSDORFER, G. G. (1983) Effects of angiotensin II on the zona glomerulosa of sodium-loaded dexamethasone-treated rats administered or not with maintenance doses of ACTH: stereology and plasma hormone concentrations. *Acta Endocrinologica,* **102,** 129–135

MAZZOCCHI, G., ROBBA, C., GOTTARDO, G., MENEGHELLI, V. and NUSSDORFER, G. G. (1982) Ultrastructure of aldosterone secreting adrenal adenomata. *Journal of Submicroscopic Cytology,* **14,** 179–185

MCDOUGALL, J. G., BUTKUS, A., COGHLAN, J. P. *et al.* (1980) Biosynthetic and morphological evidence for inhibition of aldosterone production following administration of ACTH to sheep. *Acta Endocrinologica,* **94,** 559–570

MELBY, J. C. and DALE, S. L. (1979) Adrenocorticosteroids in experimental and human hypertension. *Journal of Endocrinology,* **81,** 93P–106P

MELBY, J. C., DALE, S. L. and GRIFFING, G. T. (1983) Nonaldosterone, adrenocortical and other hormonal steroids in experimental and human hypertension. In *Hypertension,* edited by J. Genest, O. Kuchel, P. Hamet and M. Cantin, pp. 349–359. New York: McGraw-Hill

MESSERLI, F. H., KUCHEL, O., NOWACZYNSKI, W. *et al.* (1976) Mineralocorticoid secretion in essential hypertension with normal and low plasma renin activity. *Circulation,* **53,** 406–410

MITCHELL, J. R. (moderator) (1977) Renin–aldosterone profiling in hypertension (NIH conference). *Annals of Internal Medicine,* **87,** 596–612

MOLHUYSEN, J. A., GERBRANDY, J., DE VRIES, L. A. *et al.* (1950) A liquorice extract with deoxycorticosterone-like action. *Lancet,* **2,** 381–386

MOORE, T. J., WILLIAMS, G. H., DLUHY, R. G., BAVLI, S. Z., HIMATHONGKAM, T. and GREENFIELD, M. (1977) Altered renin–angiotensin–aldosterone relationships in normal renin essential hypertension. *Circulation Research,* **41,** 167–171

NEW, M. I., OBERFIELD, S. E., CAREY, R., GREIG, F., ULICK, S. and LEVINE, L. S. (1982) A genetic defect in cortisol metabolism as the basis for the syndrome of apparent mineralocorticoid excess. In *Endocrinology of Hypertension,* edited by F. Mantero, E. G. Biglieri and C. R. W. Edwards, pp. 85–101. London and New York: Academic Press

NICHOLLS, M. G., RAMSAY, L. E., BODDY, K., FRASER, R., MORTON, J. J. and ROBERTSON, J. I. S. (1979) Mineralocorticoid-induced blood pressure, electrolyte, and hormone changes, and reversal with spironolactone, in healthy men. *Metabolism,* **28,** 584–593

OGILVIE, R. I., PIAFSKY, K. M. and RUEDY, J. (1978) Antihypertensive responses to spironolactone in normal renin hypertension. *Clinical Pharmacology and Therapeutics,* **24,** 526–530

OLIVER, J. T., BIRMINGHAM, M. K., BARTOVA, A., LI, M. P. and CHAN, T. H. (1973) Hypertensive action of 18-hydroxydeoxycorticosterone. *Science,* **182,** 1249–1250

PERERA, G. A., KNOWLTON, A. I. and LOWELL, A. (1944) Effect of deoxycorticosterone acetate on the blood pressure of man. *Journal of the American Medical Association,* **125,** 1030–1035

PERRONE, R. D., BENGELE, H. H., DALE, S. L., MELBY, J. C. and ALEXANDER, E. A. (1982) Mineralocorticoid activity of 19-nor-DOC and 19-OH-DOC in adrenalectomized rats. *American Journal of Physiology,* **242,** E305–E308

PERRONE, R. D., SCHWARTZ, J. H., BENGELE, H. H., DALE, S. L., MELBY, J. C. and ALEXANDER, E. A. (1981) Mineralocorticoid activity of 19-nor-DOC and 19-OH-DOC in toad bladder. *American Journal of Physiology,* **241,** E406–E409

POWELL-JACKSON, J. D., CALIN, A., FRASER, R. *et al.* (1974) Excess deoxycorticosterone secretion from adrenocortical carcinoma. *British Journal of Medicine,* **2,** 32–33

PRIESTLEY, J. T., FERRIS, D. O., REMINE, W. H. and WOOLNER, L. B. (1968) Primary aldosteronism: surgical management and pathologic findings. *Mayo Clinic Proceedings,* **43,** 761–775

RAPP, J. P. and DAHL, L. K. (1972) Possible role of 18-hydroxydeoxycorticosterone in hypertension. *Nature,* **237,** 338–339

RE, R. N., SANCHO, J., KLIMAN, B. and HABER, E. (1978) The characterization of low renin hypertension by plasma renin activity and plasma aldosterone concentration. *Journal of Clinical Endocrinology and Metabolism,* **46,** 189–195

REIDBORD, H. and FISHER, E. (1969) Aldosteronoma and non-functioning adrenal cortical adenoma. Comparative ultrastructure study. *Archives of Pathology,* **88,** 155–161

ROSEMBERG, E., DEMANY, M. and BUDNITZ, E. (1962) Effects of administration of large amounts of d-aldosterone in normal subjects and in patients with Sheehan's syndrome. *Journal of Clinical Endocrinology and Metabolism,* **22,** 465–480

ROVNER, D. R., CONN, J. W., KNOPF, R. F., COHEN, E. L. and HSUEH, M. T. Y. (1965) Nature of renal escape from the sodium-retaining effect of aldosterone in primary aldosteronism and in normal subjects. *Journal of Clinical Endocrinology and Metabolism,* **22,** 53–64

SAFAR, M. E., SIMON, A. CH., DARD, S. A. *et al.* (1982) Aldosterone in sustained essential hypertension. *Clinical Endocrinology,* **16,** 77–88

SALTI, I. S., STIEFEL, M., RUSE, J. L. and LAIDLAW, J. C. (1969a) Non-tumorous 'primary' aldosteronism. I. Type relieved by glucocorticoid (glucocorticoid-remediable aldosteronism). *Canadian Medical Association Journal,* **101,** 1–10

SALTI, I. S., RUSE, J. L., STIEFEL, M., DELARUE, N. C. and LAIDLAW, J. C. (1969b) Non-tumorous 'primary' aldosteronism. II. Type not relieved by glucocorticoid. *Canadian Medical Association Journal,* **101,** 11–16

SCHAMBELAN, M., SLATON, P. E. and BIGLIERI, E. G. (1971) Mineralocorticoid production in hyperadrenocorticism. Role in pathogenesis of hypokalemic alkalosis. *American Journal of Medicine,* **51,** 299–303

SCHAMBELAN, M., BRUST, N. L., CHANG, B. C. F., SLATER, K. L. and BIGLIERI, E. G. (1976) Circadian rhythm and effect of posture on plasma aldosterone concentration in primary aldosteronism. *Journal of Clinical Endocrinology and Metabolism,* **43,** 115–131

SCOGGINS, B. A., COGHLAN, J. P., DENTON, D. A., MASON, R. T. and WHITWORTH, J. A. (1982) A review of mechanisms involved in the production of steroid-induced hypertension with particular reference to ACTH dependent hypertension. In *Endocrinology of Hypertension,* edited by F. Mantero, E. C. Biglieri and C. R. W. Edwards, pp. 41–67. London and New York: Academic Press

SEALEY, J. E. and LARAGH, J. H. (1973) Searching out low-renin patients: limitations and some commonly used methods. *American Journal of Medicine,* **55,** 303–314

SEKIHARA, H. (1983) 19-Hydroxyandrostenedione: evidence for a new class of sodium-retaining and hypertensinogenic steroids. *Endocrinology,* **113,** 1141–1148

SEKIHARA, H., HOLLIFIELD, J. W., ISLAND, D. P., SLATON, P. E. and LIDDLE, G. W. (1979) Evidence for heterogeneity of mineralocorticoids in urine of patients with low renin essential hypertension. *Journal of Clinical Endocrinology and Metabolism,* **48,** 143–147

SEKIHARA, H., SENNETT, J. A., LIDDLE, G. W., MCKENNA, T. J. and YARBRO, L. R. (1976) Plasma 16β-hydroxydehydroepiandrosterone in normal and pathological conditions in man. *Journal of Clinical Endocrinology and Metabolism,* **43,** 1078–1084

SENNETT, J. A., YARBRO, L. R., SLATON, P. E., HOLLIFIELD, J. W. AND LIDDLE, G. W. (1977) The C-19-mineralocorticoids in hypertension. *Circulation Research,* **41,** 260–264

SHADE, R. E. and GRIM, C. E. (1975) Suppression of renin and aldosterone by small amounts of DOCA in normal man. *Journal of Clinical Endocrinology and Metabolism,* **40,** 652–658

SIGG, C., SCHWEIZER-CAGIANUT, M. and HEDINGER, C. (1983) Zur Morphologie der Nebennieren beim primären Hyperaldosteronismus. *Schweizerische Medizinische Wochenschrift,* **113,** 357–367

SOMMERS, S. C. and TERZAKIS, J. A. (1970) Ultrastructural study of aldosterone-secreting cells of the adrenal cortex. *American Journal of Clinical Pathology,* **54,** 303–310

SONINO, N., LEVINE, L. S. and NEW, M. (1981) Mineralocorticoid and metabolic response to metyrapone in normotensive children and children with dexamethasone-suppressible and primary aldosteronism. *Acta Endocrinologica,* **98,** 87–94

SPARK, R. F. (1972) Low renin hypertension and the adrenal cortex. *New England Journal of Medicine,* **287,** 343–349

SPARK, R. F., DALE, S. L., KAHN, P. C. and MELBY, J. C. (1969) Activation of aldosterone secretion in primary aldosteronism. *Journal of Clinical Investigation,* **48,** 96–104

SPARK, R. F. and MELBY, J. C. (1968) Aldosteronism in hypertension: the spironolactone response test. *Annals of Internal Medicine,* **69,** 685–691

STOCKIGT, J. R. (1979) Mineralocorticoid excess. In *The Adrenal Gland,* edited by V. H. T. James, pp. 197–241. New York: Raven Press

STREETEN, D. H. P., TOMYCZ, N. and ANDERSON, JR, G. H. (1979) Reliability of screening methods for the diagnosis of primary aldosteronism. *American Journal of Medicine,* **67,** 403–413

SUNDSFJORD, J. A. and OEDEGAARD, A. E. (1977) Renin levels and spironolactone treatment in general practice: similar blood pressure lowering effect of spironolactone in low and normal renin patients. *European Journal of Clinical Investigation,* **7,** 389–392

SUTHERLAND, D. J. A., RUSE, J. L. and LAIDLAW, J. C. (1966) Hypertension, increased aldosterone secretion and low plasma renin activity relieved by dexamethasone. *Canadian Medical Association Journal,* **95,** 1109–1119

TAN, S.-Y., GENEL, M., FORMAN, B. H. and MULROW, P. J. (1977) Steroid profile in a case of adrenal carcinoma with severe hypertension. *American Journal of Clinical Pathology,* **67,** 591–593

TAN, S.-Y. and MULROW, P. J. (1979) Low renin essential hypertension: failure to demonstrate excess 11-deoxycorticosterone production. *Journal of Clinical Endocrinology and Metabolism,* **49,** 790–793

TAYLOR, A. A., MITCHELL, J. R., BARTTER, F. C. *et al.* (1978) Effect of aminoglutethimide on blood pressure and steroid secretion in patients with low renin essential hypertension. *Journal of Clinical Investigation,* **62,** 162–168

THOMAS, C. J., GOMEZ-SANCHEZ, C. E., MARVER, D. and GOMEZ-SANCHEZ, E. P. (1983) 19-Oxo-deoxycorticosterone binds to the renal mineralocorticoid receptor and produces sodium retention but not potassium excretion. *Endocrinology,* **113,** 517–522

TRONO, D., CEREDA, J. M. and FAVRE, L. (1983) Pseudo-syndrome de Conn par intoxication au pastis sans alcool. *Schweizerische Medizinische Wochenschrift,* **113,** 1092–1095

TUCK, M. L., WILLIAMS, G. H., DLUHY, R. G., GREENFIELD, M. and MOORE, T. J. (1976) A delayed suppression of the renin–angiotensin–aldosterone axis following saline infusion in human hypertension. *Circulation Research,* **39,** 711–717

TURPIE, A. G. and THOMSON, T. J. (1965) Carbenoxolone sodium in the treatment of gastric ulcer with special reference to side effects. *Gut,* **6,** 591–594

ULICK, S. (1976) Adrenocortical factors in hypertension. *American Journal of Cardiology,* **38,** 814–824

VALLOTTON, M. B. (1980) L'aldostéronisme primaire: problèmes de classification. *Revue Française d'Endocrinologie Clinique,* **21,** 407–416

VALLOTTON, M. B., CAPPONI, A. M., GRILLET, C. *et al.* (1981) Characterization of angiotensin-receptors on bovine adrenal fasciculata cells. *Proceedings of the National Academy of Sciences of the United States of America,* **78,** 592–596

VAUGHAN, E. D. JR, LARAGH, J. H., GAVRAS, I. *et al.* (1973) Volume factor in low and normal renin essential hypertension: treatment with either spironolactone or chlorthalidone. *American Journal of Cardiology,* **32,** 523–532

VECSEI, P., ABDELHAMID, S., HAACK, D., LICHTWALD, K., LEWICKA, S. and VON MITTELSTADT, G. (1982) Increased excretion of 18-hydroxycorticosterone in patients with adrenal adenomas and hypertension. *Clinical and Experimental Hypertension,* **4,** 1759–1770

VETTER, H., SIEBENSCHEIN, R., STUDER, A. *et al.* (1978) Primary aldosteronism: inability to differentiate unilateral from bilateral adrenal lesions by various routine clinical and laboratory data and by peripheral plasma aldosterone. *Acta Endocrinologica,* **89,** 710–725

WALLACH, L., NYARAI, L. and DAWSON, K. G. (1975) Stimulated renin: a screening test for hypertension. *Annals of Internal Medicine,* **82,** 27–34

WEBER, M. A., PURDY, R. E. and DRAYER, J. I. M. (1983) Interactions of mineralocorticoids and pressor agents in vascular smooth muscle. *Hypertension,* **5** (Suppl. 1), 41–46

WEIDMANN, P., DEMYTTENAERE-BURSZTEIN, S., MAXWELL, M. H. and DELIMA, J. (1975) Effect of ageing on plasma renin and aldosterone in normal man. *Kidney International,* **8,** 325–333

WEINBERGER, M. H. (1983) Primary aldosteronism. In *Hypertension,* edited by J. Genest, O. Kuchel, P. Hamet and M. Cantin, pp. 922–939. New York: McGraw-Hill

WEINBERGER, M. H., GRIM, C. E., HOLLIFIELD, J. W. *et al.* (1979) Primary aldosteronism: diagnosis, localization and treatment. *Annals of Internal Medicine,* **90,** 386–395

WENTING, G. J., MAN IN'T VELD, A. J., VERHOEVEN, R. P., DERKX, F. H. and SCHALENKAMP, M. A. D. H. (1977) Volume–pressure relationships during development of mineralocorticoid hypertension in man. *Circulation Research,* **40** (Suppl. 1), 163–170

WILLIAMS, G. H., BRALEY, M. S. and UNDERWOOD, R. H. (1976) The regulation of plasma 18-hydroxy-11-deoxycorticosterone in man. *Journal of Clinical Investigation,* **58,** 221–229

WILLIAMS, G. H. and DLUHY, R. G. (1977) Regulation of renin–angiotensin–aldosterone axis in hypertension. In *Hypertension,* 1st Edn, edited by J. Genest, E. Koiw and O. Kuchel, pp. 292–311. New York: McGraw-Hill

WISGERHOF, M. and BROWN, R. D. (1978) Increased adrenal sensitivity to angiotensin II in low-renin essential hypertension. *Journal of Clinical Investigation,* **61,** 1456–1462

WISGERHOF, M., CARPENTER, P. C. and BROWN, R. D. (1978) Increased adrenal sensitivity to angiotensin II in idiopathic hyperaldosteronism. *Journal of Clinical Endocrinology and Metabolism,* **47,** 938–943

WITZGALL, H., THAYIL, G. and WEBER, P. C. (1982) Rapid increase of mineralocorticoids after furosemide in low-renin essential hypertension: evidence for 18-hydroxycorticosterone to be a better marker than aldosterone. *Klinische Wochenschrift,* **60,** 847–852

WOODS, J. W., LIDDLE, G. W., STANT, E. G., MICHELAKIS, A. and BRILL, A. B. (1969) Effect of an adrenal inhibitor in hypertensive patients with suppressed renin. *Archives of Internal Medicine,* **123,** 366–370

9
Biochemical investigation of adrenocortical dysfunction

J. G. Ratcliffe

INTRODUCTION

The biochemical investigation of adrenal disease has not changed radically in the last decade although useful advances have been made in methods of assessing cortisol status. The complex measurement of cortisol secretion rate and non-specific methods for cortisol and its metabolites, e.g. 17-hydroxycorticosteroids (17-OHCS), 17-oxogenic steroids (17-OGS), fluorogenic corticosteroids, have now largely been superseded by simple non-extraction (direct) radioimmunoassays for cortisol, applicable to serum, urine and saliva. Simplified assays for 11-deoxycortisol provide an alternative end point to 17-OHCS in the metyrapone test. Unfortunately, similar progress has not been made with ACTH assays; few laboratories offer ACTH assays with adequate sensitivity, specificity and precision for confident interpretation, except perhaps in the investigation of adrenal insufficiency. The development of reliable ACTH assays for subnormal values is an important goal for the future. Diagnostic rewards from assays for ACTH-related peptides may also be anticipated following recent elucidation of how ACTH is processed from its precursor pro-opiocortin (POC) by the pituitary and by extrapituitary tumours (*Figure 9.1*). The diagnostic role of administered corticotropin-releasing factor (CRF) has yet to be established, as has the measurement of CRF in peripheral blood.

Clinically, the major problem in accurate diagnosis of adrenal disease is encountered in suspected Cushing's syndrome and this will be discussed in detail. In contrast, the investigation of suspected adrenal failure is relatively non-controversial and will only be considered briefly in this section. Congenital adrenal hyperplasia is discussed in Chapter 6.

Suspected adrenal hypofunction is investigated initially by assessing basal cortisol status and adrenal reserve using the short ACTH (tetracosactrin, ACTH 1–24) stimulation test. The diagnosis of adrenal insufficiency is confirmed by subnormal basal cortisol values (plasma or salivary cortisol at 0900 hours or 24-hour urinary free cortisol) and impaired incremental cortisol reponse to ACTH. The differential diagnosis between primary and secondary failure is usually clinically obvious. It may be confirmed by a single plasma ACTH, or lipotropin (LPH) measurement, which is clearly elevated in primary failure and normal or undetectable in secondary failure. If ACTH assays are not available, the prolonged (three-day) ACTH test is a more complex alternative. This involves assessing the change in cortisol responses to two short ACTH tests, the first performed before and the second after daily injections of depot ACTH, given intramuscularly for three days. Occasionally, it is necessary to exclude secondary adrenal insufficiency when basal cortisol and/or the short ACTH test is normal. In such circumstances the insulin hypoglycaemia test is valuable and allows concomitant assessment of other anterior

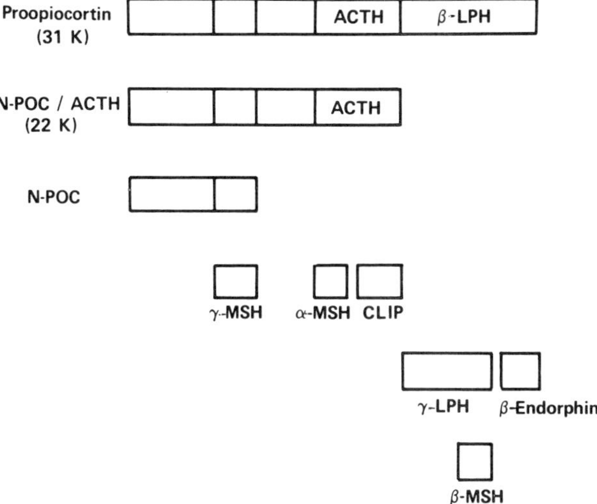

Figure 9.1 Diagrammatic representation of structural relationships between pro-opiocortin and ACTH-related peptides to which it may give rise. Major peptides in normal pituitary and/or plasma include glycosylated *N*-terminal pro-opiocortin (*N*-POC), ACTH, β-lipotropin (β-LPH), γ-LPH and β-endorphin. Other peptides, e.g. 22K fragment, γ-MSH, α-MSH, corticotropin-like intermediate lobe peptide (CLIP; 18–39 ACTH) and β-MSH, may occur in pathological conditions, especially the ectopic ACTH syndrome

pituitary hormone reserve. This test also provides the most valuable method for assessing the clinical need for additional corticosteroids in patients receiving, or having recently (within the previous three months) received steroids, who are to undergo a major stress such as surgery.

CUSHING'S SYNDROME

Overview of diagnostic problems

The interpretation of diagnostic tests must take into account other potential causes of abnormal cortisol status, of which obesity, alcohol abuse and severe depression have been documented. Rarely, the expression of Cushing's syndrome may be intermittent, with cycles of abnormality varying between hours and months (Brown *et al.*, 1973; Schteingart and McKenzie, 1980). Intermittent hypercortisolism has been reported with all major causes of Cushing's syndrome and may give rise to diagnostic confusion. Unfortunately, it is perhaps the exception rather than the rule for all biochemical tests to behave as predicted for a particular aetiology of Cushing's syndrome. Indeed, only the unwary would rely on biochemical tests alone (Aron *et al.*, 1981). Confusion is most likely to arise when an extrapituitary tumour source of ACTH is not apparent from imaging or other localization studies and a mistaken diagnosis of pituitary-dependent Cushing's is made. The underlying neoplasm may only declare itself months or years later and is usually a small bronchial carcinoid or thymoma (Howlett *et al.*, 1984).

Until the mechanism of Cushing's disease is better understood there will continue to be problems in interpreting dynamic biochemical tests. The following pituitary pathologies have been described in Cushing's disease (McNicol, 1981):

(1) Tumour of corticotrophs in the anterior lobe, with atrophy of non-tumorous corticotrophs. This is compatible with Harvey Cushing's original view, and is found in a variable proportion of cases (ranging from 26–80% in different series).
(2) Normal architecture of anterior pituitary, with no evident tumour. This implies that corticotrophs have a functional defect or that excessive ACTH secretion can occur without morphological evidence of hyperplasia as a result of hypothalamic stimulation.
(3) Hyperplasia of corticotrophs with or without a tumour in either the anterior pituitary or the zone between the anterior and posterior pituitary (intermediate lobe-like cells). Anterior lobe hyperplasia implies stimulation by hypothalamic releasing factors, while hyperplasia and/or nodules in the intermediate zone may imply a different pathogenesis since these cells may be controlled by neural factors and show relative resistance to classical corticosteroid negative feedback. Indeed, in some cases argyrophil nerve fibres have been demonstrated to course freely through adenoma tissue and end apparently around individual adenoma cells (Lamberts, de Lange and Stefanko, 1982). Several potential neurotransmitters may be involved – catecholamines (noradrenaline and dopamine), serotonin, acetylcholine, γ-aminobutyric acid (GABA), vasopressin and opioid and other peptides. Their role in controlling ACTH secretion of either anterior or intermediate lobe corticotrophs is uncertain although there is evidence that bromocriptine (a dopamine agonist) and cyproheptadine (a serotonin antagonist) can inhibit ACTH secretion, at least in the short term in some cases of Cushing's disease (Lamberts *et al.*, 1977; Krieger, Amorosa and Linick, 1975) and that valproate (a GABA agonist) can reduce ACTH levels in some cases of Nelson's syndrome (Dornhurst *et al.*, 1983).

It has also been reported that thyrotropin-releasing factor (TRF) and/or luteinizing hormone releasing factor (LH-RF) stimulate ACTH and cortisol responses acutely in a variable proportion of patients with Cushing's disease (Pieters *et al.*, 1982b). In this context, TRF can stimulate ACTH and cortisol release in pregnant women, suggesting that functioning intermediate lobe tissue is required for the effect. However, patients with Cushing's disease who respond to one or both releasing factors do not differ from the non-responders in dexamethasone suppressibility or basal or TRF-stimulated prolactin levels, and bromocriptine is equally ineffective in reducing cortisol levels in the two groups. These data do not therefore support the notion of an intermediate lobe origin of Cushing's disease in patients with a paradoxical response to releasing hormones.

Although pathology of the intermediate lobe cells may occur in a subset of patients with Cushing's disease, present data are too conflicting to allow an assessment of its significance, prevalence, and biochemical correlates.

There is also evidence that anterior and intermediate lobe corticotrophs differ in their patterns of production of ACTH-related peptides. The anterior lobe secretes mainly ACTH, *N*-terminal pro-opiocortin (*N*-POC) and β-LPH, whereas intermediate lobe cells secrete α-melanocyte stimulating hormone (α-MSH), corticotropin-like intermediate lobe peptide (CLIP) and β-endorphin, but relatively little ACTH (Lowry *et al.,* 1980). Thus a pathognomonic feature of intermediate lobe cells is the presence of α-MSH. This must be identified specifically as being *N*-terminally acetylated and *C*-terminally amidated. It is not sufficient to demonstrate *N*-terminal ACTH activity alone since several ACTH-related peptides possess this structural feature.

The distinction of these variants of Cushing's disease is of considerable practical importance since selective microadenomectomy may only be appropriate in patients with discrete anterior lobe tumours (Gold, 1979). Coincident anterior lobe hyperplasia may point to the need for radical hypophysectomy, while hyperplasia of the intermediate lobe may be more appropriately treated by neuropharmacological drug therapy. Thus a major goal must be the accurate diagnosis of the particular pathology of Cushing's disease. Biochemical investigations may have considerable potential in identifying the nature of the lesion.

Measurement of basal pituitary adrenal function

Cortisol secretion rate (CSR)

The CSR is measured by administering a known amount of radiolabeled cortisol and measuring the specific activity of a unique metabolite (usually tetrahydrocortisol or tetrahydrocortisone) in the urine (Cope and Black, 1958). It is thus a measure of adrenocortical activity and does not provide information on tissue exposure to cortisol, which is better gained from urinary free cortisol or serial plasma cortisol determinations. The CSR is increased in obesity, although urinary free cortisol excretion is essentially normal; it is not increased by oestrogens, which may raise free cortisol levels, and varies with thyroid activity. Determination of CSR is technically complex and time-consuming, and it remains purely a research procedure.

Corticosteroids

Assays of cortisol in urine provide an integrated summary of the amount of steroid excreted in a given time, and the 24-hour urinary free cortisol measurement is particularly valuable in establishing the diagnosis. Plasma steroid assays are less helpful, as frequent samples taken at short intervals (e.g. 15 min) are required to provide a clear impression of integrated steroid secretion. For clinical purposes this is impractical although a single plasma cortisol determination taken without stress in the late evening (2300–2400 hours) has some diagnostic value. Multiple sampling using saliva is more feasible and provides a useful index of the non-protein bound fraction in plasma.

URINE ASSAYS

Urine for corticosteroid measurements should be collected without preservative and stored at 4 °C until assay. Many assays have been used to assess urinary corticosteroid excretion. Each measures a different group of metabolites. The Porter-Silber method measures about half the cortisol metabolites (mainly tetrahydrocortisol and tetrahydrocortisone) and the results are referred to as 17-hydroxycorticosteroids (17-OHCS) or Porter-Silber chromogens. The 17-oxogenic steroid (17-OGS) assay measures 80–90% of the total metabolic products of cortisol, and also metabolites such as pregnanetriol. The 17-oxosteroid (17-OS) assay measures metabolites of adrenal and testicular androgens in addition to those of corticosteroids. All these methods lack specificity and diagnostic precision (Rudd, 1983).

The best single diagnostic test remains the measurement of urinary free cortisol, which provides a good index of the integrated level of plasma non-protein bound hormone (Burke and Beardwell, 1973). There are, however, methodological pitfalls. Urinary cortisol excretion is only about 200 nmol (72.5 pg) per 24 hours, which represents less than 1% of the cortisol secretion rate, whereas excretion of 17-OGS or 17-OS is about 40 000 nmol (11 500 μg)per 24 hours. Fluorimetric methods (which measure 11-hydroxycorticosteroids) give higher values than saturation analysis techniques, and as little as 20% of fluorogenic corticosteroids in urine may be cortisol. Various compounds interfere with the fluorimetric assay, notably spironolactone, trilostane and tetracyclines. However, in the absence of drug interference such assays can give good diagnostic discrimination between obese or hirsute women and patients with Cushing's syndrome. Competitive protein binding and radioimmunoassay (RIA) methods are more widely used although specificity may only be achieved by preliminary solvent extraction. More recently, good specificity has been obtained with carefully selected antisera which allow simple direct assay without prior extraction. Antisera produced to cortisol-3-CMO immunogen together with iodinated tracer ([^{125}I]-histamine-3-CMO cortisol) provide the basis for a satisfactory clinical assay (Riad-Fahmy, Read and Hughes, 1983). Urinary cortisol directly reflects the unbound fraction, which represents about 5% of total plasma cortisol at normal plasma concentrations. Unbound cortisol is ultrafiltered at the glomerulus and most is reabsorbed in the renal tubules along with other low-molecular weight solutes, leaving a small portion to be excreted. As total plasma cortisol concentration rises, the unbound fraction increases in proportion, until the binding capacity of cortisol-binding globulin (CBG) is exceeded at plasma cortisol levels of more than 20 μg/dl (550 nmol/l). Thereafter there is a disproportionate increase in the unbound fraction. For example, a doubling of plasma cortisol from 20 to 40 μg/dl (550–1100 nmol/l) results in a more than five-fold increase in free cortisol. There is an excellent correlation between plasma unbound cortisol and urinary free cortisol in Cushing's syndrome. The urinary free cortisol excretion is also independent of several factors which alter excretion of cortisol metabolites by changing the kinetics of metabolism and/or conjugation (e.g. obesity, pregnancy, drugs). The main practical drawback of the test is the need for accurate 24-hour collection, which may be difficult to achieve as an outpatient. Two simpler urine tests have therefore been proposed as screening tests for Cushing's syndrome.

(1) Determination of the free cortisol:creatinine ratio in an early morning specimen of urine (Walker, 1979). However, the additional measurement of creatinine adds to the analytical imprecision of the ratio.

(2) Measurement of the total amount of cortisol in urine passed during the night and immediately on rising (Mattingly and Tyler, 1976). Fluorimetric corticosteroids have been used successfully for this test although specificity can be improved by assaying cortisol by RIA.

Both tests overcome the problem of the incomplete 24-hour urine collection and take advantage of the fact that in normal subjects cortisol excretion is minimal during most of the night, while in Cushing's syndrome high cortisol excretion persists. Several specimens can be taken on an outpatient basis when the diagnosis is equivocal or there is evidence of periodicity in cortisol excretion.

In reviewing the value of urinary corticosteroid measurements, Crapo (1979) confirmed the diagnostic superiority of 24-hour urinary free cortisol assays over 24-hour 17-OHCS or 17-OGS measurements. In the 15 separate studies reviewed, only 3.3% of 479 lean, obese or chronically ill controls had elevated levels, while 5.6% of 248 patients with Cushing's syndrome had normal levels. This high resolution is perhaps surprising since the 'between-batch' imprecision of urinary cortisol assays is often between 10 and 20% (coefficient of variation), and presumably reflects the great discriminatory power of the test. In this context, Levell (1980) has demonstrated the rather marginal effects of changes in analytical imprecision of urinary 24-hour 11-OHCS assays on diagnostic discrimination. There is more limited evidence that the early morning cortisol:creatinine ratio, overnight cortisol or fluorogenic corticosteroid estimation or even the ratio of 24-hour 17-OHCS to creatinine provide such good diagnostic discrimination. All these tests can be used in ambulatory subjects so that hospitalization is not essential.

The major unconjugated urinary metabolite of cortisol is 6β-hydroxycortisol, which comprises about 1–2% of the total daily cortisol secretion. Direct radioimmunoassay of 6β-hydroxycortisol in a 24-hour specimen of urine has been reported to give excellent diagnostic discrimination of hypercortisolaemic states, with levels at least ten times higher than normal in Cushing's syndrome. Urinary 6β-hydroxycortisol discriminated better than either urinary free cortisol or 17-OHCS, with no false negatives or positives, in adolescents with Cushing's syndrome (Saenger and Peterson, 1984). The marked increase in 6β-hydroxycortisol is due apparently to dose-dependent induction of 6β-hydroxylase by cortisol (Voccia *et al.*, 1979). This test merits further assessment although it should be noted that several hormones and drugs can also induce 6β-hydroxylation and hence increase excretion of the metabolite.

PLASMA OR SERUM ASSAYS

Since cortisol is relatively stable in blood, serum specimens are preferred. As with urine, a range of techniques is available for measuring serum corticosteroids, e.g. colorimetry for 17-OHCS (Porter-Silber chromogens), fluorimetry for 11-OHCS and saturation analysis (competitive protein binding (CPB), radioimmunoassay). Single estimations may be misleading but serial measurements are useful in dynamic tests. Dynamic tests also have the advantage that all samples can be measured in a single assay, thus reducing analytical imprecision. The fluorimetric assay is convenient, rapid, cheap and automatable, but only 50–60% of fluorescence is due to cortisol, and drug interference is common. However, prednisolone and dexamethasone do not interfere in doses used diagnostically or therapeutically, which is an important advantage over many immunoassays.

CPB methods, which require preliminary extraction, have now been superseded by radioimmunoassay. Antisera with better specificities have been developed using immunogens coupled in the 3-position, and γ-emitting labels (e.g. selenium-75, iodine-125) are available commercially. More convenient and reliable separation procedures (e.g. double antibody solid phase) have replaced charcoal. Finally, by employing displacing agents (pH4 ± 8-anilino-naphthalene sulphonic acid), cortisol can be assayed without prior extraction (Ratcliffe, 1983). The precision of such assays is less than 10% between assay over the clinically important range. However, there remain problems of specificity in that most radioimmunoassays significantly overestimate the true value determined by gas chromatography-mass spectrometry or high-performance liquid chromatography. There may be cross-reaction with polar metabolites and conjugates (e.g. deoxycortisol) and some synthetic steroids (e.g. prednisolone, dexamethasone). As with other non-extraction assays, the control of matrix effects requires careful selection of hormone free serum for making up cortisol standards. Blood bank serum from which endogenous cortisol has been removed by charcoal is often satisfactory.

Several non-isotopic immunoassays for cortisol have been described (using enzyme, fluorescent or chemiluminescent labels) but so far their performance has been less robust than that of RIA, and signal detection requires additional steps and in some cases special equipment. Non-isotopic immunoassays usually require preliminary serum extraction because of problems in signal generation in the presence of serum components. However, certain non-isotopic immunoassays do not require physical separation of free and bound moieties, and fluorescence polarization immunoassays have been described which are both homogeneous and direct (Kobayashi *et al.*, 1979). In principle, such assays would be suitable for automation. Further developments can be expected to make these alternative methods attractive for the clinical chemistry laboratory.

SALIVARY ASSAYS

The recent developments in radioimmunoassay techniques have been applied to assays for cortisol in mixed saliva, which has practical advantages. Collection is not invasive and allows multiple sampling by the patient, both factors of potential importance in assessing circadian rhythms in detail. Direct assays in saliva are feasible since protein binding is much less of a problem than in serum (Walker *et al.*, 1978). Saliva concentrations of cortisol reflect, but are not identical with, free plasma levels and the concentrations are high relative to other steroids such as testosterone and oestradiol (Peters *et al.*, 1982). As expected, salivary levels have a similar relationship to plasma cortisol levels as urinary free cortisol concentrations; when the binding capacity of serum corticosteroid-binding globulin (CBG) is exceeded, there is a disproportionate increase in salivary cortisol concentration. In the investigation of Cushing's syndrome by salivary cortisol assays, serial determinations and dynamic tests may have greater value than single basal assessment of cortisol status by single specimens. They may have a special diagnostic role when renal handling of cortisol is abnormal.

ACTH and related peptides

There is little evidence that plasma ACTH assays are useful in establishing the diagnosis of Cushing's syndrome, although it has been reported that ACTH levels in Cushing's disease are higher than in normal subjects if specimens are taken at

0900 hours rather than earlier in the morning (Horrocks and London, 1982). They do, however, have an important role in the differential diagnosis. Normal or elevated levels distinguish ACTH dependent (pituitary-driven or ectopic) from ACTH independent (adrenal adenoma or carcinoma) causes (Rees and Lowry, 1979). In practice, this simple distinction has often been blurred by application of poor ACTH assays. This has led to reports of low normal basal ACTH levels in Cushing's syndrome due to adrenal tumours, and undetectable levels in the pituitary-driven variety. These technical problems are compounded when scrupulous attention is not paid to sample collection and handling. ACTH is generally believed to be unstable in blood and to adsorb readily to glass surfaces. It is recommended that specimens are collected into cooled plastic tubes containing lithium heparin, centrifuged immediately at 4 °C, and the plasma decanted into a second polystyrene tube, snap-frozen in dry ice and stored at −20 °C until assay. Haemolyzed or thawed specimens are usually considered to be unsuitable for ACTH analysis.

However, recent work using both sensitive bio- and immunoassays has challenged this view, which was based largely on examination of the stability of exogenous ACTH by the Lipscomb-Nelson bioassay in the hypophysectomized rat. Endogenous ACTH may be much more stable than was previously thought, with little loss of endogenous ACTH activity in whole blood for up to two hours at room temperature (A. Lambert, personal communication). Re-examination of the specimen handling requirements is clearly needed since some flexibility in the protocol for blood separation would be convenient and encourage the development of more reliable commercial plasma ACTH assays.

For the last 15 years ACTH has been measured in plasma for clinical purposes almost exclusively by radioimmunoassay. The ACTH assay is relatively difficult, and most assays, including those employing commercial kits, are poorly validated. None are able to distinguish subnormal levels and most cannot even measure low normal levels (below 10–20 ng/l; 2.2–4.4 pmol/l) reliably. Near the limits of detection, non-specific factors (often of high molecular weight and poorly characterized) become intrusive and assays lack parallelism with standard 1–39 ACTH (Nicholson *et al.*, 1984). Furthermore, molecular species other than 1–39 ACTH may cross-react.

In attempts to circumvent these problems prior extraction of relatively large volumes of plasma (1–5 ml) has been employed. Several simple procedures are available, based on adsorption of ACTH to talc, silicic acid or leached silica glass, and subsequent elution and evaporation. Experience with leached silica glass as adsorbent over 15 years in the author's laboratory has shown it to have the advantage of removing non-specific interference and reducing tracer damage by proteolytic enzymes in plasma (Ratcliffe and Edwards, 1971). It is also relatively selective in extracting 1–39 ACTH and discriminates against both high molecular weight precursors and fragments. These advantages are bought at the expense of greater technical complexity, some loss of assay precision and reduced sample capacity (less than 20 samples per assay). Nevertheless, the improved specificity at low normal levels is of great help in distinguishing ACTH-dependent from adrenal causes of Cushing's syndrome. Extraction of 5 ml plasma allows a sensitivity of 1–5 ng/l (0.22–1.1 pmol/l) to be achieved reliably. In Cushing's disease ACTH levels are seldom less than 20 ng/l (4.4 pmol/l) whereas in adrenal causes of the syndrome they are undetectable (less than 1 ng/l; 0.22 pmol/l). Unfortunately, ACTH is a poor immunogen, and few antisera with the necessary avidity and specificity are available. ACTH tracer of high specific activity and immunoreactivity is also essential, and this can be achieved by conventional methods of iodination (chloramine T, iodogen) although the iodinated peptide must be purified subsequently on silicates, gel chromatography or high performance chromatography. Bound and free hormone may be separated by double antibody or other methods after an assay incubation time of at least 24 hours.

Standardization should employ a generally available material such as MRC 74/555. The potency of this ACTH preparation has recently been revised to 25 μg per ampoule from its original designation in 1974 of about 11.6 μg per ampoule. This, together with the variety of 'in house' standards employed, makes direct comparison of numerical values between laboratories difficult.

The overall quality of ACTH radioimmunoassay is also difficult to assess since external quality assessment schemes rarely cover this analyte because of the perceived problems of stability. Evidently it behoves clinicians and laboratory workers to validate any ACTH assay thoroughly before diagnostic use. In inexperienced hands any ACTH assay may cause diagnostic confusion and it is better to have no ACTH assay than one of uncertain validity. On the other hand, a sound ACTH assay provides valuable diagnostic information and reduces the need for dynamic tests. The development of two-site labelled antibody assays for ACTH employing monoclonal antibodies offers the potential for greater reliability.

Because of these difficulties and our improved understanding of the biosynthesis of ACTH, attention has been directed to the possible diagnostic use of ACTH-related peptides. The anterior pituitary processes the precursor molecule into *N*-terminal pro-opiocortin (1–76) (*N*-POC), ACTH and β-LPH, which are secreted in equimolar amounts (*see Figure 9.1*). In animals with a well-defined intermediate lobe the processing of these peptides proceeds further. Thus, ACTH is split into α-MSH and CLIP, and β-LPH into γ-LPH and β-endorphin. In man there is no well-defined intermediate lobe except in the fetus and possibly in pregnancy (Pieters *et al.*, 1982), but cells adjacent to the posterior lobe and scattered cells in the anterior pituitary of intermediate lobe origin may synthesize a similar profile of ACTH-related peptides.

Several assays for β-MSH-related peptides, the lipotropins, have been described which measure β- and/or γ-LPH. Preliminary plasma extraction is usually required. These assays appear to have similar diagnostic discrimination to ACTH in Cushing's syndrome, and in Cushing's disease the plasma LPH:ACTH ratio is approximately unity while in ectopic ACTH syndrome it tends to be increased, possibly reflecting the greater stability of circulating LPH and its slower renal clearance (Gilkes, Rees and Besser, 1977). The lipotropins are stable in blood or plasma for at least 24 hours at ambient temperature so that the LPH assay may be a convenient clinical tool for assessing ACTH status indirectly (Gray and Ratcliffe, 1979).

N-POC 1–76 circulates in higher concentrations than ACTH and shows similar changes to ACTH and LPH in pituitary–adrenal disorders. It has thus been proposed as another indirect marker for corticotroph function (Chan, Seidah and Chretien, 1983; Hale *et al.*, 1984). Again, the greater stability in blood and technical simplicity of the plasma assay suggest that it may become useful diagnostically when reagents are more widely available. The biological role of *N*-POC 1–76 is unclear although there is *in vitro* evidence that it may potentiate ACTH-mediated adrenal steroidogenesis (Al-Dujaili *et al.*, 1981). If *N*-POC has a biological role its diagnostic role may be further enhanced. There have also been suggestions that portions of *N*-POC (*N*-POC 1–28, *N*-POC 2–59) stimulate adrenal DNA synthesis *in vitro* and mitosis *in vivo* and thus may be involved in the control of adrenal growth (Estivariz *et al.*, 1982). There is almost no information on circulating levels of α-MSH and CLIP in Cushing's disease although these would clearly be of interest in patients with abnormal function of intermediate lobe cells.

So far, no molecular forms of ACTH-related peptides have been shown to be unique to either pituitary or ectopic sources, although different proportions of the peptides are commonly found. Thus, peptides with higher molecular weight than ACTH, e.g. POC, POC/ACTH (22K fragment), are more abundant in ectopic sources but also occur in patients with large, locally invasive pituitary tumours (Ratter *et al.*, 1983). Characteristically, tumours associated with the ectopic ACTH syndrome also secrete α-MSH, CLIP and β-endorphins, a pattern typically found in the intermediate lobe cells. These possible differences in the profile of ACTH-related peptides have not yet been exploited diagnostically.

Dynamic tests

Suppression tests

A characteristic feature of pituitary adrenal function in Cushing's syndrome is resistance to suppression by exogenous corticosteroids. This feature forms the basis of several variants of the dexamethasone test described by Liddle over 20 years ago (Liddle, 1960). Dexamethasone was originally chosen because it has little sodium-retaining activity and neither the parent drug nor its metabolites interfere with urinary 17-OHCS measurement. However, apparent failure of suppression may occur in patients who do not have Cushing's syndrome and are receiving drugs such as diphenylhydantoin (phenytoin) and carbamazepine, which increase hepatic metabolism of dexamethasone. Individuals with a high alcohol intake may also fail to suppress normally after dexamethasone, either as a result of enzyme induction or alcohol-induced Cushing's syndrome, or both. In addition, patients with severe depression are abnormally resistant to suppression. This feature has subsequently been used as a diagnostic criterion for endogenous depression, using a modified form of the test in which plasma cortisol is determined at 1600 hours following oral administration of dexamethasone 1 mg the previous midnight. Conversely, patients with Cushing's syndrome may suppress normally on the low dose, possibly due to impaired metabolic clearance of dexamethasone with high circulating levels (false negatives) (Caro *et al.*, 1978).

As originally described, the full dexamethasone test requires complete 24-hour urine collections for six consecutive days. After two control days, dexamethasone 0.5 mg is given every six hours for two days (low dose), followed by 2 mg every six hours for the final two days (high dose). Using a cut-off level of 4 mg (14 μmol) 17-OHCS or less per day on the second day for normal suppression, the test has excellent predictive value, with virtually no false positives and only 6.4% false negatives in the six series reviewed by Crapo (1979) There is less uniformity in cut-off levels when alternative end points are used: for urinary 17-OGS it lies between 3 and 7 mg/day (10–24 μmol/day); for urinary free cortisol it varies with the method of assay but is about 110 nmol/day (40 μg) by RIA; and for serum cortisol taken 48 hours after starting dexamethasone it is about 60 nmol/l (2.2μg/dl) by RIA. With these end points there are perhaps more false positives than with the original test.

There is some evidence that doses of dexamethasone other than 2 mg/day may provide better diagnostic discrimination. Thus, Streeten *et al.* (1969) reported that administration of a total dose of 20 μg/kg body weight per day divided into 6-hourly doses provided complete separation of Cushing's syndrome and controls, using an upper limit of normal for urinary 17-OHCS of 1 mg/g (390 nmol/mol) creatinine per day on the second day of the test. The simultaneous measurement of plasma cortisol and dexamethasone levels has also been advocated. The timing of drug administration may warrant attention: it is conventional to start and end the test at 0900 hours but greater reliability is probably achieved by choosing midday so that the problem of having to give a dose at 0300 hours in the morning is avoided. It is perhaps surprising that the test is still widely used in much the same form as originally described, and re-examination of the optimal way to perform the suppression test in suspected Cushing's syndrome is clearly overdue.

Many laboratories now use an overnight suppression test instead of the low dose test. Dexamethasone 1–2 mg is given between 2300 and 2400 hours and plasma cortisol measured between 0800 and 0900 hours in the morning. In normal subjects, the plasma fluorogenic cortisol is less than 200 nmol/l (7.2 μg/dl), compared to a value of less than 100 nmol/l (3.6 μg/dl) by radioimmunoassay. Several protocols differing in detail have been used successfully, provided each laboratory determines criteria for the cut-off value for plasma cortisol. The dexamethasone dose should be at least 1 mg and the test should be carried out in outpatients rather than hospitalized subjects or chronically ill patients (Connolly *et al.,* 1968). This reduces the proportion of false positive results due to stress or accelerated dexamethasone metabolism by drugs. In addition, oestrogens may reduce cortisol metabolism and increase binding to CBG, thus giving spurious positive responses. The main advantage of the test is that there are few false negatives in patients with Cushing's syndrome (1.9% in the 12 studies reviewed by Crapo in 1979), and a normal suppression virtually rules out the diagnosis. With the above provisos it can be recommended as a simple and effective screening test.

The high dose dexamethasone test is still widely employed in the differential diagnosis of Cushing's syndrome. Suppression of urinary 17-OHCS, 11-OHCS or free cortisol to below 50% of baseline values is usually seen in patients with Cushing's disease but not in those with adrenal tumours or the ectopic ACTH syndrome. However, the criterion of 50% suppression should not be applied rigidly since some patients with Cushing's disease may only suppress to 50–60% of basal values. It is useful to assess urinary creatinine excretion to ensure equivalence of urine collection. There are a few patients with Cushing's disease who do not suppress to at least 60% of basal values on 8 mg of dexamethasone. Patients with nodular adrenal hyperplasia are particularly resistant. Smals *et al.* (1984), in a study of 13 patients with macronodular adrenal hyperplasia, suggested that this is a later stage of diffuse hyperplasia in which greater autonomy has developed in response to long-standing stimulation by endogenous ACTH. This is also indicated by the lesser stimulation of cortisol by exogenous ACTH, which is intermediate between that in bilateral diffuse hyperplasia and that in classical adrenal adenoma. Despite the frequency of incomplete dexamethasone suppression, an adequate metyrapone response is often found, suggesting ACTH dependence. Overall, it seems ACTH-dependent hyperplasia coexists with autonomy of the macronodules. This is compatible with low, variable and sometimes undetectable basal ACTH levels, and emphasizes the requirement for sensitive and reliable ACTH assays for the differential diagnosis.

Occasionally, patients with adrenal tumours or ectopic ACTH secretion appear to suppress or even increase steroid levels on 8 mg dexamethasone. Some of these are due to spontaneous cyclical changes over days or months in cortisol production. Reproducibility of testing is then important for differential diagnosis. Rarely, patients with pituitary-dependent Cushing's disease are resistant to suppression by dexamethasone in doses up to 32 mg per day but show the usual degree of suppressibility seen in Cushing's disease to exogenous cortisol (Carey, 1980). The reason for this is unknown but loss of feedback receptors for dexamethasone in the presence of intact receptors for cortisol may be responsible (*see* Chapter 4).

An overnight high dose dexamethasone suppression test has also been suggested as a more rapid and reliable alternative to the standard two-day test for the differential diagnosis of Cushing's syndrome (Aron *et al.*, 1980). Dexamethasone 8 mg given orally at 2300 hours at night suppressed plasma cortisol, measured at 0700–0800 hours the following morning, to below 50% of baseline values in 95% of patients with Cushing's disease but not in patients with adrenal carcinoma or ectopic ACTH syndrome. Again, it cannot be recommended generally until wider experience is available.

Metyrapone

The metyrapone test was introduced by Liddle *et al.* (1959) to assess feedback control of ACTH. In normal subjects given metyrapone 750 mg four-hourly for 24 hours the urinary excretion of 17-OGS at least doubles during the day of administration of the drug or, more usually, during the subsequent 24 hours. Typically, in Cushing's disease the mean rise in 17-OGS is threefold, whereas in Cushing's syndrome due to an adrenal tumour or ectopic ACTH syndrome there is no increase. In several studies reviewed by Crapo (1979), 98% of patients with Cushing's disease had positive metyrapone tests, but a significant proportion of patients with an adrenal tumour and ectopic ACTH syndrome (16% and 46% respectively) also responded. Only 57% of patients with nodular adrenal hyperplasia responded. Thus, while a lack of response provides evidence against Cushing's disease, a positive response is more difficult to interpret. False negative tests may occur in patients taking diphenylhydantoin or oestrogens because of increased hepatic metabolism of metyrapone induced by these drugs.

In addition to inhibiting adrenal 11β-hydroxylation, metyrapone also inhibits cortisol synthesis at an early stage in the pathway. There is a delay of 4–6 hours between the fall in cortisol level and the subsequent increase in plasma ACTH level, although as expected 11-deoxycortisol concentration rises as that of cortisol declines. The drug should be taken with a snack or milk to reduce symptoms of epigastric discomfort and requires medical supervision with an awareness that glucocorticoid insufficiency may be precipitated particularly in patients with adrenal tumours.

Because the standard test requires four days (two control days, one day on drug, one day after drug) of 24-hour urine collection several alternatives have been suggested.

(1) Administration of metyrapone as in the standard test, with determination of serum 11-deoxycortisol at 0800 hours, after 24 hours on the drug. Such a one-day test has been reported to be more accurate than the high dose

dexamethasone test in distinguishing Cushing's disease from adrenal tumour if deoxycortisol is assayed by RIA after chromatography and a cut-off level of 10 μg/dl (289 nmol/l) is chosen (Sindler, Griffing and Melby, 1983). Similar good diagnostic discrimination was found by Perry *et al.* (1984) using a direct (non-extraction) RIA for deoxycortisol and a cut-off level of 1050 nmol/l (36.4 μg/dl). Nevertheless, occasional patients with ectopic ACTH syndrome showed positive responses. Serum 11-deoxycortisol was a better end point than plasma ACTH for differential diagnosis, and the assay was simpler and quicker as no special sample preparation was required. However, there remain problems of specificity with direct serum 11-deoxycortisol assays. Cortisol levels should be monitored by fluorimetry initially to ensure adequate adrenal inhibition.

(2) Single-dose metyrapone test, giving 30 mg/kg (a total of approximately 2 g) by mouth at 2300–2400 hours and measuring serum 11-deoxycortisol at 0800 hours in the morning (Jubiz *et al.*, 1970). The rationale is that the pituitary adrenal axis is most sensitive to declining cortisol levels between 0200 and 1000 hours, but experience in the differential diagnosis of Cushing's syndrome is too limited to recommend its general use.

Corticotrophin releasing factor (CRF)

The isolation and subsequent synthesis of ovine CRF has opened up the possibility of its use in the investigation of hypothalamic pituitary pathophysiology (Vale *et al.,* 1981). The 41-amino acid peptide is a potent stimulator of ACTH and glucocorticoid secretion in several species. In man the effective dose range is between 0.03 and 30 μg/kg body weight (De Bold *et al.,* 1983). When it is given intravenously, ACTH levels rise within a few minutes of injection to a peak at 15–45 min and remain elevated for four to eight hours. The release of ACTH and cortisol is biphasic with 0.3 and 30 μg/kg doses, suggesting initial secretion of a readily-releasable pool of ACTH followed later by secretion of a second pool of newly synthesized and/or processed peptide. The disappearance curve of immunoreactive (IR) CRF from plasma is bi-exponential, with a half-life of 11.6 min for the fast and 73 min for the slow component (Schulte *et al.,* 1984). CRF is specific for ACTH and does not affect levels of prolactin, growth hormone, thyrotropin or gonadotropins. It is not yet known whether shorter analogues of CRF-41 retain biological activity. Human CRF differs from that in sheep in seven of 41 amino acids. It is metabolized more rapidly than ovine CRF and, in the dose range 0.5–5 μg/kg body weight, produces brief monophasic pulses of ACTH and cortisol which more closely resemble those occurring spontaneously or under stress (Schuermeyer *et al.,* 1984).

CRF-41 is under intensive scrutiny as a test of hypothalamic pituitary function. Patients with apparent ACTH deficiency often show an adequate ACTH response to CRF, suggesting that the principal defect is hypothalamic (Lytras *et al.,* 1984). The ACTH response to CRF in Cushing's disease is of considerable interest in elucidating its pathophysiology and a possible role in the differential diagnosis of Cushing's syndrome has been sought. On the basis of current hypotheses of the pathogenesis of Cushing's disease the following responses might be predicted:

(1) With generalized hyperplasia or normal architecture of corticotrophs in the anterior lobe, exogenous CRF might be expected to give an exaggerated

ACTH and cortisol response. This assumes that the anterior lobe corticotrophs are under increased endogenous CRF drive.

(2) In Cushing's disease associated with tumours of anterior lobe corticotrophs and atrophy of non-tumorous corticotrophs, exogenous CRF might give impaired, normal or exaggerated ACTH responses, depending on the number and affinity of CRF receptors on the tumour cells. This assumes that endogenous CRF secretion is decreased so that non-tumorous corticotrophs are suppressed and unresponsive to a single dose of CRF.

(3) In Cushing's disease associated with hyperplasia or tumour of intermediate lobe cells, exogenous CRF might give an impaired ACTH response since these cells appear to be primarily under neural rather than CRF-mediated vascular control. However, it is not clear whether the normal corticotrophs of the anterior lobe would be capable of responding to exogenous CRF.

In Cushing's syndrome due to an adrenal tumour, an impaired ACTH response would be predicted, as pituitary corticotrophs are suppressed by the decreased endogenous CRF. In the ectopic ACTH syndrome a variety of responses might be predicted: if the extrapituitary tumour lacks CRF receptors and does not secrete CRF, then pituitary corticotrophs would be suppressed and no response to exogenous CRF would be expected. However, some tumours are capable of producing CRF-like substances leading to hyperplasia of pituitary corticotrophs, and here the response might be exaggerated. It would seem unlikely, therefore, that the interpretation of CRF tests in Cushing's syndrome will be simple. Preliminary clinical investigations in Cushing's syndrome using CRF doses of 1 μg/kg or 100–500 μg i.v. have suggested that responses are variable (Nakahara *et al.*, 1983; Chrousos *et al.*, 1984). Most patients with Cushing's disease have shown an exaggerated incremental and peak ACTH response but impaired and normal responses have also been reported, with no apparent relation to duration or severity of disease or to basal cortisol concentrations. For example, a patient with Cushing's disease due to bilateral nodular hyperplasia with low and non-suppressible ACTH levels showed no ACTH response to 200 μg CRF (Pieters *et al.*, 1983). A patient with Cushing's disease with biochemical findings suggestive of an intermediate lobe adenoma (subnormal response to metyrapone, suppression with bromocriptine) had only a modest response to CRF. Clearly, further work correlating pituitary histopathology and responses to conventional dynamic tests with responses to CRF are required. Most patients with ectopic ACTH syndrome have shown no response to CRF but there has been at least one exception (Lytras *et al.*, 1984). It will be important to establish whether such responders also respond to metyrapone and/or dexamethasone. In patients with adrenal tumours, the response was impaired or absent, in accordance with theoretical predictions. So far, therefore, it does not seem that this type of CRF test will provide clear-cut diagnostic discrimination in Cushing's syndrome. As Orth (1984) has emphasized, to be of major benefit the CRF test must replace, and not merely add to, the expensive and time-consuming inpatient tests now performed.

It is also clear that the role of other substances with ACTH-releasing activity must be assessed with CRF. For example, arginine vasopressin (VP) stimulates ACTH release, although with much less potency than CRF. There is strong evidence from *in vitro* studies that vasopressin (and maybe other factors) acts synergistically with CRF (Gillies, Lenton and Lowry, 1982). Both CRF and VP are localized together in neurones of the paraventricular nucleus, which is functionally related to the control of ACTH secretion in the rat. Both peptides can reach the hypophyseal portal system via the parvocellular division of the neurohypophyseal tract at the median eminence. Concentrations of CRF and VP in portal blood are much higher than in peripheral blood and a direct interaction can occur at pituitary level. Recently, it has been shown that in normal females VP 10 IU i.m. augmented the ACTH response to CRF (1 μg/kg i.v.) by a factor of five (Liu *et al.*, 1983). These considerations may be important in the pathogenesis of Cushing's disease and in devising the most discriminating differential diagnostic tests in Cushing's syndrome.

The role of CRF in tests employing stimuli thought to act at hypothalamic level (glucagon, insulin hypoglycaemia) also remains uncertain. In most patients with Cushing's disease, the ACTH response to hypoglycaemia, although not to CRF, is impaired. This could mean that secretion of CRF is inhibited by chronic glucocorticoid excess or that mediators other than CRF are required for ACTH release.

Finally, the direct measurement of CRF in peripheral blood may in theory provide valuable information on the pathophysiology of Cushing's disease. Thus far, radioimmunoassays of ovine CRF do not detect human hypothalamic CRF, so that it is not yet known whether peripheral venous concentrations can be used as indices of pituitary portal venous levels; however, they may be of value in detecting cases of ectopic CRF secretion.

Other tests

A variety of other tests have been used in the investigation of Cushing's syndrome. Lysine vasopressin and pyrogen both have unpleasant side-effects and should not be used for routine diagnostic purposes. The insulin hypoglycaemia test is still useful occasionally in helping to distinguish Cushing's syndrome from situations mimicking the syndrome. A normal cortisol response occurs typically only in the latter. Unfortunately, occasional anomalous responses diminish its diagnostic value, although it may still be useful in evaluating patients with severe depression and elevated corticosteroid levels which fail to suppress with dexamethasone. In a review of nine studies involving 74 patients with Cushing's syndrome, Crapo (1979) reported that 18% of patients had a normal cortisol response to hypoglycaemia. Furthermore, normal ACTH responses return after bilateral adrenalectomy in patients with Cushing's disease. ACTH stimulation tests have little value in the differential diagnosis of Cushing's syndrome. Patients with Cushing's disease tend to have exaggerated cortisol responses to pharmacological doses of exogenous ACTH, whereas some patients with adrenal tumours have impaired responses, but the overlap is considerable.

Selective venous sampling with ACTH measurements

Because of the importance and occasional difficulty in distinguishing Cushing's disease from ectopic ACTH syndrome, selective venous catheterization with ACTH measurements has been used to localize the source of ACTH (Findling *et al.*, 1981; Drury *et al.*, 1982). Identification of a pituitary source requires sampling as near to the pituitary as possible, e.g. at the inferior petrosal sinus, with simultaneous sampling of peripheral venous blood as a control since ACTH secretion from pituitary and ectopic sources may be intermittent and pulsatile. This is discussed further in Chapter 7.

CAT scanning may be of value in locating occult ACTH-producing tumours, for example in the thymus, where conventional mediastinal radiographic tomography may be misleading because of mediastinal adiposity, or when searching for small tumours of the lung or pancreas, or an adrenal phaeochromocytoma.

Alcohol-induced (pseudo) Cushing's syndrome

Physical and biochemical features of Cushing's syndrome can occur in patients with chronic alcohol abuse which disappear as ethanol is eliminated (Rees *et al.*, 1977). The biochemical abnormalities described include markedly elevated plasma corticosteroids, loss of circadian rhythm of cortisol production, resistance to overnight and low-dose dexamethasone and elevated urine 17-OGS and 17-OS. The cortisol response to insulin hypoglycaemia is usually normal but is impaired in 20% of chronic alcoholics. Liver function tests are variable but evidence of liver damage is usually slight (Lamberts *et al.*, 1979).

The underlying mechanism is unclear; alcohol-mediated ACTH secretion and/or impaired metabolism of cortisol by the liver have been proposed. ACTH levels are either normal or suppressed. Occasionally, a prolonged half-life of cortisol is found. In some patients, alcohol ingestion can cause severe hyperlipidaemia and consequent interference in the fluorogenic corticosteroid assay. The main problem is to recognize the possibility that alcohol can cause florid signs of Cushing's syndrome. A specimen for blood ethanol is recommended. The clinical and biochemical abnormalities resolve within 3–4 weeks of stopping alcohol intake.

Strategy for diagnosis

Clearly, all the available tests have their problems. In view of the expense of prolonged investigation in hospital and the need for a rapid diagnosis, many centres have moved towards using a small number of the most discriminating tests, some of which can be performed on an outpatient basis. The following represents a personal view.

Outpatient tests

(1) The overnight dexamethasone suppression test is very useful. Normal suppression of plasma cortisol at 0900 hours, after dexamethasone 1.5 mg orally at 2300–2400 hours the previous night, virtually rules out the diagnosis of Cushing's syndrome. Patients who have inadequate suppression should be further investigated. The cut-off plasma cortisol level should be determined locally, but is between 100 and 200 nmol/l (3.6 and 7.2 μg/dl). Alternatively, the overnight urinary free cortisol test can be used for screening outpatients.

(2) Urinary free cortisol determination is the best single test to establish the diagnosis in patients with abnormal screening tests. Ideally several (at least three) 24-hour urine collections should be made, as undue reliance on a single

specimen may be misleading. A specific precise radioimmunoassay for cortisol is recommended, with strict quality control using urine pools, although CPB and fluorimetric methods have also been used successfully.

If the diagnosis remains in doubt, serial determination of urinary free cortisols is indicated, ideally while the patient is not receiving any drugs likely to interfere with cortisol metabolism. It may be necessary to study a patient over a period of several months, repeating the test at regular intervals, if intermittent Cushing's syndrome is suspected. Only when the diagnosis of Cushing's syndrome is securely established are further tests indicated.

Inpatient tests

(1) Differential diagnosis between ACTH-dependent and independent causes is best made by plasma ACTH measurements on at least three specimens, including one at 2300 hours. A sensitive assay (detection limit 10 ng/l; 2.2 pmol/l) for ACTH is required for this and it must be specific for ACTH within the low normal range. Few commercial kits are reliable for this purpose since most show non-specific plasma effects at low levels. We have therefore continued to extract ACTH from plasma before assay, a step which increases the effective sensitivity and removes non-specific interfering factors. The plasma ACTH assay should clearly identify patients with an adrenal tumour (ACTH levels below 10 ng/l; 2.2 pmol/l) but will not distinguish Cushing's disease from the ectopic ACTH syndrome unless the level is very high (more than 300 ng/l; 66 pmol/l). It is also unusual for ACTH levels in the ectopic ACTH syndrome to be in the low normal range, e.g. 20–50 ng/l (4.4–11 pmol/l).

(2) High dose dexamethasone suppression (two control days, two test days) is useful to distinguish Cushing's disease from the ectopic ACTH syndrome. As previously discussed, anomalous results occur occasionally, particularly in Cushing's disease due to nodular adrenal hyperplasia and ectopic ACTH secretion due to a bronchial carcinoid or thymoma. In order to shorten the period of investigation, specimens for basal ACTH and also plasma cortisol are taken during the control days of the dexamethasone test.

Other tests have a role under certain circumstances. In patients with increased urinary free cortisol and possible pseudo-Cushing's syndrome, the diagnosis can be aided by the insulin hypoglycaemia test and the differential diagnosis between Cushing's disease and ectopic ACTH syndrome by the metyrapone test, using serum deoxycortisol and ACTH measurement at 0800 hours after metyrapone at midnight. Rarely, selective venous sampling may help.

The demonstration of hypokalaemic alkalosis remains a valuable sign of ectopic ACTH production so that plasma electrolytes should be measured early in the diagnostic protocol.

Acknowledgement

The author is grateful to Mrs Pauline Bullock for excellent secretarial assistance.

References

AL-DUJAILI, E. A. S., HOPE, J., ESTIVARIZ, F. E., LOWRY, P. J. and EDWARDS, C. R. W. (1981) Circulating human pituitary pro-γ-melanotropin enhances the adrenal response to ACTH. *Nature,* **291,** 156–158

ARON, D. C., FINDLING, J. W., TYRRELL, J. B., FITZGERALD, P. A. and FORSHAM, P. H. (1980) The overnight high dose dexamethasone suppression test: a rapid method for differential diagnosis in Cushing's syndrome. *Hormone Research,* **13,** 334

ARON, D. C., TYRRELL, J. B., FITZGERALD, P. A., FINDLING, J. W. and FORSHAM, P. H. (1981) Cushing's syndrome: problems in diagnosis. *Medicine,* **60,** 25–35

BROWN, R. D., VAN LOON, G. R., ORTH, D. N. and LIDDLE, G. W. (1973) Cushing's disease with periodic hormonogenesis: one explanation for paradoxical response to dexamethasone. *Journal of Clinical Endocrinology and Metabolism,* **36,** 445–451

BURKE, C. W. and BEARDWELL, C. G. (1973) Cushing's syndrome: an evaluation of the clinical usefulness of urinary free cortisol and other urinary steroid measurements in diagnosis. *Quarterly Journal of Medicine,* New Series **42,** 175–204

CAREY, R. M. (1980) Suppression of ACTH by cortisol in dexamethasone-non-suppressible Cushing's disease. *New England Journal of Medicine,* **302,** 275–279

CARO, J. F., MEIKLE, A. W., CHECK, J. H. and COHEN, S. N. (1978) 'Normal suppression' to dexamethasone in Cushing's disease: an expression of decreased metabolic clearance for dexamethasone. *Journal of Clinical Endocrinology and Metabolism,* **47,** 667–670

CHAN, J. S. D., SEIDAH, N. G. and CHRETIEN, M. (1983) Measurement of *N*-terminal (1–76) of human pro-opiomelanocortin in human plasma: correlation with adrenocorticotropin. *Journal of Clinical Endocrinology and Metabolism,* **56,** 791–796

CHROUSOS, G. P., SCHULTE, H. M., OLDFIELD, E. H., GOLD, P. W., CUTLER, G. B. and LORIAUX, D. L. (1984) The corticotropin releasing factor stimulation test. *New England Journal of Medicine,* **310,** 622–626

CONNOLLY, C. K., GORE, M. B. R., STANLEY, N. and WILLS, M. R. (1968) Single dose dexamethasone suppression in normal subjects and hospital patients. *British Medical Journal,* **2,** 665–667

COPE, C. L. and BLACK, E. G. (1958) The production rate of cortisol in man. *British Medical Journal,* **1,** 1020–1024

CRAPO, L. (1979) Cushing's syndrome: a review of diagnostic tests. *Metabolism,* **28,** 955–977

DEBOLD, C. R., DECHERNEY, G. S., JACKSON, R. V. *et al.* (1983) Effect of synthetic ovine corticotropin releasing factor: prolonged duration of action and biphasic response of plasma adrenocorticotropin and cortisol. *Journal of Clinical Endocrinology and Metabolism,* **57,** 294–298

DORNHURST, A., JENKINS, J. S., LAMBERTS, S. W. J. *et al.* (1983) The evaluation of sodium valproate in the treatment of Nelson's syndrome. *Journal of Clinical Endocrinology and Metabolism,* **56,** 985–991

DRURY, P. L., RATTER, S., TOMLIN, S. *et al.* (1982) Experience with selective venous sampling in diagnosis of ACTH-dependent Cushing's syndrome. *British Medical Journal,* **284,** 9–12

ESTIVARIZ, F. E., ITURRIZA, F., MCLEAN, C., HOPE, J. and LOWRY, P. J. (1982) Stimulation of adrenal mitogenesis by *N*-terminal pro-opiocortin peptides. *Nature,* **297,** 419–421

FINDLING, J. W., ARON, D. C., TYRRELL, J. B. *et al.* (1981) Selective venous sampling for ACTH in Cushing's syndrome. *Annals of Internal Medicine,* **94,** 647–652

GILKES, J. J. H., REES, L. H. and BESSER, G. M. (1977) Plasma immunoreactive corticotrophin and lipotrophin in Cushing's syndrome and Addison's disease. *British Medical Journal,* **2,** 996–998

GILLIES, G. E., LENTON, E. A. and LOWRY, P. J. (1982) Corticotropin releasing activity of the new CRF is potentiated several times by vasopressin. *Nature,* **299,** 355–357

GOLD, E. M. (1979) The Cushing syndromes: changing views of diagnosis and treatment. *Annals of Internal Medicine,* **90,** 829–844

GRAY, C. E. and RATCLIFFE, J. G. (1979) Clinical evaluation of a radioimmunoassay for β-MSH related peptides (lipotrophins) in human plasma. *Clinical Endocrinology,* **10,** 163–172

HALE, A. C., RATTER, S. J., TOMLIN, S. J., LYTRAS, N., BESSER, G. M. and REES, L. H. (1984) Measurement of immunoreactive γ-MSH in human plasma. *Clinical Endocrinology,* **21,** 139–148

HORROCKS, P. M. and LONDON, D. R. (1982) Diagnostic value of 9 a.m. plasma adrenocorticotrophic hormone concentrations in Cushing's disease. *British Medical Journal,* **285,** 1302–1303

HOWLETT, T. A., DRURY, P. L., REES, L. H. and BESSER, G. M. (1984) Cushing's syndrome due to ectopic ACTH production: 15 years' experience. *Third Joint Meeting of British Endocrine Societies (Edinburgh),* March 1984, abstract 81

JUBIZ, W., MEIKLE, A. W., WEST, C. D. and TYLER, F. H. (1970) Single dose metyrapone test. *Archives of Internal Medicine,* **125,** 472–474

KOBAYASHI, Y., MIYAI, K., TSUBOTA, N. and WATANABE, F. (1979) Direct fluorescence polarization immunoassay of serum cortisol. *Steroids,* **34,** 829–834

KRIEGER, D. T., AMOROSA, L. and LINICK, F. (1975) Cyproheptadine-induced remission of Cushing's disease. *New England Journal of Medicine,* **293,** 893–896

LAMBERTS, S. W. J., KLIJN, J. G. M., DE JONG, F. H. and BIRKENHÄGER, J. C. (1979) Hormone secretion in alcohol-induced pseudo-Cushing's syndrome. *Journal of the American Medical Association,* **242,** 1640–1643

LAMBERTS, S. W. J., DE LANGE, S. A. and STEFANKO, S. Z. (1982) Adrenocorticotropin secreting pituitary adenomas originate from the anterior or the intermediate lobe in Cushing's disease: differences in the regulation of hormone secretion. *Journal of Clinical Endocrinology and Metabolism,* **54,** 286–291

LAMBERTS, S. W. J., TIMMERMANS, H. A. T., DE JONG, F. H. and BIRKENHÄGER, J. C. (1977) The role of dopaminergic depletion in the pathogenesis of Cushing's disease and the possible consequences for medical therapy. *Clinical Endocrinology,* **7,** 185–193

LEVELL, M. J. (1980) Effects of analytical error on the clinical discrimination of the urinary 11-hydroxycorticosteroid assay. *Annals of Clinical Biochemistry,* **17,** 237–240

LIDDLE, G. W. (1960) Tests of pituitary–adrenal suppressibility in the diagnosis of Cushing's syndrome. *Journal of Clinical Endocrinology and Metabolism,* **20,** 1539–1560

LIDDLE, G. W., ESTEP, H. L., KENDALL, J. W., WILLIAMS, W. C. and TOWNES, A. S. (1959) Clinical application of a new test of pituitary reserve. *Journal of Clinical Endocrinology and Metabolism,* **19,** 875–894

LIU, J. H., MUSE, K., CONTRERAS, P. *et al.* (1983) Augmentation of ACTH-releasing activity of synthetic corticotropin releasing factor by vasopressin in women. *Journal of Clinical Endocrinology and Metabolism,* **57,** 1087–1089

LOWRY, P. J., GILLIES, G., HOPE, J. and JACKSON, S. (1980) Structure and biosynthesis of peptides related to corticotrophin and lipotrophin. *Hormone Research,* **13,** 201–210

LYTRAS, N., GROSSMAN, A., PERRY, L. *et al.* (1984) Corticotrophin releasing factor: responses in normal subjects and patients with diseases of the hypothalamus and pituitary. *Clinical Endocrinology,* **20,** 71–84

MATTINGLY, D. and TYLER, C. (1976) Overnight urinary 11-hydroxycorticosteroid estimations in diagnosis of Cushing's syndrome. *British Medical Journal,* **3,** 668–669

MCNICOL, A. M. (1981) Patterns of corticotropic cells in the adult human pituitary in Cushing's disease. *Diagnostic Histopathology,* **4,** 335–341

NAKAHARA, M., SHIBASAKI, T., SHIZUME, K. *et al.* (1983) Corticotropin releasing factor test in normal subjects and patients with hypothalamic–pituitary–adrenal disorders. *Journal of Clinical Endocrinology and Metabolism,* **57,** 963–968

NICHOLSON, W. E., DAVIS, D. R., SHERRELL, B. J. and ORTH, D. N. (1984) Rapid radioimmunoassay for corticotropin in unextracted human plasma. *Clinical Chemistry,* **30,** 259–265

ORTH, D. N. (1984) The old and the new in Cushing's syndrome. *New England Journal of Medicine,* **310,** 649–651

PERRY, L. A., DRURY, P. L., REES, L. H. and BESSER, G. M. (1984) Measurement of 11-deoxycortisol during the metyrapone test in the differential diagnosis of Cushing's syndrome. *Third Joint Meeting of British Endocrine Societies (Edinburgh),* March 1984, abstract 184

PETERS, J. R., WALKER, R. F., RIAD-FAHMY, D. and HALL, R. (1982) Salivary cortisol assays for assessing pituitary–adrenal reserve. *Clinical Endocrinology,* **17,** 583–592

PIETERS, G. F. F. M., HERMUS, A. R. M. M., SMALS, A. G. H., BARTELINK, A. K. M., BENRAAD, T. J. and KLOPPENBORG, P. W. C. (1983) Responsiveness of the hypophyseal–adrenocortical axis to corticotropin releasing factor in pituitary dependent Cushing's disease. *Journal of Clinical Endocrinology and Metabolism,* **57,** 513–516

PIETERS, G. F. F. M., SMALS, A. G. H., GOVERDE, H. J. M. and KLOPPENBORG, P. W. C. (1982a) Paradoxical responsiveness of adrenocorticotropin and cortisol to thyrotropin releasing hormone in pregnant women: evidence for intermediate lobe activity? *Journal of Clinical Endocrinology and Metabolism,* **55,** 387–389

PIETERS, G. F. F. M., SMALS, A. G. H., GOVERDE, H. J. M., PESMAN, G. J., MEYER, E. and KLOPPENBORG, P. W. C. (1982b) Adrenocorticotropin and cortisol responsiveness to thyrotropin releasing hormone and luteinizing hormone releasing hormone discloses two subsets of patients with Cushing's disease. *Journal of Clinical Endocrinology and Metabolism,* **55,** 1188–1197

RATCLIFFE, W. A. (1983) Direct (non-extraction) serum assays for steroids. In *Immunoassays for Clinical Chemistry,* 2nd Edn., edited by W. M. Hunter and J. E. T. Corrie, pp. 401–409. Edinburgh: Churchill Livingstone

RATCLIFFE, J. G. and EDWARDS, C. R. W. (1971) The extraction of adrenocorticotrophin and arginine vasopressin from human plasma by porous glass. In *Radioimmunoassay Methods,* edited by K. E. Kirkham and W. M. Hunter, pp. 502–512. Edinburgh: Livingstone

RATTER, S. J., GILLIES, G., HOPE, J. *et al.* (1983) Pro-opiocortin related peptides in human pituitary and ectopic ACTH secreting tumours. *Clinical Endocrinology,* **18,** 211–218

REES, L. H., BESSER, G. M., JEFFCOATE, W. J., GOLDIE, D. J. and MARKS, V. (1977) Alcohol-induced pseudo-Cushing's syndrome. *Lancet*, **1**, 726–728

REES, L. H. and LOWRY, P. J. (1979) Adrenocorticotrophin and lipotrophin. In *Hormones in Blood*, 3rd Edn., vol. 3, edited by C. H. Gray and V. H. T. James, pp. 130–179. London: Academic Press

RIAD-FAHMY, D., READ, G. F. and HUGHES, I. A. (1983) Corticosteroids. In *Hormones in Blood*, 3rd Edn., vol. 4, edited by C. H. Gray and V. H. T. James, pp. 285–315. London: Academic Press

RUDD, B. T. (1983) Urinary 17-oxogenic and 17-oxosteroids, a case for deletion from the clinical chemistry repertoire. *Annals of Clinical Biochemistry*, **20**, 65–71

SAENGER, P. and PETERSON, R. E. (1984) Usefulness of urinary 6β-hydroxycortisol excretion in the diagnosis of Cushing's syndrome. *Seventh International Congress of Endocrinology*, abstract 2215. Amsterdam: Excerpta Medica

SCHTEINGART, D. E. and MCKENZIE, A. K. (1980) Twelve hour cycles of adrenocorticotropin and cortisol secretion in Cushing's disease. *Journal of Clinical Endocrinology and Metabolism*, **51**, 1195–1198

SCHUERMAYER, T. H., AVGERINOS, P. C., GOLD, P. W. *et al.* (1984) Human corticotropin releasing factor: dose-response and time-course of ACTH and cortisol in man. *Seventh International Congress of Endocrinology*, abstract 2068. Amsterdam: Excerpta Medica

SCHULTE, H. M., CHROUSOS, G. P., BOOTH, J. D. *et al.* (1984) Corticotropin releasing factor: pharmacokinetics in man. *Journal of Clinical Endocrinology and Metabolism*, **58**, 192–196

SINDLER, B. H., GRIFFING, G. T. and MELBY, J. C. (1983) The superiority of the metyrapone test versus the high dose dexamethasone test in the differential diagnosis of Cushing's syndrome. *American Journal of Medicine*, **74**, 657–662

SMALS, A. G. H., PIETERS, G. F. F. M., VAN HAELST, U. J. G. and KLOPPENBORG, P. W. C. (1984) Macronodular adrenocortical hyperplasia in long standing Cushing's disease. *Journal of Clinical Endocrinology and Metabolism*, **58**, 25–31

STREETEN, D. H. P., STEVENSON, C. T., DALAKOS, T. G. *et al.* (1969) The diagnosis of hypercortisolism. *Journal of Clinical Endocrinology and Metabolism*, **29**, 1191–1211

VALE, W., SPIESS, J., RIVIER, C. and RIVIER, J. (1981) Characterization of 41-residue ovine hypothalamic peptide that stimulates secretion of corticotrophin and β-endorphin. *Science*, **213**, 1394–1397

VOCCIA, E., SAENGER, P., PETERSON, R. E. *et al.* (1979) 6β-Hydroxycortisol excretion in hypercortisolemic states. *Journal of Clinical Endocrinology and Metabolism*, **48**, 467–471

WALKER, M. S. (1979) Screening for Cushing's syndrome using early morning urine samples. *Annals of Clinical Biochemistry*, **16**, 86–88

WALKER, R. F., RIAD-FAHMY, D. and READ, G. F. (1978) Adrenal status assessed by direct radioimmunoassay of cortisol in whole saliva or parotid fluid. *Clinical Chemistry*, **24**, 1460–1463

10
Interactions between adrenal cortex and medulla

C. Weinkove and D. C. Anderson

INTRODUCTION

There have been some excellent reviews in recent years on selected aspects of interactions between the adrenal cortex and medulla (Pohorecky and Wurtman, 1971; Carballeira and Fishman, 1980; Ungar and Phillips, 1983). However, most have been restricted to a single aspect of this relationship, for example the effect of glucocorticoids on catecholamine synthesis and secretion.

We attempt here to examine a wider range of established and potential sites of interaction. As well as the direct interactions between the adrenal cortex and medulla and their secretory products, we have considered their interactions on target tissue and organs and feedback control of their secretion.

It is evident that physiological and biochemical responses to the products of the two components of the adrenal gland are integrated in a complex manner. We have tried to identify some of the more important biological consequences of these interactions in both health and disease in man, but do not claim that this review is comprehensive. When considering interactions in peripheral tissues, we have tried to distinguish between catecholamines of local neural origin (mainly norepinephrine) and circulating catecholamines from the adrenal medulla (predominantly epinephrine). We are mainly concerned here with the latter and must stress the major difference in the mechanism of action of adrenal steroids and catecholamines. In general, catecholamines activate adenylate cyclase within minutes, while adrenal steroids promote new protein synthesis and take hours to days to produce their effects.

ANATOMICAL RELATIONSHIPS BETWEEN ADRENAL CORTEX AND MEDULLA

Embryology

Although the concentric arrangement of adrenal cortex surrounding the medulla is found only in eutherian mammals among vertebrate classes, varying degrees of intermingling of the respective components are evident in protherian mammals, as well as in birds, reptiles and amphibians (Carballeira and Fishman, 1980). In the early stages of embryonic growth, immature cortical cells, of mesodermal origin, coalesce into compact masses on either side of the aorta. Meanwhile immature chromaffin cells (phaeochromoblasts), of ectodermal origin, detach themselves from the sympathetic primordia, penetrate the cortical buds, and coalesce centrally to form the medulla. As they mature, they attain intense chromoaffinity. The later development and zonation of the cortex is considered in Chapter 2.

Little is known of what attracts the adrenal medullary precursor cells to the primitive cortex. However, there is little doubt that glucocorticoids play an important part in medullary cell maturation. Thus, culture of fetal rat sympathetic ganglia cells at a sufficiently early stage, in the presence of high levels of glucocorticoids, has been shown to produce permanent changes in their appearance and probably in their catecholamine content. Neonatal rats possess para-aortic nests of chromaffin positive cells (the organ of Zuckerkandl) which rapidly atrophy after birth (Lempinen, 1966). These, however, can be maintained by administration of high doses of glucocorticoids in the neonatal period (Eränkö, Lempinen and Räisänen, 1966, 1967). Further support for a role of glucocorticoids in shaping the future of the phaeochromoblasts comes from experiments in which sympathetic tissue from animals of various ages has been cultured with and without glucocorticoids. Although changes such as the induction of the enzyme phenylethanolamine *N*-methyltransferase (*see below*) occur later, it appears that only early stem cells (phaeochromoblasts) can be induced to have the appearance and features typical of adrenal medullary cells (Coupland, 1965).

Mature anatomy

Figure 10.1 shows in stylized form some of the important relationships between the mammalian adrenal cortex, medulla and vasculature, first studied in detail by Symington and colleagues (Symington, 1969). The adrenal glands each receive arterial blood from three arteries, which give rise to a network of coalescing small arteries over the surface of the gland. Straight capillaries arise from this arterial plexus and descend through the zona fasciculata to the reticularis, where they form a further plexus. Where there is a medulla (in the head and body of the gland) these capillaries provide the major blood supply for it, although some arterioles descend from the capsular plexus to supply the medulla directly.

From the medullary capillaries, venules rich in elastic fibres drain into the central vein, which is largely enveloped by a cuff of cortical tissue of unknown function (*Figure 10.1*). The central vein is also unusual in that its wall contains bundles of mainly longitudinal smooth muscle fibres. These are pierced by the medullary veins as they join the central vein. Contraction of the longitudinal muscle bundles may occlude the medullary veins and lead to damming up of blood in the medulla and adjacent cortex, exposing the inner zone to high concentrations of catecholamines. Periodic contraction and relaxation of the longitudinal muscle bundles might lead to pulsatile hormonal discharge as the venules fill and collapse.

Some workers (Symington, 1969; Neville and O'Hare, 1979) report little if any innervation of the cortex, although Dallman, Engeland and McBride (1977) supported some earlier findings of adrenocortical innervation (*see also* Chapter 1).

EFFECT OF THE ADRENAL CORTEX ON CATECHOLAMINE SYNTHESIS

Catecholamine pathway

The main steps involved in catecholamine biosynthesis are indicated in *Figure 10.2*. Starting with the amino acid tyrosine, there are four enzyme steps: tyrosine hydroxylase (TH), dihydroxyphenylalanine (DOPA) decarboxylase, dopamine β-hydroxylase (DBH), and phenylethanolamine *N*-methyltransferase (PNMT). These enzymes are found in the central nervous and peripheral sympathetic

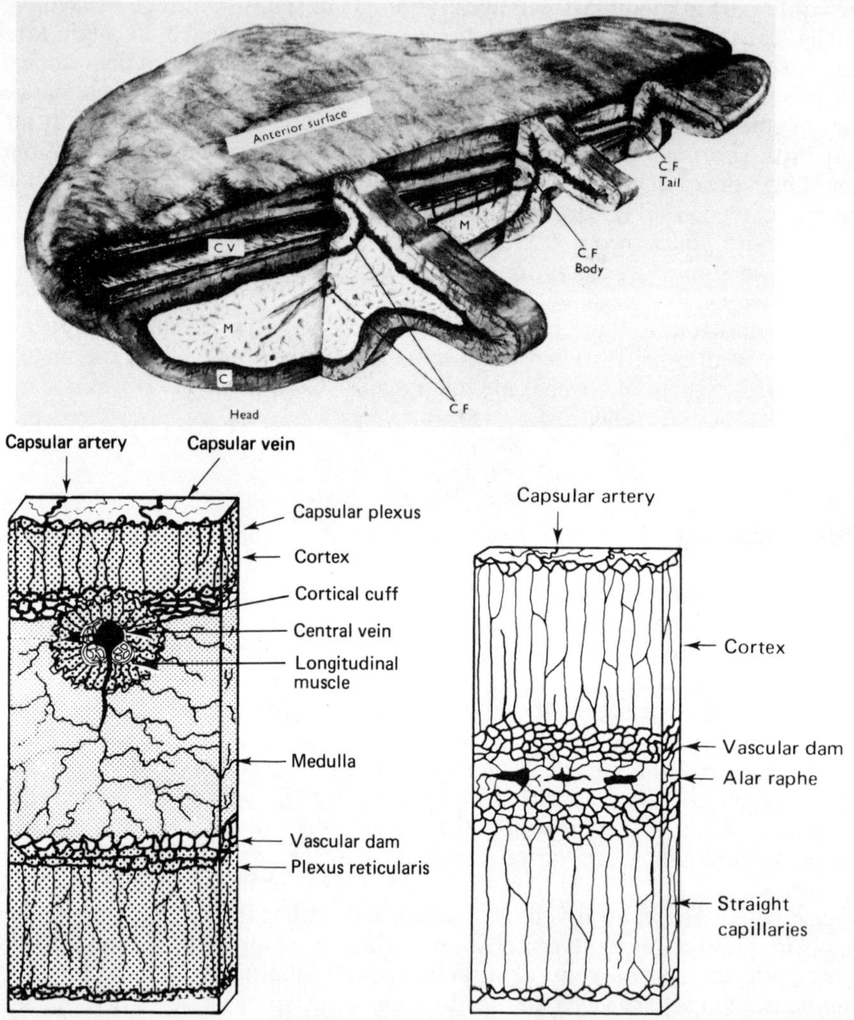

Figure 10.1 Anatomy of the normal human adrenal gland. Shown in cross-section are the capillaries from the capsule draining through the cortex (C) into a deeper plexus, before passing through the medulla (M) and entering the central vein (CV) between the bundles of longitudinal muscle. (*Bottom left*) Diagrammatic representation of structure in head. (*Bottom right*) Structure in tail, where there is no head. (Modified from Symington, 1969, and reproduced by permission of the Author and Publishers, *Journal of Endocrinology*)

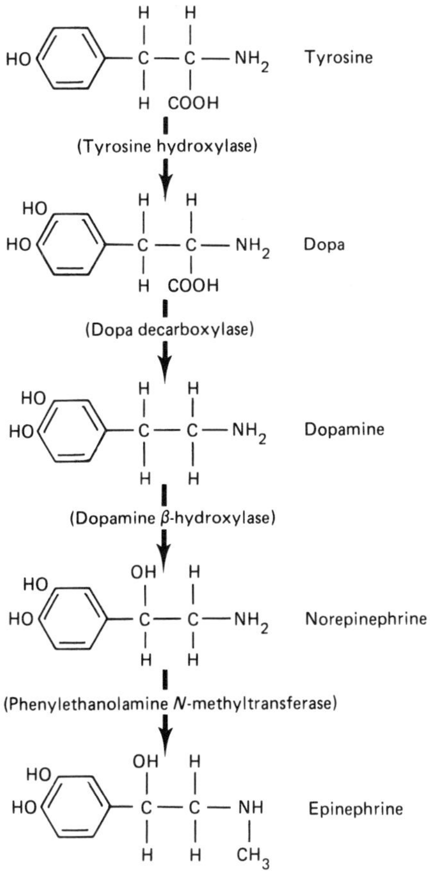

Tyrosine

(Tyrosine hydroxylase)

Dopa

(Dopa decarboxylase)

Dopamine

(Dopamine β-hydroxylase)

Norepinephrine

(Phenylethanolamine *N*-methyltransferase)

Epinephrine

Figure 10.2 The pathway and enzymes concerned in the biosynthesis of adrenal catecholamines

nervous systems, except for PNMT, which is largely restricted to the adrenal medulla, the major if not the sole source of epinephrine in blood.

There is good evidence that ACTH exerts significant effects on two of these enzymes, namely TH (which is rate-limiting for the whole pathway) and DBH (Weinshilboum and Axelrod, 1970), while PNMT activity depends on high local concentrations of corticosteroids (Axelrod, 1977).

After its formation by dopa decarboxylase, dopamine is actively transported by an ATP-dependent mechanism to the chromaffin storage granule, where it is acted upon probably in the granule by DBH. This is a copper-containing mixed-function oxidase which requires ascorbic acid and oxygen. The dependence of this enzyme on ascorbic acid may be of importance in cortico-medullary interactions (*see later*). The final enzyme, PNMT, is cystolic and shifts a methyl group from the donor *S*-adenosylmethionine to the amino nitrogen of norepinephrine to form epinephrine. This enzyme requires oxygen, and is sensitive to product inhibition.

Effects of glucocorticoids

That the adrenal cortex influences the nature of medullary secretions was suggested by the observation that the ratio of epinephrine to norepinephrine in the adrenal depended on the amount of cortex surrounding the medulla (Shepherd and West, 1951). Wurtman and Axelrod (1965) then showed that PNMT activity was markedly reduced in rats two to three weeks after hypophysectomy. This deficiency could be prevented or corrected by physiological doses of ACTH, or pharmacological doses of glucocorticoids. The latter will, of course, mimic the physiologically high steroid concentrations normally found in the adrenal medulla.

Later experiments with cultures of bovine adrenal medullary cells have confirmed these observations. Using their *in vitro* system, Hersey and Distefano (1979) concluded that cortisol maintained PNMT activity at a maximum level and that PNMT activity did not apparently fluctuate with variations in glucocorticoid concentration *in vivo*. Glucocorticoids can only initiate PNMT activity in rat extra-adrenal chromaffin tissue during the first five days of postnatal development (Ciaranello, Jacobowitz and Axelrod, 1973), and some other role must be sought for adrenal steroids once PNMT activity is established. Ciaranello (1978) has produced evidence that adrenal steroids produce their major effect on PNMT activity by reducing enzyme degradation. Apparently, the presence of the methyl donor cofactor *S*-adenosylmethionine prevents PNMT degradation, and adrenal steroids may be necessary for its synthesis. Ungar and Phillips (1983), however, concluded that glucocorticoids have other effects, including a direct influence on epinephrine output from the adrenal gland as well as induction of PNMT. Glucocorticoids also have a more general tropic effect on the medullary cells (Mikulášková-Rochová and Linèt, 1967), with effects on other catecholamine biosynthetic enzymes, such as tyrosine hydroxylase (Tischler *et al.,* 1983) and dopamine β-hydroxylase (Snider *et al.,* 1977).

The teleological argument has often been put forward that the medulla is inside the cortex because PNMT activity, and hence epinephrine synthesis, demands high concentrations of glucocorticoids. However, Hersey and Distefano (1979) have shown that with prolonged culture of bovine medullary cells, PNMT activity became less dependent on glucocorticoids. It is possible therefore that the sensitivity of PNMT to glucocorticoids is 'tuned down' *in vivo* so that the response occurs over the intra-adrenal rather than the systemic concentration range.

In the frog, PNMT of similar substrate specificity to the mammalian adrenal enzyme is not confined to the adrenal, but has a more general neurotransmitter role and is not dependent on glucocorticoids. Furthermore, although the adrenal medulla in mammals is the principal source of circulating epinephrine, specific nerve cells in the brain, for example in the locus coeruleus, also synthesize this catecholamine (Hökfelt *et al.,* 1973; Saavedra *et al.,* 1974). PNMT synthesis in such locations is responsive to levels of glucocorticoids found in the systemic circulation. It could be argued that the dependence of PNMT on glucocorticoids over the high concentration range seen within the adrenal may have evolved in order to link epinephrine synthesis to cortisol release. Thus, acute stress leads to epinephrine release from storage granules, and a virtually simultaneous pulse of cortisol secretion under the action of ACTH. This would then induce synthesis of more PNMT and accelerate epinephrine replacement.

Effect of ascorbic acid

Ascorbic acid has fascinating roles in both the cortex and the medulla. It is present in high concentration in the cortex, and is released from adrenal cortical cells upon stimulation by ACTH, acting through cyclic adenosine monophosphate (cAMP). It has long been held to be important in steroidogenesis, ascorbic acid being known to inhibit 21-hydroxylase activity (Cooper and Rosenthal, 1962). This effect can be reversed by epinephrine and other antioxidants (Kitabchi, 1967a), with ascorbate exerting a 'braking influence' on steroid biosynthesis (Kitabchi, 1967b).

Ascorbic acid is specifically taken up by adrenomedullary cells in culture by a sodium-dependent carrier transport mechanism (Diliberto, Heckman and Daniels, 1983). The ascorbate transport system in both adrenal cortex and medulla requires calcium ions, and can concentrate the vitamin 60- to 100-fold within the cell. The concentration is particularly high in the chromaffin vesicle, and probably another cytosolic compartment (Diliberto, Heckman and Daniels, 1983). Release with a variety of secretagogues was calcium-dependent, and 1.5 times greater than the content of the granules alone (Levine *et al.*, 1983).

The only known ascorbate-dependent reaction in chromaffin cells is dopamine β-hydroxylation. Regulation of the ascorbate concentration within chromaffin vesicles requires a system for the accumulation and concentration of ascorbate (*see above*), and a system for the reduction of the semidehydroascorbate generated by dopamine β-hydroxylation. Semidehydroascorbate may be reduced by cytochrome b-561 of the vesicle membrane, which is maintained in reduced form by cytosolic ascorbate and mitochondrial semidehydroascorbate reductase (Diliberto, Heckman and Daniels, 1983).

In the adrenal medulla ascorbate is released from chromaffin vesicles and cells by a nicotinic-receptor mediated action of acetylcholine. Glucocorticoids released from the adrenal cortex inhibit ascorbic acid transport into medullary cells (Levine *et al.*, 1983) so that much ascorbic acid secreted by the cortex apparently bypasses the medulla. The significance of ascorbate concentration and release by the adrenal cortex on medullary function is therefore still obscure. In addition to the effect on the enzyme DBH, it is possible that high ascorbate concentration in the medullary cells is important in preventing oxidation of epinephrine to adrenochrome, which would inhibit 11β-hydroxylation within the medulla (Sweat and Bryson, 1965; Carballeira and Fishman, 1980).

The clinical significance of ascorbate deficiency or excess on steroid synthesis is uncertain. Liakakos *et al.* (1975) measured plasma cortisol concentrations in healthy children before and after an injection of depot synthetic ACTH. After five days of continuous ascorbic acid administration (1 g three times a day), the basal cortisol concentration had not changed but the cortisol response to synthetic ACTH was blunted. This would provide some support for an inhibitory role for ascorbate in steroidogenesis.

EFFECTS OF THE ADRENAL MEDULLA ON CORTICAL FUNCTION

Possible cooperation of the medulla on cortisol biosynthesis

Although most workers have assumed that the cortex alone is responsible for steroid biosynthesis, work by Carballeira and Venning (1964) suggests that the medulla may have the ability to complete the process of glucocorticoid biosynthesis. They incubated slices of medullary phaeochromocytoma, free of adrenocortical contamination, with C-21 and other steroid precursors and found that the medulla converted steroids on the pathway from pregnenolone onwards to cortisol. Further studies are needed, since this might indicate an important role for the medulla in preventing steroid precursors from flooding into the circulation during periods of stress.

Local influence of medullary catecholamines on cortical function

It has been shown that addition of epinephrine to adrenocortical mitochondria inhibits cortisol synthesis by reducing 11β-hydroxylase activity (Sweat and Bryson, 1965). Addition of reducing agents such as ascorbic acid prevented this effect, suggesting that inhibition was caused by an oxidation product of epinephrine, probably adrenochrome. Conceivably, therefore, exposure to adrenochrome might contribute to the low level of 11β-hydroxylase activity in the zona reticularis.

It is possible to remove the adrenal medulla in rats and allow the adrenal cortex to regenerate without any medullary tissue. Torda and Kvetňanský (1983) have demonstrated that these medullectomized animals maintained normal basal plasma corticosteroid concentrations, but had a reduced steroid response to immobilization stress. Soliman and Kolta (1981) showed that, while epinephrine had no effect on intact adrenal glands *in vitro,* it stimulated steroidogenesis and corticosterone release in the regenerating medullectomized rat adrenal gland. They offered no explanation of the mechanism of action, but epinephrine may have acted by reversing ascorbic acid inhibition of 21-hydroxylase activity (Cooper and Rosenthal, 1962; Kitabchi, 1967a).

Sandberg *et al.* (1953) found that epinephrine infusion produced no change in basal corticosteroid secretion or metabolism in their human volunteers, although it did reduce the corticosteroid response to ACTH. Segre *et al.* (1966) agreed that epinephrine infusion did not effect the metabolic clearance of cortisol, but reported an increased cortisol production rate in normal men and women after epinephrine infusion. Animal experiments have not resolved these contradictory findings. Cushman, Alter and Hilton (1966) could not demonstrate any change in basal cortisol release when they perfused isolated dog adrenal pouches with catecholamines. They and others did not comment on the fact that catecholamine infusion inhibited ACTH stimulated cortisol release in their experimental model. Since in this experiment no reducing agent was used to prevent catecholamine oxidation, it is possible that they were examining the effect of adrenochromes which inhibited the adrenocortical response to ACTH.

In man, the interplay between adrenochrome, epinephrine and ascorbic acid might act as a local regulatory mechanism for the biosynthesis of the biologically active 11-oxygenated steroids in the adrenal gland. Interestingly, the adrenal medulla has also been shown to contain ACTH, albeit in low concentrations (Evans *et al.,* 1983), providing it with the agents to both stimulate (through ACTH and epinephrine) and inhibit (via adrenochrome formation) steroid biosynthesis. The association between phaeochromocytoma and Cushing's syndrome is discussed later.

Systemic effects of adrenal catecholamines on hypothalamo–pituitary–adrenal function

It is difficult to distinguish the effects on the pituitary of catecholamines from the median eminence (arriving via the hypophyseal portal blood supply) from those of adrenal catecholamines delivered by the systemic circulation. It seems possible, however, that circulating epinephrine levels in conditions of acute severe stress are high enough at 10^{-8} mol/l (Little *et al.*, 1984) to have a significant effect on the pituitary gland. In fact, epinephrine injection has been used in man and animals as a test for an intact pituitary–adrenal axis, with the fall in circulating eosinophils being used as a sensitive indicator (Sandberg *et al.*, 1953).

Giguère, Côté and Labrie (1982) have recently shown that cultured rat pituitary cells possess α-receptors, and that stimulation of these receptors by an epinephrine concentration of 3×10^{-8} mol/l leads to direct release of ACTH into the culture medium. Pre-incubation of pituitary cells with dexamethasone or other glucocorticoid analogues, but not with sex steroids, inhibited this response. This is therefore opposite to the stimulatory effect exhibited by glucocorticoids on β-receptor function in other systems (*see below*).

In contrast to the direct effect on pituitary ACTH release, norepinephrine has been shown to inhibit corticotropin release from the rat hypothalamus (Jones and Hillhouse, 1977), while epinephrine stimulates corticotropin release from the intermediate lobe (Tilders, Berkenbosch and Smelik, 1980). It appears, therefore, that the catecholamines exert different effects on various parts of the hypothalamic–pituitary axis. Catecholamine infusion studies in man are difficult to interpret. Wilcox *et al.* (1975) reported that dopamine infusion caused a rise in plasma ACTH and cortisol concentrations; norepinephrine infusion had an opposite effect. Few *et al.* (1980) found an early fall followed by a late rise in plasma cortisol concentration after norepinephrine infusion. They attributed the late rise in plasma ACTH and cortisol to the metabolic effects of the norepinephrine infusion and not to a direct hypothalamic–pituitary action. Müller-Hess *et al.* (1974) reported that epinephrine infusion reduced the cortisol and growth hormone response to insulin-induced hypoglycaemia; however, blood glucose concentrations did not fall as low or as rapidly in the epinephrine-infused volunteers.

The pathophysiological importance of the different effects of catecholamines in the control of ACTH and cortisol secretion is further complicated by the finding that adrenal steroids may themselves influence brain catecholamine metabolism. Rastogi and Singhal (1978) have reported that bilateral adrenalectomy produced an increase in rat brain tyrosine hydroxylase and monoamine oxidase activity, with an increase in catecholamine turnover. These changes could be reversed by corticosterone treatment of the adrenalectomized animals, suggesting that the interaction between catecholamines and adrenal steroids extends well beyond the confines of the adrenal gland.

INTEGRATION OF CORTEX AND MEDULLA IN THE RESPONSE TO STRESS

The concept of stress is vague and has been substantially modified since the days of Selye and the 'general adaptation syndrome' (Selye, 1970). In a large population study of healthy individuals, Summers *et al.* (1983) found a strong statistical correlation between the urinary excretion of steroid and catecholamine metabolites. Under conditions of stress this simple relationship does not always hold, and adrenal medullary secretion may be separated from adrenocortical effects. Thus, Goldstein *et al.* (1982) showed that the surgical and psychological stress caused by dental extraction increased plasma catecholamine concentration, with no significant change in plasma cortisol, and Melsom *et al.* (1976) were able to demonstrate that, following myocardial infarction, diazepam reduced the plasma epinephrine but not the plasma cortisol concentration. Mannelli *et al.* (1982) measured intraoperative adrenal vein plasma catecholamine and cortisol concentrations, and showed that the cortex and medulla responded differently to surgical stress. We confine our discussion here to a consideration of the role of the cortex and medulla in glucose homeostasis and the response to hypotension.

Glucose homeostasis

It is well recognized that glucocorticoids and epinephrine both have diabetogenic (anti-insulin) actions and play an important counter-regulatory effect in elevating blood glucose in response to hypoglycaemia. Their release in response to stress is probably of importance in making glucose available for muscular work, so producing stress-hyperglycaemia.

Response to hypoglycaemia

Hypoglycaemia is a potent stimulus to epinephrine and norepinephrine secretion by the adrenal medulla (Goldfien *et al.*, 1961), acting via a central nervous system response mediated by the splanchnic nerves. This leads to tachycardia and a widened pulse pressure. The epinephrine response to hypoglycaemia (blood glucose levels below 2.5 mmol/l), which is initiated within a few minutes, is augmented by the β_1-receptor blocker metoprolol (Hansson *et al.*, 1977), diminished by the non-selective β-receptor blocking agent penbutolol (Hansson and Hökfelt, 1976) and increased by propranolol (Nilsson, Karlberg and Soderberg, 1980).

Various mechanisms have been proposed to explain the influence of β-blockade on epinephrine effects. The augmented epinephrine response may be due to its reduced clearance during β-blockade, or in some way to peripheral β-blockade itself, or it may be due to an action of the β-blocker directly on the central nervous system (Nilsson, Karlberg and Soderberg, 1980). Some workers (Lauridsen, Christensen and Lyngsøe, 1983) report that β-receptor blockade (non-selective) or β_1-blockade (selective) prolongs the hypoglycaemic response to insulin, but this is not a consistent finding (Nilsson, Karlberg and Soderberg, 1980).

Glucocorticoids maintain the blood glucose in the fasting state by maintaining the supply of amino acids from peripheral tissues and by stimulating hepatic gluconeogenesis. They also stimulate lipolysis, acting with growth hormone to promote the release of fatty acids and glycerol. These mechanisms appear to require protein synthesis (Exton *et al.*, 1972), and although the ACTH and consequent cortisol response to hypoglycaemia is rapid it is unlikely that glucocorticoids exert a major effect early enough to participate in the prompt restoration of blood glucose during insulin-induced hypoglycaemia.

Nevertheless, there is strong experimental evidence, for example from studies in adrenalectomized rats, that the hepatic gluconeogenic and glycogenolytic, as well as the peripheral lipolytic responses to epinephrine and glucagon, depend on the presence of glucocorticoids (Exton *et al.*, 1972). It appears that this so-called 'permissive' action of adrenal steroids involves the maintenance of a normal intracellular environment so that the sensitivity of physiological processes to cyclic AMP is preserved (*see later*).

Stress hormones and hyperglycaemia

A variety of different forms of stress may be associated with hyperglycaemia, which is not reproduced by the separate administration to normal individuals of cortisol, epinephrine or glucagon but appears to depend on their combined effects.

Epinephrine alone produces hyperglycaemia both by stimulation of hepatic glucose production, which is transient, and by inhibition of glucose utilization, which is sustained (Sacca *et al.*, 1980). Norepinephrine (with its predominantly α-adrenergic action) has little effect, while isoprenaline (isoproterenol, a selective β-adrenergic stimulating drug) causes sustained activation of glucose production followed by a proportional elevation of glucose uptake; plasma glucagon levels do not change in response to either. The mechanism underlying the difference in response to isoproterenol and epinephrine is unclear. The observation that isoproterenol increased glucose output in the presence of two- to threefold elevation of peripheral insulin levels demonstrates that activation of hepatic β-receptors has a potent anti-insulin effect.

Infusion of cortisol in dogs or man leads to a transient small rise in blood glucose by reducing glucose utilization without influencing its production (Eigler, Sacca and Sherwin, 1979; Sacca *et al.*, 1981; Shamoon, Hendler and Sherwin, 1981). Glucagon infusion results in a transient rise in glucose production and clearance. When epinephrine and glucagon are infused together their effects are additive, with the inhibitory effect of epinephrine on glucose clearance predominating. Cortisol markedly accentuates the effects of both glucagon and epinephrine when infused simultaneously.

Studies in both dogs (Eigler, Sacca and Sherwin, 1979) and man (Shamoon, Hendler and Sherwin, 1981) show that simultaneous elevation of all three hormones rapidly results in a diabetic-like state, characterized by a sustained glucose elevation which is far greater than the sum of the responses to the three hormones separately. Under these conditions there is glucose over-production with impaired glucose utilization, and plasma glucose levels are in excess of 12 mmol/l.

It appears, therefore, that stress-hyperglycaemia results from a complex interaction of these three hormones. Such a synergistic effect is not observed on ketone body production, which is negatively correlated with insulin production (Batstone *et al.*, 1976). In mild diabetics, particularly the elderly who become ill for any other reason, there is the potential for setting up a vicious cycle. The stress resulting from fluid depletion may stimulate production of these three hormones and so further impair glucose tolerance and precipitate a rapidly deteriorating hyperosmolar state.

Response to hypotension

In contrast to hypoglycaemia, hypotension acts through baroreceptors in the carotid body and aortic arch to cause a generalized sympathetic activation affecting mainly the vasculature. As a consequence, peripheral norepinephrine concentrations rise without a comparable change in plasma epinephrine concentration. Similarly, cold stress principally stimulates norepinephrine release (Cryer, 1980).

Epinephrine deficiency in hypopituitarism

Hypophysectomy has been shown to cause a decline in the activity of the medullary enzymes tyrosine hydroxylase (Mueller, Thoenen and Axelrod, 1970), dopa-decarboxylase, dopamine β-hydroxylase (Weinshilboum and Axelrod, 1970) and PNMT (Wurtman and Axelrod, 1965). This is reflected in a parallel decline in epinephrine production. Thus, Rudman *et al.* (1981) showed that ACTH-deficient children suffered an 80–90% reduction in plasma concentration of epinephrine at rest and after exercise, but no change in the circulating levels of norepinephrine. They found this to be related to ACTH (and secondary cortisol) deficiency, and not to deficiency of any other pituitary hormone. Not only are epinephrine concentrations low basally, they also fail to rise in response to potent stimuli of epinephrine release such as hypoglycaemia. Experimental suppression of adrenal steroidogenesis has been shown to reduce adrenal medullary activity. Badder *et al.* (1980) studied the effect of exogenous steroid treatment with and without aminoglutethimide and showed a significant reduction in the epinephrine response to hypovolaemic stress in beagle dogs. Similarly, treating patients with cortisol (20 mg/day) reduced urinary epinephrine excretion during an insulin tolerance test (Duckworth, Masi and Kitabchi, 1974).

Tietze *et al.* (1972) described a group of children with adrenal medullary hyporesponsiveness who were noted to have little or no clinical response to insulin-induced hypoglycaemia. Adrenal medullary deficiency in a hypopituitary diabetic patient would mean that loss of consciousness could occur without the customary premonitory symptoms and signs. Presumably the fact that hypopituitary patients cope without catecholamine replacement depends on the presence of a predominantly extra-adrenal source of circulating norepinephrine which is unaffected by lack of ACTH. However, there may well be circumstances outside everyday existence when epinephrine deficiency in hypopituitarism or after glucocorticoid suppression would be a serious physiological disadvantage.

ADRENAL CORTEX AND MEDULLA, RENAL FUNCTION AND ELECTROLYTE HOMEOSTASIS

Catecholamines and the renin–angiotensin–aldosterone system

It is well-recognized that a fall in renal perfusion pressure stimulates the renin–angiotensin system, increases aldosterone secretion from the zona glomerulosa and thus sodium and water retention in the distal renal tubules, and that glucocorticoids are essential for normal water excretion (Goldberg, 1981; Ganong, 1983a). The influence of the adrenal medulla and the sympathetic nervous system on renal function and the renin–angiotensin–aldosterone system is less well appreciated. This is surprising, since an effect of the sympathetic nervous system on salt and water excretion was recognized by Claude Bernard as early as 1859, as quoted by Güllner (1983). Catecholamines, released at sympathetic nerve endings or secreted from the adrenal medulla directly into the circulation, may influence renal function in a number of ways. They are known to change cardiac output and vascular resistance, to stimulate renin release via a β-effect on the juxtaglomerular apparatus (Young and Landsberg, 1977), and to stimulate aldosterone synthesis, both directly and indirectly, by changing sodium and water absorption in the proximal tubule (Güllner, 1983). Finally, they may stimulate antidiuretic hormone (ADH) release from the posterior pituitary gland.

ACTH, potassium, sodium and angiotensin II are well known to affect aldosterone release, and each may be directly or indirectly influenced by catecholamines. It is difficult in the intact animal to distinguish some of these direct and indirect effects. Isolated organ and tissue experiments suffer the opposite disadvantage in that they bear little relationship to the behaviour of intact animals.

It has been known for some time that increases in dietary sodium or saline infusions increase the urinary excretion of dopamine while decreasing that of norepinephrine (Alexander *et al.*, 1974), suggesting a possible role for both of these catecholamines in sodium excretion. The mechanism of the natriuretic action of dopamine appears to be multiple, involving an increase in renal plasma flow and glomerular filtration rate, as well as inhibition of the tubular transport of sodium (*see* review by Lee, 1982).

Dopamine has also been shown to have a direct inhibitory effect on angiotensin-stimulated aldosterone biosynthesis in bovine adrenal cells (McKenna *et al.*, 1979). In normal man, Sowers *et al.* (1981) have demonstrated that dopamine receptors impose a maximum tonic inhibition on aldosterone secretion, since the addition of L-dopa caused no further fall in aldosterone concentration. Metoclopramide, a competitive antagonist of dopamine in the central nervous and cardiovascular systems, was shown to stimulate aldosterone release independently of any changes in ACTH, electrolytes or the renin–angiotensin system. Later, Sowers *et al.* (1982) demonstrated that dopamine reduced the activity of 18-hydroxylase, a key enzyme in the pathway to aldosterone synthesis. It would seem from the *in vitro* experiments of McKenna *et al.* (1979) that the concentrations of dopamine required to inhibit aldosterone synthesis maximally are considerably higher than those normally found in the circulation (micromolar compared to nanomolar). They suggested that either circulating conjugated dopamine might be deconjugated in the adrenal cortex or, because of significant blood flow from medulla to cortex, adrenocortical dopamine concentrations might be higher than those in the general circulation. This is another indication to re-examine the corticomedullary circulation.

Some of the effects of sympathetic stimulation on renin release and renal tubular sodium reabsorption are probably mediated via the renal nerves, rather than by adrenal medullary epinephrine secretion (*see* review by Güllner, 1983). Many studies, however, have been performed on anaesthetized animals and their significance for normal physiology is uncertain. Lifshitz (1978) prepared dogs with one innervated and one denervated kidney and a split bladder. After recovery from surgery experiments were carried out in the conscious animals. The denervated kidney was able to absorb sodium as efficiently as the contralateral innervated kidney in response to either volume expansion or haemorrhage, suggesting that circulating catecholamines from the adrenal medulla are involved.

Johnson and Barger (1981) studied the effects of venous and arterial infusions of epinephrine and norepinephrine on renal water and electrolyte handling in conscious trained dogs. They have shown that, like neurally released norepinephrine, circulating norepinephrine causes sodium and potassium retention by a direct action on tubular transport mechanisms independent of any change in glomerular filtration rate and renal plasma blood flow. Intrarenal norepinephrine had no direct effect on renal reabsorption of water, while systemic norepinephrine increased the urinary flow rate, supporting previous reports that norepinephrine acts by suppressing ADH release. The physiological significance of these direct humoral effects is uncertain, since the concentrations of norepinephrine necessary to elicit them are found only under severe stress which would also cause increased sympathetic neural activity. On the other hand, a modest rise in circulating epinephrine concentration was shown to have potent effects on renal water and electrolyte handling. In contrast to norepinephrine, the antinatriuretic action was more potent when epinephrine was infused systemically than when it was given intrarenally. This was probably due to increased renin secretion and the subsequent generation of angiotensin II. Angiotensin II can reduce sodium excretion directly without altering glomerular filtration rate or renal plasma flow. Alternatively, epinephrine may have increased renal nerve activity, with the same effect on renal tubular mechanisms. Either way, the work of Johnson and Barger (1981) points to some physiological role for circulating epinephrine in renal sodium conservation.

Dopamine appears to hold an unique place amongst the catecholamines affecting renal function. In man, 98% of circulating dopamine is sulphoconjugated (Snider and Kuchel, 1983), compared to 80% of norepinephrine which is conjugated (Goldberg and Weder, 1980). Furthermore, the amount of free dopamine excreted far exceeds the glomerular filtration rate (*see* review by Lee, 1982).

The current view is that dopamine is synthesized locally from circulating L-dopa and acts on three types of vascular receptor (α, β- and dopamine-specific). α-Receptor stimulation leads to vasoconstriction and β-receptor stimulation to vasodilation. The pharmacological effects of dopamine are therefore complex, producing natriuresis by increasing renal blood flow and glomerular filtration rate, as well as by a direct inhibition of tubular transport of sodium. Urinary dopamine excretion is increased by sodium chloride loading and decreased by a sodium bicarbonate load. Frusemide (furosemide), which inhibits chloride transport in the loop of Henle, also increases dopamine excretion. This suggests that dopamine excretion is sensitive to the chloride load presented to the macula densa and may play a physiological role in electrolyte and fluid balance.

Catecholamines, glucocorticoids and ADH

A further point of interaction between adrenocortical and adrenomedullary function centres around the release of antidiuretic hormone (vasopressin), which is influenced by hypothalamic osmoreceptors as well as other stimuli, such as pain, emotion and (through stretch and baroreceptors) hypovolaemia. While α-stimulation produces a diuresis, both ADH and adrenergic β-stimulation increase intracellular cyclic AMP and cause water retention.

Schrier *et al.* (1972) studied the mechanism of the antidiuretic action of β-adrenergic stimulation in anaesthetized dogs. Under controlled conditions (with no change in glomerular filtration rate or renal vascular resistance), isoproterenol had little direct renal effect. The antidiuretic action was independent of renal innervation but required an intact posterior pituitary gland. In subsequent similar experiments they found that the diuretic action of α-adrenergic stimulation was due to a suppression of vasopressin release from the posterior pituitary (Schrier, Berl and Harbottle, 1973).

Their results were supported by McDonald *et al.* (1977), who compared the effects of isoproterenol (β-stimulant) with those of norepinephrine (predominantly an α-stimulant) on renal water excretion in normal Sprague-Dawley rats and in rats with congenital diabetes insipidus, and found that the lack of ADH in the latter prevented the expected responses to α- and β-stimulation, despite similar changes in blood pressure and glomerular filtration rate in both groups of animals. Further support for the work of Schrier and co-workers has come from Johnson and Barger (1981). Thus, adrenergic stimulation or depression of baroreceptors would act via central mechanisms causing or inhibiting ADH release and so affecting free water clearance.

Physiological studies on the renin–angiotensin system in man

In view of the many factors which influence renin release (Goldberg and Weder, 1980), it is not surprising that Lightman *et al.* (1981), in diurnal studies on 10 normal volunteers, could find no correlation between circadian changes in plasma norepinephrine concentration and plasma renin activity, aldosterone or cortisol. Curiously, they could not even demonstrate a correlation between plasma renin activity and aldosterone concentration.

Henrich *et al.* (1977) have demonstrated that changes in plasma aldosterone concentration can be dissociated from those of plasma renin activity by altering plasma potassium concentration. They studied the effects of isokalaemic and hypokalaemic ultrafiltration dialysis in eight patients with end-stage renal failure. All had raised plasma catecholamine concentrations, and similar blood pressures and electrolyte concentrations before dialysis and the same weight loss after dialysis. Plasma renin activities rose in both groups of patients after dialysis, but while the plasma aldosterone concentration rose in the isokalaemic group it fell significantly in those made mildly hypokalaemic. The latter also showed a significant drop in plasma catecholamine concentration after dialysis. It appears that circulating catecholamines had some effect on aldosterone secretion, either directly or via changes in serum potassium, which was independent of plasma renin activity.

Vetter *et al.* (1976) studied renin and aldosterone secretion in seven patients with phaeochromocytoma before and after α-adrenergic blockade with phentolamine. All seven patients had raised urinary excretion of norepinephrine, but only those with additional excess excretion of epinephrine (or in one case of dopamine) showed elevated plasma renin activity and aldosterone concentration, which dropped with α-adrenergic blockade. They interpreted their findings to indicate that β-adrenergic stimulation (via epinephrine) was the prime cause of the raised renin activity, but this does not explain the reduction in renin and aldosterone caused by α-blockade in these patients.

Adrenal hormones and potassium and phosphate homeostasis

The adrenocortical hormones cortisol and aldosterone play an essential role in maintaining the plasma potassium concentration within narrow limits, as shown, for example, in the hyperkalaemia of Addison's disease and the hypokalaemia of Conn's or severe Cushing's syndromes. The role of the adrenal medulla in potassium homeostasis is less well known.

The plasma potassium concentration is affected by both internal and external potassium balance. Catecholamines appear to influence the internal distribution of potassium rather than directly affecting its renal excretion. The other factors affecting internal balance include acid–base status, body fluid tonicity, and insulin as well as mineralocorticoids (Cox, 1981).

Bia, Tyler and DeFronzo (1982) studied the effects of chronic (7–10 days) adrenal insufficiency on extrarenal potassium tolerance by infusing potassium into rats after acute nephrectomy. They demonstrated impaired potassium tolerance in the glucocorticoid-replaced adrenalectomized rats which could not be attributed to acidaemia, hypotension, changes in plasma glucose or insulin concentration, or potassium retention prior to study. Chronic aldosterone or acute epinephrine replacement completely corrected the abnormality. They argue for a role of both hormones in extrarenal potassium disposal. Rosa *et al.* (1980) studied potassium tolerance in healthy volunteers with and without β-receptor blockade (propranolol) or stimulation (epinephrine), and were able to confirm earlier animal work which had suggested that the hypokalaemic effect of epinephrine was independent of insulin, aldosterone and renal potassium excretion, and that it involved some direct cellular action of catecholamines on β-receptors. Using specific β-receptor blockers it has been demonstrated that the potassium-lowering effect is mediated by β_2-receptors (Editorial, 1983; Struthers and Reid, 1984).

It is interesting to note that plasma epinephrine concentrations will be raised in clinical conditions which are associated with a rise in plasma potassium (e.g. acidaemia and hypertonicity) and can therefore have an important protective function.

Body and co-workers (1983) have also demonstrated that epinephrine, infused to provide physiological plasma concentrations, induced dose-dependent decrements in plasma phosphate concentration. This fall was independent of any changes in serum ionized calcium, immunoreactive parathyroid hormone or immunoreactive calcitonin. This epinephrine effect, mediated probably via β-receptors, could explain the hypophosphataemia which occurs in several clinical states such as sepsis, acute myocardial infarction and toxic shock. It may account for the association between hypokalaemia and hypophosphataemia (Anderson, Peters and Stewart, 1969) and serve to provide the cells with an essential ingredient to cope with the hypermetabolism associated with increased epinephrine release.

The renin–angiotensin system in panhypopituitarism

In addition to the more obvious effects of disordered ADH secretion from the posterior gland, abnormalities of the anterior pituitary will also disturb renal function, by mechanisms involving the adrenal cortex and medulla. Bakiri, Benmiloud and Vallotton (1983) measured plasma renin substrate (PRS) and plasma renin activity (PRA) in 20 females with Sheehan's syndrome before and after therapy with cortisol, and again after treatment with cortisol and thyroxine. They concluded that panhypopituitarism altered the renin–angiotensin system, producing low basal levels of PRS and PRA, with subnormal responses to hypovolaemic stimulation. Glucocorticoid treatment alone did not improve the hyporeninism, which responded, however, to combined treatment with cortisol and thyroxine. Successful treatment was not associated with a rise in PRS, and the authors concluded that these patients had a diminished sensitivity of renin secretion to hypovolaemia and that thyroxine increased the responsiveness of receptor cells to adrenergic stimulation. Stiles *et al.* (1981) have shown that experimental hypothyroidism reduces both the number of rat reticulocyte β-receptors and their binding affinity for specific agonists. The combined deficiencies of ACTH and thyroid-stimulating hormone (TSH) in panhypopituitary patients could therefore reduce both the epinephrine response to hypovolaemia and the sensitivity of the juxtaglomerular cells to β-stimulation.

The renin–angiotensin system in the Shy-Drager syndrome

The complexity of the renin–angiotensin system and its relationship to sympathetic nervous system activity is illustrated by recent work on patients with the Shy-Drager syndrome. This disorder is characterized by progressive degeneration of the central nervous system, resulting in significant autonomic and extrapyramidal dysfunction, with severe postural hypotension as a prominent and incapacitating feature (*see* review by Moskowitz, 1977). Sasaki *et al.* (1983) have shown that in normal individuals intravenous frusemide induces a rise in both plasma renin activity and aldosterone concentration; yet most of their Shy-Drager patients responded only with a rise in the latter. They suggested that the fall in blood pressure in these patients had stimulated ACTH release and so caused aldosterone secretion. Their claim was supported by the absence of other known stimuli of aldosterone release, such as changes in serum sodium or potassium concentration, and by the demonstration that the aldosterone response in these patients could be blocked by pretreatment with dexamethasone.

In this disorder the rise in ACTH following hypovolaemic stress appeared to compensate for the absence of renin release. This underlines the advantage of having more than one mechanism of controlling aldosterone release, and illustrates yet another site of interaction between adrenocortical and sympathetic nervous activity.

ENDOGENOUS OPIATES AND THE ADRENAL CORTEX AND MEDULLA

It is now well-recognized that the natural opiate β-endorphin is derived from the same precursor, pro-opiomelanocortin, as ACTH and that it is secreted by the anterior pituitary gland in parallel with ACTH, in equimolar concentrations (Mains and Eipper, 1978). The purpose of this co-secretion is unknown, although the secretion of a natural analgesic in times of stress might be of protective value. There is circumstantial evidence to link the control of endogenous opiate and ACTH release. Thus, morphine blocks stress-induced increases in ACTH secretion in animals, and naloxone, a morphine antagonist which readily crosses the blood–brain barrier, stimulates a rise in cortisol (probably via ACTH) as well as in luteinizing hormone and follicle-stimulating hormone (FSH) (Moreley *et al.*, 1980).

The other endogenous opiates may be derived from proenkephalin or preneoendorphin dynorphin. Natural opiates derived from the former (and possibly the latter) precursor are present in very high concentrations in the bovine and human adrenal medulla. Enkephalins have been shown to be stored with epinephrine in chromaffin cells and released with epinephrine during stress (Viveros *et al.*, 1979; Boarder, Erdelyi and Barchas, 1982). Furthermore, β-endorphin immunoreactivity has been reported in human adrenal glands, along with significant amounts of ACTH and α-MSH (Evans *et al.*, 1983).

GLUCOCORTICOIDS AND ADRENAL CATECHOLAMINES: MECHANISMS OF TARGET ORGAN INTERACTIONS

There is extensive evidence that glucocorticoids modulate the response of target tissues to catecholamines. Although they do not produce the reactions by themselves, small amounts of glucocorticoids must be present for a number of metabolic reactions to occur. Catecholamines require the 'permissive action' of glucocorticoids in order to exert their calorigenic, lipolytic, pressor and bronchodilator effects (Ganong, 1983b).

Some of these effects occur at very high concentrations and are shared by other steroids, again at supraphysiological concentrations. For example, corticosteroids have long been known to potentiate the vascular effects of catecholamines (Ramey and Goldstein, 1957; Raab, 1959; Schmid, Eckstein and Abboud, 1967). Yard and Kadowitz (1972), in experiments on perfused hindquarters of dog and cat, showed that hydrocortisone enhanced the response to epinephrine, but not to norepinephrine or to nerve stimulation. This effect was observed only at a dose of 10 mg/kg and not at 1 or 100 mg/kg. Furthermore, it did not occur with the potent synthetic glucocorticoid dexamethasone, which makes it unlikely that it is mediated via 'classical' glucocorticoid receptors (but *see* Chapter 4).

Many mechanisms have been proposed to explain corticosteroid enhancement of catecholamine action. These vary both with the animal species and with the catecholamine action studied. For example, Salt (1972) reported that various steroids (including corticosterone, testosterone, progesterone and 17β-oestradiol) produced a dose-dependent inhibition of norepinephrine uptake by target tissue ($uptake_2$) in the isolated perfused rat heart, while Nicol and Rae (1972) demonstrated similar steroidal inhibition of epinephrine and norepinephrine uptake by arterial smooth muscle. Failure of the peripheral tissue to take up and metabolize catecholamines would result in prolonged exposure to its stimulant action and could explain the enhancing effect of adrenal steroids. However, the non-specific nature of the receptors, the high concentrations of the steroids required, and the fact that the degree of inhibition of $uptake_2$ was not related to the biological potency of the steroid derivative, make this an unlikely explanation of all steroid-potentiating effects.

As well as stimulating PNMT synthesis, glucocorticoids have been shown to inhibit monoamine oxidase (MAO) and catecholamine *O*-methyltransferase (COMT) activity in the adrenal glands of the rat (Parvez and Parvez, 1972). Conversely, adrenalectomy causes a rise in the excretion of catecholamine metabolites (Avakian and Callingham, 1968) and increased MAO activity in the rat heart (Westfall and Osada, 1969). Steroid inhibition of COMT and MAO activity in target tissue would enhance catecholamine action.

Corticosteroids are also believed to enhance tissue and organ sensitivity to the action of catecholamines by increasing both the number of β-adrenergic receptors and their binding affinity. Davies and Lefkowitz (1981) have described the receptor complex as consisting of at least three components: the receptor, a nucleotide regulatory (N) site, and a catalytic (C) moiety. Kinetic studies using labelled ligands suggest that these complexes exist in two interconvertible forms, a high-affinity and a low-affinity state. The high-affinity state was associated with greater adenylate cyclase activity. Using β-adrenergic receptors isolated from human neutrophil membranes, these authors showed that corticosteroids increased the ratio of the high- to the low-affinity state complexes. Because the steroid effect was demonstrable both *in vivo* and *in vitro*, it was not due to any alteration in the neutrophil cell population. Although the precise mechanism of steroid modulation of catecholamine receptor binding is unknown, the authors favoured some steroid-induced change in the N-site.

If glucocorticoids exerted their permissive effect solely by changing the catecholamine-receptor binding characteristics, one would expect steroid deficiency to be accompanied by decreased cellular cAMP. This is not the case. Exton *et al.* (1972) found that, despite normal or increased cAMP accumulation, glucagon and epinephrine stimulation of glucose synthesis from lactate was markedly reduced in isolated perfused livers from fasted adrenal-deficient rats. In response to epinephrine, cyclic AMP was increased in the fat pads of these animals, although free fatty acid and glycerol release remained impaired. The authors concluded that, in the absence of glucocorticoids, gluconeogenesis, glycogenolysis and lipolysis became less sensitive to the action of cAMP. They suggested that the 'permissive' action of adrenal steroids might involve the maintenance of a normal intracellular ionic environment.

Cortisol pretreatment produced different effects on the adenylate cyclase system of pig skin (Iizuka *et al.*, 1980). In short-term experiments, cortisol added singly or in combination with other stimulators of adenylate cyclase in the skin (epinephrine or histamine) had no effect on cAMP accumulation. However, preincubation with cortisol for more than 6 hours led to a greater accumulation of cAMP following epinephrine stimulation. Cortisol appeared to protect the adenylate cyclase system without causing a decrease in phosphodiesterase activity.

Using isolated rat epididymal fat cells, Lamberts *et al.* (1975) found that preincubation with dexamethasone, or treating rats with cortisol, enhanced epinephrine-stimulated lipolysis and increased cAMP-dependent protein kinase activity in cell homogenates. Potentiation was not limited to epinephrine but affected the lipolytic response to glucagon as well. Since total phosphodiesterase activity was unchanged, the authors attributed the potentiating action to the induction of protein kinase by glucocorticoids.

In summary, corticosteroids could potentiate the action of catecholamines by acting at many possible sites. Of these, the direct steroid effect on the number and binding affinity of the adrenergic receptors would appear the most simple. Unfortunately, changes in the catecholamine binding status do not always correspond to biological activity. Soman, Shamoon and Sherwin (1980) found that epinephrine infusions in humans produced only transient increases in lipolysis and hepatic glucose production. They had hoped to demonstrate downregulation of β-adrenergic receptors, but found instead that the loss of responsiveness was not associated with parallel changes in β-adrenergic binding to lymphocytes.

It has also been shown that steroids enhance cAMP accumulation in response to adrenergic stimulation, protect the epinephrine-adenylate cyclase system (Iizuka *et al.*, 1980), increase protein kinase activity in response to cAMP (Lamberts *et al.*, 1975), and change the intracellular ionic environment to improve catecholamine responsiveness (Bennett and Gardiner, 1978). Steroids also act by inhibiting norepinephrine uptake$_2$ (Salt, 1972) and degradation by COMT and MAO (Avakian and Callingham, 1968; Westfall and Osada, 1969; Parvez and Parvez, 1972). To this formidable list we must add the effect of ACTH and related peptides in potentiating catecholamine effects on cardiac muscle directly (Bassett, Strand and Cairncross, 1978). In fact, these authors found the potentiating effect of ACTH to be greater than that observed with either cortisol or corticosterone.

The embarrassing number of mechanisms used to explain the potentiating effect of adrenal steroids (and pituitary peptides) on catecholamine action may be a reflection of the different species and tissues studied. Despite our uncertainty as to the 'true' mode of corticosteroid action, it is important to examine some of the biological consequences of the peripheral interaction of these adrenal hormones.

CLINICAL IMPLICATIONS OF THE POTENTIATING EFFECT OF CORTICOSTEROIDS ON ADRENERGIC ACTION

Lungs

Glucocorticoids are known to be important in promoting surfactant production, which prepares the fetus for breathing air. Changes in the amniotic fluid lecithin/sphingomyelin (L/S) ratios and phosphatidylglycerol percentages are used to predict the respiratory distress syndrome (RDS) (Gluck and Kulovich, 1973; Hallman *et al.*, 1976), and glucocorticoids given to the mother help maturation of the immature fetal lung. β-Adrenergic receptors have been identified in the fetal lung, and babies born to mothers given β-sympathomimetic agents show a decreased incidence of RDS. Recently, Divers *et al.* (1982) analyzed amniotic fluid and found a positive correlation between cortisol, catecholamines and their metabolites, and L/S ratios. It appears that cortisol and epinephrine complement each other in producing fetal lung maturation, perhaps by inducing phosphatidylglycerol synthesis.

It is tempting to draw an analogy with the situation in the adult lung, where glucocorticoids and β-agonists have related and complementary therapeutic effects in bronchial asthma. Epinephrine is believed to act directly on bronchial smooth muscle β-receptors and/or indirectly via mast cells, where they prevent histamine release. Wheezing and sudden deaths in asthmatics are known to have a peak incidence during the early hours of the morning. Barnes *et al.* (1980) have studied the relationship of changes in peak flow to changes in circulating epinephrine, cyclic AMP and histamine concentrations. Epinephrine and cAMP concentrations were closely correlated, with the lowest levels recorded at 0400 hours, when histamine concentration was at its peak. The night-time drop in plasma cortisol concentration preceded these changes by four hours. It has therefore been suggested that nocturnal wheezing in asthmatics may be due to low epinephrine concentration at a time when β-receptor number or responsiveness is at its lowest, because of the preceding fall in plasma cortisol concentration.

Cardiovascular system

The occurrence of sudden deaths in the 1960s of asthmatics treated with non-specific β-receptor agonists such as isoprenaline is a graphic illustration of the importance of the cardiac effects of catecholamines. The β_1-adrenergic effect is particularly cardiotoxic.

Glucocorticoids and mineralocorticoids appear to modify the cardiac responses to catecholamines. Thus, in animals that are either chronically stressed or pretreated with prednisone or desoxycorticosterone acetate, the pure β-agonist isoproterenol has greatly increased toxicity. Similarly, aminophylline, which increases intracellular cAMP concentrations, is also more toxic to the myocardium after exposure to glucocorticoids. This increase in cardiotoxicity is not seen with epinephrine, which possesses both α- and β-adrenergic activity. However, simultaneous blockade with the α-antagonist phenoxybenzamine eliminates this advantage of the naturally occurring catecholamine over isoproterenol (Green, Guideri and Lehr, 1980).

It is obvious that these effects of steroids and catecholamines are of great importance after myocardial infarction, when myocardial excitability is greatly enhanced in the ischaemic area. In the management of phaeochromocytoma it is accepted practice that α-blockade precedes β-blockade, because of the risk of hypertensive crises provoked by the α-agonist action of the norepinephrine. At the myocardial level, however, the α-adrenergic effect is cardioprotective and the β-adrenergic effect arrhythmogenic.

Catecholamines and steroids in severe shock

We have already discussed the potentiating effects of cortisol on epinephrine synthesis, release and action. However, the physiological significance of steroid secretion in the response to trauma and its pharmacological value in the management of shock is controversial (Editorial, 1977). Animal experiments appear to indicate that steroids reduce mortality mainly when given before injury, as shown with fludrocortisone in normal rats (Da Vanzo, 1964) and corticosterone in steroid-depleted rats (Barton and Little, 1978). This is consistent with the time steroids take to exert their receptor-mediated effects.

No one will deny the importance of steroid replacement in patients with adrenocortical insufficiency. However, there is uncertainty as to the value of pharmacological doses of steroids in injured patients. In one study, administration of steroids to unselected acutely shocked patients did not confer any clinical advantage (Shubin and Weil, 1966). In a different study, patients with acute septic shock given an intravenous bolus of steroids (dexamethasone or methylprednisolone) did significantly better than controls given saline and conventional antibiotic therapy (Schumer, 1976). Steroid administration caused an impressive reduction in mortality (from 38.4% to 10.4%) in this selected group of patients. The mechanism of steroid action under these conditions is uncertain. Schumer felt that it was probably immunological – inhibiting the release of the 'autocoids' histamine and serotonin.

After major trauma the steroid effects on catecholamine sensitivity would occur too late to be of any benefit to the victim, and another role must be found for the high plasma cortisol concentration. Steroids might stimulate tyrosine hydroxylase activity to replenish catecholamine stores, as demonstrated in phaeochromocytoma cell cultures (Tischler *et al.*, 1983). Munck, Guyre and Holbrook (1984) have argued that steroid secretion following trauma has an anti-inflammatory role, protecting the patient from a possible autoimmune reaction triggered by the excess of cellular destruction.

Phaeochromocytoma and adrenocortical function

Considering the many ways in which the adrenal cortex and medulla can each enhance the other's activity it is not surprising to find a definite (though uncommon) association between phaeochromocytoma and Cushing's syndrome

(Williams *et al.*, 1960; Steiner, Goodman and Powers, 1968). The nature of this association has not been resolved, though there are interesting clues.

ACTH, which has been found in the normal human adrenal medulla (Evans *et al.*, 1983), may arise from phaeochromocytoma or associated thyroid carcinoma. Manger and Gifford (1977) reported ACTH in the serum of two of their patients with phaeochromocytoma, which disappeared postoperatively. Hutchinson, Evans and Davidson (1958) described adrenocortical hyperplasia in addition to a phaeochromocytoma in the surgical specimen from a patient with Cushing's syndrome.

Adrenal medullary tissue may be able to complete glucocorticoid biosynthesis from pregnenolone onwards (Carballeira and Venning, 1964), and a large amount of cortisol was found in an extract of one phaeochromocytoma (Ramsay and Langlands, 1962). In one series of eight patients with phaeochromocytoma, two had increased cortisol secretion (Vetter *et al.*, 1974).

Apart from Cushing's syndrome, phaeochromocytomas have been found in association with other adrenal gland abnormalities. A malignant phaeochromocytoma has simulated an adrenocortical carcinoma with clinical features of Addison's disease (McGavack *et al.*, 1942), and adrenal virilism has regressed after removal of a phaeochromocytoma (Neff *et al.*, 1942). These early case reports were based on the clinical features and lack complete adrenocortical function tests. If the cortico–medullary interactions reviewed earlier in this chapter are so powerful, one would expect the association to be based on stronger evidence. However, apart from isolated (and unsubstantiated) reports of associated Addison's disease (e.g. MacKeith, 1944), adrenocortical activity is always almost normal in patients with phaeochromocytoma (Hume, 1960; Ramsay and Langlands, 1962). Perhaps adrenal medullary tumours distort the complicated vascular connections between cortex and medulla, resulting in loss of some of the normal interactions. However, if a phaeochromocytoma should itself produce cortisol, this may lead to suppression of normal adrenocortical activity and the danger of postoperative hypotension (Mulrow, Cohn and Yesner, 1959).

Some phaeochromocytomas appear to retain their sensitivity to cortisol and ACTH, and acute and sometimes fatal hypertensive crises may follow use of the latter for diagnostic purposes (Moorhead *et al.*, 1966). The mechanism of ACTH and cortisol stimulation of catecholamine secretion and release has already been discussed (*see above*), and the danger of using the ACTH stimulation and metyrapone test in any patient who might possibly have a phaeochromocytoma should be stressed (Steiner, Goodman and Powers, 1968; Critchley, West and Waite, 1974).

The clinical endocrinologist should be aware of the positive (though rare) association between phaeochromocytoma and Cushing's syndrome, and use a simple screening test for phaeochromocytoma, such as the measurement of urinary methoxycatecholamines (Pisano, 1960), before embarking on the various steroid stimulation and suppression tests.

SUMMARY AND CONCLUSIONS

There are a wide range of interactions between the adrenal cortex and medulla and their products, both physical and functional. The interpretation of these interactions is complicated by the fact that the adrenal medullary hormones, notably norepinephrine, also double as central neurotransmitters and that, like the steroid hormones, they affect many tissues. In the whole animal it is difficult to distinguish primary from secondary effects.

It appears to be well-established that glucocorticoids, and other products of the ACTH-stimulated cortex, such as ascorbic acid, have important effects on epinephrine synthesis and release. A direct influence of adrenal medullary hormones on cortical function is less well established and, unless the direction of adrenal blood flow can be shown to be reversed, appears to be less biologically significant.

Epinephrine may augment ACTH release under some conditions of 'stress', and cortisol generally increases the target organ sensitivity to the β-adrenergic effects of epinephrine. Such interactions may be important in understanding fetal lung maturation, bronchial asthma and the management of cardiogenic shock. Complex interactions also exist between epinephrine and the renin–angiotensin–aldosterone system, with a recently appreciated role for epinephrine in internal potassium homeostasis.

References

ALEXANDER, R. W., GILL, J. R. JR, YAMABE, H., LOVENBERG, W. and KEISER, H. R. (1974) Effects of dietary sodium and of acute saline infusion on the interrelationship between dopamine excretion and adrenergic activity in man. *Journal of Clinical Investigation,* **54,** 194–200

ANDERSON, D. C., PETERS, T. J. and STEWART, W. K. (1969) Association of hypokalaemia and hypophosphataemia. *British Medical Journal,* **4,** 402–403

AVAKIAN, V. M. and CALLINGHAM, B. A. (1968) An effect of adrenalectomy upon catecholamine metabolism. *British Journal of Pharmacology and Chemotherapy,* **33,** 211P–212P

AXELROD, J. (1977) Catecholamines: effects of ACTH and adrenal corticoids. *Annals of the New York Academy of Sciences,* **297,** 275–283

BADDER, E. M., SANTEN, R., SAMOJLIK, E. and HARRISON, T. S. (1980) Adrenal medullary epinephrine secretion: effects of cortisol alone and combined with aminoglutethimide. *Journal of Laboratory and Clinical Medicine,* **96,** 815–821

BAKIRI, F., BENMILOUD, M. and VALLOTTON, M. B. (1983) The renin–angiotensin system in panhypopituitarism: dynamic studies and therapeutic effects in Sheehan's syndrome. *Journal of Clinical Endocrinology and Metabolism,* **56,** 1042–1047

BARNES, P., FITZGERALD, G., BROWN, M. and DOLLERY, M. (1980) Nocturnal asthma and changes in circulating epinephrine, histamine and cortisol. *New England Journal of Medicine,* **303,** 263–303

BARTON, R. N. and LITTLE, R. A. (1978) Effects of inhibition of adrenal steroidogenesis on compensation of fluid loss and on survival after limb ischaemia in the rat. *Journal of Endocrinology,* **76,** 293–302

BASSETT, J. R., STRAND, F. L. and CAIRNCROSS, K. D. (1978) Glucocorticoids, adrenocorticotrophic hormone and related polypeptides on myocardial sensitivity to noradrenaline. *European Journal of Pharmacology,* **49,** 243–249

BATSTONE, G. F., HINKS, L., WHITEFOOT, R., BLOOM, S., LAING, J. E. and ALBERTI, K. G. M. M. (1976) Hormonal changes after thermal injury. *Journal of Endocrinology,* **68,** 38P–39P

BENNETT, T. and GARDINER, S. M. (1978) Corticosteroid involvement in the changes in noradrenergic responsiveness of tissues from rats made hypertensive by short-term isolation. *British Journal of Pharmacology,* **64,** 129–136

BERNARD, C. (1859) *Leçons sur les Propriétés Physiologique des Liquides de l'Organisme,* pp. 172–173. Paris: Ballière

BIA, M. J., TYLER, K. A. and DEFRONZO, R. A. (1982) Regulation of extrarenal potassium homeostasis by adrenal hormones in rats. *American Journal of Physiology,* **242,** F641–F648

BOARDER, M. R., ERDELYI, E. and BARCHAS, J. D. (1982) Opioid peptides in human plasma: evidence for multiple forms. *Journal of Clinical Endocrinology and Metabolism,* **54,** 715–720

BODY, J.-J., CRYER, P. E., OFFORD, K. P. and HEATH, H. (1983) Epinephrine as a hypophosphataemic hormone in man. Physiological effects of circulating epinephrine on plasma calcium, magnesium, phosphorus, parathyroid hormone and calcitonin. *Journal of Clinical Investigation,* **71,** 572–578

CARBALLEIRA, A. and FISHMAN, L. M. (1980) The adrenal functional unit: a hypothesis. *Perspectives in Biology and Medicine,* **23,** 573–597

CARBALLEIRA, A. and VENNING, E. H. (1964) Conversion of steroids by chromaffin tissue. I. Studies with a pheochromocytoma. *Steroids,* **4,** 329–350

CIARANELLO, R. D. (1978) Regulation of phenylethanolamine *N*-methyltransferase. *Biochemical Pharmacology,* **27,** 1895–1897

CIARANELLO, R. D., JACOBOWITZ, D. and AXELROD, J. (1973) Effect of dexamethasone on phenylethanolamine *N*-methyltransferase in chromaffin tissue of the neonatal rat. *Journal of Neurochemistry,* **20,** 797–805

COOPER, D. Y. and ROSENTHAL, O. (1962) Action of noradrenaline and ascorbic acid on C-21 hydroxylation of steroids by adrenocortical microsomes. *Archives of Biochemistry and Biophysics,* **96,** 331–335

COUPLAND, R. E. (1965) *The Natural History of the Chromaffin Cell.* London: Longmans Green

COX, M. (1981) Potassium homeostasis. *Medical Clinics of North America,* **65,** 363–384

CRITCHLEY, J. A. J. H., WEST, C. P. and WAITE, J. (1974) Dangers of corticotrophin in phaeochromocytoma. *Lancet,* **2,** 782

CRYER, P. E. (1980) Physiology and pathophysiology of the human sympathoadrenal neuroendocrine system. *New England Journal of Medicine,* **303,** 436–444

CUSHMAN, P., ALTER, S. and HILTON, J. G. (1966) Cortisol secretion by the dog adrenal: effects of cyclic adenosine monophosphate, dichloroisoproterenol, dihydroergotamine and adrenaline. *Journal of Endocrinology,* **34,** 271–272

DALLMAN, M. F., ENGELAND, W. C. and MCBRIDE, M. H. (1977) The neural regulation of compensatory adrenal growth. *Annals of the New York Academy of Sciences,* **297,** 373–392

DA VANZO, J. P. (1964) Effect of 2-methyl-9-alpha-fluorohydrocortisone on survival of tourniquet-shocked rats. *Archives Internationales de Pharmacodynamie et de Thérapie,* **150,** 442–446

DAVIES, A. O. and LEFKOWITZ, R. J. (1981) Agonist-promoted high affinity state of the β-adrenergic receptor in human neutrophils: modulation by corticosteroids. *Journal of Clinical Endocrinology and Metabolism,* **53,** 703–708

DILIBERTO, E. J. JR, HECKMAN, G. D. and DANIELS, A. J. (1983) Characterization of ascorbic acid transport by adrenomedullary chromaffin cells. *Journal of Biological Chemistry,* **258,** 12886–12894

DIVERS, W., BABAKNIA, A., HOPPER, B. R., WILKES, M. M. and YEN, S. S. C. (1982) Fetal lung maturation: amniotic fluid catecholamines, phospholipids and cortisol. *American Journal of Obstetrics and Gynecology,* **142,** 440–444

DUCKWORTH, W. C., MASI, A. T. and KITABCHI, A. E. (1974) Effect of low-dose cortisol therapy on epinephrine excretion in man. *Journal of Clinical Endocrinology and Metabolism,* **39,** 750–753

EDITORIAL (1977) The big shot. *Lancet,* **1,** 633–634

EDITORIAL (1983) Adrenaline and potassium: everything in flux. *Lancet,* **2,** 1401–1403

EIGLER, N., SACCA, L. and SHERWIN, R. S. (1979) Synergistic interactions of physiologic increments of glucagon, epinephrine, and cortisol in the dog. *Journal of Clinical Investigation,* **63,** 114–123

ERÄNKÖ, O., LEMPINEN, M. and RÄISÄNEN, L. (1966) Adrenaline and noradrenaline in the organ of Zuckerkandl and adrenals of newborn rats treated with hydrocortisone. *Acta Physiologica Scandinavica,* **66,** 253–254

ERÄNKÖ, O., LEMPINEN, M. and RÄISÄNEN, L. (1967) Effect of hydrocortisone administration *in utero* on the adrenaline and noradrenaline content of extra-adrenal chromaffin tissue in the rat. *Acta Physiologica Scandinavica,* **69,** 255–256

EVANS, C. J., ERDELYI, E., WEBER, E. and BARCHAS, J. D. (1983) Identification of pro-opiomelanocortin-derived peptides in the human adrenal medulla. *Science,* **221,** 957–960

EXTON, J. H., FRIEDMANN, N., WONG, E. H.-A., BRINEAUX, J. P., CORBIN, J. D. and PARK, C. R. (1972) Interaction of glucocorticoids with glucagon and epinephrine in the control of gluconeogenesis and glycogenolysis in liver and of lipolysis in adipose tissue. *Journal of Biological Chemistry,* **247,** 3579–3588

FEW, J. D., GAWELL, M. J., IMMS, F. J. and TIPTAFT, E. M. (1980) The influence of the infusion of noradrenaline on plasma cortisol levels in man. *Journal of Physiology,* **309,** 375–389

GANONG, W. F. (1983a) *Review of Medical Physiology,* 11th edn., pp. 301–302. Los Altos: Lange Medical Publications

GANONG, W. F. (1983b) *Review of Medical Physiology,* 11th edn., p. 300. Los Altos: Lange Medical Publications

GIGUÈRE, V., CÔTÉ, J. and LABRIE, F. (1982) Specific inhibition by glucocorticoids of the α_1-adrenergic stimulation of adrenocorticotropin release in rat anterior pituitary cells. *Endocrinology,* **110,** 1225–1230

GLUCK, L. and KULOVICH, M. V. (1973) Lecithin/sphingomyelin ratios in amniotic fluid in normal and abnormal pregnancy. *American Journal of Obstetrics and Gynecology,* **115,** 539–546

GOLDBERG, M. (1981) Hyponatremia. *Medical Clinics of North America,* **65,** 251–269

GOLDBERG, L. I. and WEDER, A. B. (1980) Connections between endogenous dopamine, dopamine receptors and sodium excretion: evidence and hypotheses. *Recent Advances in Clinical Pharmacology,* **2,** 149–166

GOLDFIEN, A., MOORE, R., ZILELI, S., HAVENS, L. L., BOLING, L. and THORN, G. W. (1961) Plasma epinephrine and norepinephrine levels during insulin-induced hypoglycaemia in man. *Journal of Clinical Endocrinology,* **21,** 296–304

GOLDSTEIN, D. S., DIONNE, R., SWEET, J. *et al.* (1982) Circulatory plasma catecholamine, cortisol, lipid, and psychological responses to a real life stress (third molar extractions): effects of diazepam sedation and of inclusion of epinephrine with the local anesthetic. *Psychosomatic Medicine,* **44,** 259–272

GREEN, M., GUIDERI, G. and LEHR, D. (1980) Role of α- and β-adrenergic activation in ventricular fibrillation. Death of corticoid-pretreated rats. *Journal of Pharmaceutical Sciences,* **69,** 441–444

GÜLLNER, H. G. (1983) Regulation of sodium and water excretion by catecholamines. *Life Sciences,* **32,** 921–925

HALLMAN, M., KULOVICH, M., KIRKPATRICK, E., SUGARMAN, R. G. and GLUCK, L. (1976) Phosphatidylinositol and phosphatidylglycerol in amniotic fluid: indices of lung maturity. *American Journal of Obstetrics and Gynecology,* **125,** 613–617

HANSSON, B.-G., DYMLING, J.-F., MANHEM, P. and HÖKFELT, B. (1977) Long-term treatment of moderate hypertension with the β_1-receptor blocking agent metoprolol. II. Effect of submaximal work and insulin-induced hypoglycaemia on plasma catecholamines and renin activity, blood pressure and pulse rate. *European Journal of Clinical Pharmacology,* **11,** 247–254

HANSSON, B.-G. and HÖKFELT, B. (1976) Long-term treatment of moderate hypertension with penbutolol (Hoe 893d). II. Effect on the response of plasma catecholamines and plasma renin activity to insulin-induced hypoglycaemia. *European Journal of Clinical Pharmacology,* **9,** 245–251

HENRICH, W. L., KATZ, F. H., MOLINOFF, P. B. and SCHRIER, R. W. (1977) Competitive effects of hypokalaemia and volume depletion on plasma renin activity, aldosterone and catecholamine concentrations in hemodialysis patients. *Kidney International,* **12,** 279–284

HERSEY, R. M. and DISTEFANO, V. (1979) Control of phenylethanolamine *N*-methyltransferase by glucocorticoids in cultured bovine adrenal medullary cells. *Journal of Pharmacology and Experimental Therapeutics,* **209,** 147–152

HÖKFELT, T., FUXE, K., GOLDSTEIN, M. and JOHANNSON, O. (1973) Evidence for adrenaline neurons in the rat brain. *Acta Physiologica Scandinavica,* **89,** 286–289

HUME, D. M. (1960) Pheochromocytoma in the adult and in the child. *American Journal of Surgery,* **99,** 458–496

HUTCHINSON, G. B., EVANS, J. A. and DAVIDSON, D. C. (1958) Pitfalls in the diagnosis of pheochromocytoma. *Annals of Internal Medicine,* **48,** 300–309

IIZUKA, H., KAMIGAKI, K., NEMOTO, O., AOYAGI, T. and MIURA, Y. (1980) Effects of hydrocortisone on the adrenaline–adenylate cyclase system of the skin. *British Journal of Dermatology,* **102,** 703–710

JOHNSON, M. D. and BARGER, A. C. (1981) Circulating catecholamines in control of renal electrolyte and water excretion. *American Journal of Physiology,* **240,** F192–F199

JONES, M. T. and HILLHOUSE, E. W. (1977) Neurotransmitter regulation of corticotropin-releasing factor *in vitro. Annals of the New York Academy of Sciences,* **297,** 536–560

KITABCHI, A. E. (1967a) Inhibition of steroid C-21 hydroxylase by ascorbate: alterations of microsomal lipids in beef adrenal cortex. *Steroids,* **10,** 567–576

KITABCHI, A. E. (1967b) Ascorbic acid in steroidogenesis. *Nature,* **215,** 1385–1386

LAMBERTS, S. W. J., TIMMERMANS, H. A. T., KRAMER-BLANKESTIJN, M and BIRKENHÄGER, J. C. (1975) The mechanism of the potentiating effect of glucocorticoids on catecholamine-induced lipolysis. *Metabolism,* **24,** 681–689

LAURIDSEN, U. B., CHRISTENSEN, N. J. and LYNGSØE, J. (1983) Effects of nonselective and β-1-selective blockade on glucose metabolism and hormone responses during insulin-induced hypoglycemia in normal man. *Journal of Clinical Endocrinology and Metabolism,* **56,** 876–882

LEE, M. R. (1982) Dopamine and the kidney. *Clinical Science,* **62,** 439–448

LEMPINEN, M. (1966) Effect of hydrocortisone on histochemically demonstrable catecholamines of the para-aortic body of the rat. *Acta Physiologica Scandinavica,* **66,** 251–252

LEVINE, M., ASHER, A., POLLARD, H. and ZINDER, O. (1983) Ascorbic acid and catecholamine secretion from cultured chromaffin cells. *Journal of Biological Chemistry,* **258,** 13111–13115

LIAKAKOS, D., DOULAS, N. L., IKKOS, D., ANOUSSAKIS, C., VLACHOS, P. and JOURAMANI, G. (1975) Inhibitory effect of ascorbic acid (vitamin C) on cortisol secretion following adrenal stimulation in children. *Clinica Chimica Acta,* **65,** 251–255

LIFSCHITZ, M. D. (1978) Lack of a role for the renal nerves in renal sodium reabsorption in conscious dogs. *Clinical Science and Molecular Medicine,* **54,** 567–572

LIGHTMAN, S. L., JAMES, V. H. T., LINSELL, C., MULLEN, P. E., PEART, W. S. and SEVER, P. S. (1981) Studies of diurnal changes in plasma renin activity, and plasma noradrenaline, aldosterone and cortisol concentrations in man. *Clinical Endocrinology,* **14,** 213–223

LITTLE, R. A., FRAYN, K. N., RANDALL, P. E., STONER, H. B. and MAYCOCK, P. F. (1984) Plasma catecholamine concentrations in states of acute stress and trauma in man. *Archives of Emergency Medicine,* (in press)

MAINS, R. E. and EIPPER, B. P. (1978) Co-ordinate synthesis of corticotrophins and endorphins by mouse pituitary tumour cells. *Journal of Biological Chemistry,* **253,** 651–655

MANGER, W. M. and GIFFORD, R. W. (Eds) (1977) *Pheochromocytoma,* p. 130. New York: Springer-Verlag

MANNELLI, M., GHERI, R. G., SELLI, C. *et al.* (1982) A study on human adrenal secretion. Measurement of epinephrine, norepinephrine, dopamine and cortisol in peripheral and adrenal venous blood under surgical stress. *Journal of Endocrinological Investigation,* **5,** 91–95

MCDONALD, K. M., KURUVILA, K. C., AISENBREY, G. A. and SCHRIER, R. W. (1977) Effect of α- and β-adrenergic stimulation on renal water excretion and medullary tissue cyclic AMP in intact and diabetes insipidus rats. *Kidney International,* **12,** 96–103

MCGAVACK, T. H., BENJAMIN, J. W., SPEER, F. D. and KLOTZ, S. (1942) Malignant pheochromocytoma of the adrenal medulla (paraganglioma): report of case simulating carcinoma of the adrenal cortex with secondary adrenal insufficiency. *Journal of Clinical Endocrinology,* **2,** 332–338

MACKEITH, R. (1944) Adrenal–sympathetic syndrome: chromaffin tissue tumour with paroxysmal hypertension. *British Heart Journal,* **6,** 1–12

MCKENNA, T. J., ISLAND, D. P., NICHOLSON, W. E. and LIDDLE, G. W. (1979) Dopamine inhibits angiotensin-stimulated aldosterone biosynthesis in bovine adrenal cells. *Journal of Clinical Investigation,* **64,** 287–291

MELSOM, M., ANDREASSEN, P., MELSOM, H., HANSEN, T., GRENDAHL, H. and HILLESTAD, L. K. (1976) Diazepam in acute myocardial infarction: clinical effects and effects on catecholamines, free fatty acids, and cortisol. *British Heart Journal,* **38,** 804–810

MIKULÂŠKOVÁ-ROCHOVÁ, J. and LINÈT, O. (1967) The influence of hypophysectomy, corticotrophin and cortisol on the adrenal medulla of the rat. *Journal of Endocrinology,* **39,** 611–612

MOORHEAD, E. L., CALDWELL, J. R., KELLY, A. R. and MORALES, A. R. (1966) The diagnosis of pheochromocytoma: analysis of 26 cases. *Journal of the American Medical Association,* **196,** 1107–1113

MORELEY, J. E., BARANETSKY, N. G., WINGERT, T. D. *et al.* (1980) Endocrine effects of naloxone-induced opiate receptor blockade. *Journal of Clinical Endocrinology and Metabolism,* **50,** 251–257

MOSKOWITZ, M. A. (1977) Diseases of the autonomic nervous system. *Clinics in Endocrinology and Metabolism,* **6,** 745–768

MUELLER, R. A., THOENEN, H. and AXELROD, J. (1970) Effect of the pituitary and ACTH on the maintenance of basal tyrosine hydroxylase activity in the rat adrenal gland. *Endocrinology,* **86,** 751–755

MÜLLER-HESS, R., GESER, G. A., JÉQUIER, E., FELBER, J. P. and VANNOLLI, A. (1974) Effects of adrenaline on insulin-induced release of growth hormone and cortisol in man. *Acta Endocrinologica,* **75,** 260–273

MULROW, P. J., COHN, G. L. and YESNER, R. (1959) Isolation of cortisol from a pheochromocytoma. *Yale Journal of Biology and Medicine,* **31,** 363–372

MUNCK, A., GUYRE, P. M. and HOLBROOK, N. J. (1984) Physiological functions of glucocorticoids in stress and their relation to pharmacological actions. *Endocrine Reviews,* **5,** 25–44

NEFF, F. C., TICE, G. M., WALKER, G. A. and OCKERBLAD, N. (1942) Adrenal tumour in female infant with hypertrichosis, hypertension, overdevelopment of external genitalia, obesity, but absence of breast enlargement. *Journal of Clinical Endocrinology,* **2,** 125–127

NEVILLE, A. M. and O'HARE, M. J. (1979) Aspects of structure, function and pathology. In *The Adrenal Gland,* edited by V. H. T. James, pp. 1–65. New York: Raven Press

NICOL, C. J. M. and RAE, R. M. (1972) Inhibition of accumulation of adrenaline and noradrenaline in arterial smooth muscle by steroids. *Proceedings of the British Pharmacological Society,* **4,** 361P–362P

NILSSON, O. R., KARLBERG, B. E. and SODERBERG, A. (1980) Plasma catecholamines and cardiovascular responses to hypoglycemia in hyperthyroidism before and during treatment with metoprolol or propranolol. *Journal of Clinical Endocrinology and Metabolism,* **50,** 906–911

PARVEZ, H. and PARVEZ, S. (1972) Control of catecholamine release and degradation by the glucocorticoids. *Experientia,* **28,** 1330–1332

PISANO, J. J. (1960) A simple analysis for normetanephrine and metanephrine in urine. *Clinica Chimica Acta,* **5,** 406–414

POHORECKY, L. A. and WURTMAN, R. J. (1971) Adrenocortical control of epinephrine synthesis. *Pharmacological Reviews,* **23,** 1–35

RAAB, W. (1959) Transmembrane cationic gradient and blood pressure regulation. Interaction of corticoids, catecholamines and electrolytes on vascular cells. *American Journal of Cardiology,* **4,** 752–774

RAMEY, E. R. and GOLDSTEIN, M. S. (1957) The adrenal cortex and the sympathetic nervous system. *Physiological Reviews,* **37,** 155–195

RAMSAY, I. D. and LANGLANDS, J. H. M. (1962) Phaeochromocytoma with hypotension and polycythaemia. *Lancet,* **2,** 126–128

RASTOGI, R. B. and SINGHAL, R. L. (1978) Evidence for the role of adrenocortical hormones in the regulation of noradrenaline and dopamine metabolism in certain brain areas. *British Journal of Pharmacology,* **62,** 131–136

ROSA, R. M., SILVA, P., YOUNG, J. B. *et al.* (1980) Adrenergic modulation of extrarenal potassium disposal. *New England Journal of Medicine,* **302,** 431–434

RUDMAN, D., MOFFITT, S. D., FERNHOFF, P. M., BLACKSTONE, R. D. and FARAJ, B. A. (1981) Epinephrine deficiency in hypocorticotropic hypopituitary children. *Journal of Clinical Endocrinology and Metabolism,* **53,** 722–729

SAAVEDRA, J. M., PALKOVITS, M., BROWNSTEIN, M. J. and AXELROD, J. (1974) Localisation of phenylethanolamine *N*-methyltransferase in the rat brain nuclei. *Nature,* **248,** 695–696

SACCA, L., CICALA, M., CORSO, G., UNGARO, B. and SHERWIN, R. S. (1981) Effect of counterregulatory hormones on kinetic response to ingested glucose in dogs. *American Journal of Physiology,* **240,** E465–E473

SACCA, L., MORRONE, G., CICALA, M., CORSO, G. and UNGARO, B. (1980) Influence of epinephrine, norepinephrine and isoproterenol on glucose homeostasis in normal man. *Journal of Clinical Endocrinology and Metabolism,* **50,** 680–684

SALT, P. J. (1972) Inhibition of noradrenaline uptake$_2$ in the isolated rat heart by steroids, clonidine and methoxylated phenylethylamines. *European Journal of Pharmacology,* **20,** 329–340

SANDBERG, A. A., NELSON, D. H., PALMER, J. G., SAMUELS, L. T. and TYLER, F. H. (1953) The effects of epinephrine on the metabolism of 17-hydroxycorticosteroids in the human. *Journal of Clinical Endocrinology and Metabolism,* **13,** 629–647

SASAKI, K., MURABAYASHI, S., BABA, T. *et al.* (1983) The mechanism of aldosterone response to furosemide test in patients with Shy-Drager syndrome. *Hormone and Metabolic Research,* **15,** 610–615

SCHMID, P. G., ECKSTEIN, J. W. and ABBOUD, F. M. (1967) Comparison of effects of deoxycorticosterone and dexamethasone on cardiovascular responses to norepinephrine. *Journal of Clinical Investigation,* **46,** 590–598

SCHRIER, R. W., BERL, T. and HARBOTTLE, J. A. (1973) Mechanism of effect of α-adrenergic stimulation with norepinephrine on renal water excretion. *Journal of Clinical Investigation,* **52,** 502–511

SCHRIER, R. W., LIEBERMAN, R., UFFERMAN, R. C. and HARBOTTLE, J. A. (1972) Mechanism of antidiuretic effect of β-adrenergic stimulation. *Journal of Clinical Investigation,* **51,** 97–111

SCHUMER, W. (1976) Steroids in the treatment of clinical septic shock. *Annals of Surgery,* **184,** 333–341

SEGRE, E. J., FRIEDRICH, E. H., DODEK, O. I. *et al.* (1966) Effects of epinephrine on the production and metabolic clearance of cortisol in normal men and women with idiopathic hirsutism. *Acta Endocrinologica,* **53,** 561–570

SELYE, H. (1970) The evolution of the stress concept. Stress and cardiovascular disease. *American Journal of Cardiology,* **26,** 289–299

SHAMOON, H., HENDLER, R. and SHERWIN, R. S. (1981) Synergistic interactions among anti-insulin hormones in the pathogenesis of stress hyperglycemia in humans. *Journal of Clinical Endocrinology and Metabolism,* **52,** 1235–1241

SHEPHERD, D. M. and WEST, G. B. (1951) Noradrenaline and the suprarenal medulla. *British Journal of Pharmacology,* **6,** 665–674

SHUBIN, H. and WEIL, M. H. (1966) Failure of corticosteroid to potentiate sympathomimetic pressor response during shock. *Journal of the American Medical Association,* **197,** 154–155

SNIDER, S. R. and KUCHEL, O. (1983) Dopamine: an important neurohormone of the sympathoadrenal system. Significance of increased peripheral dopamine release for the human stress response and hypertension. *Endocrine Reviews,* **4,** 291–309

SNIDER, S. R., SAHAR, D., PRASAD, A. L. N. and FAHN, S. (1977) Changes in adrenal dopamine concentration after metyrapone or ACTH administration: implications for the *in vivo* regulation of dopamine β-hydroxylase by glucocorticoids. *Life Sciences,* **20,** 1077–1086

SOLIMAN, K. F. A. and KOLTA, M. G. (1981) *In vitro* response of the regenerating adrenal gland to epinephrine. *Research Communications in Chemical Pathology and Pharmacology,* **32,** 373–376

SOMAN, V. R., SHAMOON, H. and SHERWIN, R. S. (1980) Effects of physiological infusion of epinephrine in normal humans: relationship between the metabolic response and β-adrenergic binding. *Journal of Clinical Endocrinology and Metabolism,* **50,** 294–297

SOWERS, J. R., BERG, G., MARTIN, V. S. and MAYES, D. M. (1982) Dopaminergic modulation of aldosterone secretion in the Rhesus monkey: evidence that dopamine affects the late pathway of aldosterone biosynthesis by inhibiting the conversion of corticosterone to 18-hydroxycorticosterone. *Endocrinology,* **110,** 1173–1177

SOWERS, J. R., BRICKMAN, A. S., SOWERS, D. K. and BERG, G. (1981) Dopaminergic modulation of aldosterone secretion in man is unaffected by glucocorticoids and angiotensin blockade. *Journal of Clinical Endocrinology and Metabolism,* **52,** 1078–1084

STEINER, A. L., GOODMAN, A. D. and POWERS, S. R. (1968) Study of a kindred with pheochromocytoma, medullary thyroid carcinoma, hyperparathyroidism and Cushing's disease: multiple endocrine neoplasia, type 2. *Medicine,* **47,** 371–409

STILES, G. L., STADEL, J. M., DE LEAN, A. and LEFKOWITZ, R. J. (1981) Hypothyroidism modulates β-adrenergic adenylate cyclase interactions in rat reticulocytes. *Journal of Clinical Investigation,* **68,** 1450–1455

STRUTHERS, A. D. and REID, J. L. (1984) Adrenaline causes hypokalaemia in man by β_2-adrenoreceptor stimulation. *Clinical Endocrinology,* **20,** 409–414

SUMMERS, K. M., HARRISON, G. A., HUME, D. A. and PALMER, C. D. (1983) Urinary hormone levels: a population study of associations between steroid and catecholamine excretion rates. *Annals of Human Biology,* **10,** 99–110

SWEAT, M. L. and BRYSON, M. J. (1965) Steroid 11β-oxygenation, ascorbic acid, epinephrine and adrenochrome interrelations in the adrenal gland. *Endocrinology,* **76,** 773–775

SYMINGTON, T. (1969) *Functional Pathology of the Human Adrenal Gland,* pp. 13–21. Edinburgh and London: Livingstone

TIETZE, H. U., ZURBRÜGG, R. P., ZUPPINGER, K. A., JOSS, E. E. and KÄSER, H. (1972) Occurrence of impaired cortisol regulation in children with hypoglycemia associated with adrenal medullary hyporesponsiveness. *Journal of Clinical Endocrinology and Metabolism,* **34,** 948–958

TILDERS, F. J. H., BERKENBOSCH, F. and SMELIK, P. G. (1980) Adrenergic mechanisms in the control of corticotropin release from the intermediate lobe. In *Catecholamines and Stress: Recent Advances,* edited by E. Usdin, R. Kvetňanský and I. J. Kopin, pp. 125–130. New York: Elsevier/North-Holland

TISCHLER, A. S., PERLMAN, R. L., MORSE, G. M. and SHEARD, B. E. (1983) Glucocorticoids increase catecholamine synthesis and storage in PC12 pheochromocytoma cell cultures. *Journal of Neurochemistry,* **40,** 364–370

TORDA, T. and KVETŇANSKÝ, R. (1983) Effect of adrenal medullectomy on the activity of COMT and MAO in adrenal cortex of control and stressed rats. *Endocrinologica Experimentalis,* **17,** 47–52

UNGAR, A. and PHILLIPS, J. H. (1983) Regulation of the adrenal medulla. *Physiological Reviews,* **63,** 787–843

VAN LOON, G. R. and SOLE, M. J. (1980) Plasma dopamine: source, regulation, and significance. *Metabolism,* **29** (Suppl. 1), 1119–1123

VETTER, H., FISCHER, N., BAYER, J. M., SCHNITZ, TH.E., WERNING, C. and VETTER, W. (1974) Aldosteron, Cortisol und Plasma-Renin-Aktivität beim Phäochromocytom. *Klinische Wochenschrift,* **52,** 719–721

VETTER, H., VETTER, W., WARNHOLZ, C. *et al.* (1976) Renin and aldosterone secretion in pheochromocytoma. *American Journal of Medicine,* **60,** 866–871

VIVEROS, O. H., DILIBERTO, E. J., HAZUM, E. and CHANG, K. J. (1979) Opiate-like materials in the adrenal medulla: evidence for storage and secretion with catecholamines. *Molecular Pharmacology,* **16,** 1101–1108

WEINSHILBOUM, R. and AXELROD, J. (1970) Dopamine-β-hydroxylase activity in the rat after hypophysectomy. *Endocrinology,* **87,** 894–899

WESTFALL, T. C. and OSADA, H. (1969) Influence of adrenalectomy on the synthesis of norepinephrine in the rat heart. *Journal of Pharmacology and Experimental Therapeutics,* **167,** 300–308

WILCOX, C. S., AMINOFF, M. J., MILLAR, J. G. B., KEENAN, J. and KREMER, M. (1975) Circulating levels of corticotrophin and cortisol after infusions of L-dopa, dopamine and noradrenaline in man. *Clinical Endocrinology,* **4,** 191–198

WILLIAMS, G. A., CROCKETT, C. L., BUTLER, W. W. S. and CRISPELL, K. R. (1960) The coexistence of pheochromocytoma and adrenocortical hyperplasia. *Journal of Clinical Endocrinology and Metabolism,* **20,** 622–631

WURTMAN, R. J. and AXELROD, J. (1965) Adrenaline synthesis: control by the pituitary and adrenal glucocorticoids. *Science (Washington),* **150,** 1464–1465

YARD, A. C. and KADOWITZ, P. J. (1972) Studies on the mechanism of hydrocortisone potentiation of vasoconstrictor responses to epinephrine in the anesthetized animal. *European Journal of Pharmacology,* **20,** 1–9

YOUNG, J. B. and LANDSBERG, L. (1977) Catecholamines and the regulation of hormone secretion. *Clinics in Endocrinology and Metabolism,* **6,** 657–695

11
Pharmacokinetics of natural and synthetic glucocorticoids

Linda E. Gustavson and Leslie Z. Benet

Several processes are interposed between the administration of a drug and the appearance of its therapeutic effect (*Figure 11.1*). The pattern of the drug concentration–time curve measurable in systemic body fluids (blood or plasma) is a function of three major pharmacokinetic variables: the bioavailability, the distribution and the clearance (or loss) of the drug. Pharmacokinetics may be defined as what the body does to the drug, as opposed to pharmacodynamics, which is what the drug does to the body. The purpose of this chapter is to review the pharmacokinetics of natural and synthetic glucocorticoids.

PRINCIPLES OF CLINICAL PHARMACOKINETICS

A large number of parameters may be used to define the pharmacokinetics of a drug or metabolite; these may vary between animal models, healthy volunteers and

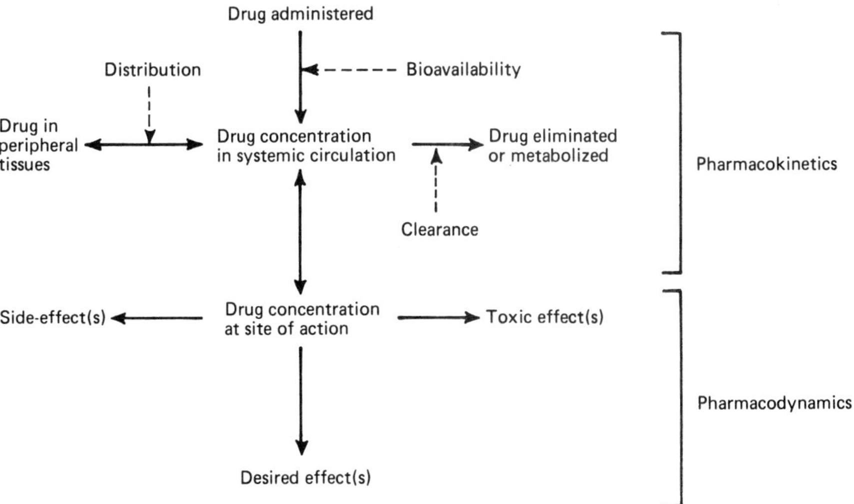

Figure 11.1 The inter-relationship between pharmacokinetics and pharmacodynamics in drug therapy

patient populations. Several comprehensive reviews of these parameters are available (Wagner, 1975; Rowland and Tozer, 1980; Gibaldi and Perrier, 1982; Benet, Massoud and Gambertoglio, 1984).

Clearance

Under steady-state conditions, if the drug concentration (c) is known, the clearance (C) in a particular patient will determine the correct dosing rate. At the simplest level, clearance of a compound is the rate of elimination by all routes relative to its concentration in a biological fluid (plasma or blood); clearance is usually further defined as blood clearance (C_B), plasma clearance (C_P) or unbound clearance (C^u), depending on the concentration measured (c_B, c_P or c^u). Clearance indicates the volume of fluid (blood or plasma) which would be cleared of a compound (unbound or total) over a given time, and is expressed as volume per unit of time.

If blood concentration is used to define clearance, the maximum clearance possible is equal to the sum of the blood flows to the various organs of elimination. For a compound eliminated solely by the liver, blood clearance is therefore limited to liver blood flow, approximately 1500 ml/min in humans. Blood clearance may be determined from plasma clearance if the blood/plasma distribution ratio or the red blood cell (RBC)/plasma concentration ratio is known:

$$\frac{C_P}{C_B} = \frac{c_B}{c_P} = 1 + \text{Ht}\,(\frac{c_{RBC}}{c_P} - 1)$$

where Ht is the hematocrit. Dividing the rate of elimination at each organ by the systemic concentration of the compound will yield the respective clearance at that organ. Added together, these separate clearances equal total systemic clearance. Other routes of elimination might include saliva or sweat, or other sites of metabolism, such as adrenal or muscle tissue. Appearance of unchanged compound in the urine represents renal clearance. Within the liver, elimination of a drug may occur via biotransformation to one or more metabolites and/or excretion of unchanged drug into the bile.

For many drugs, elimination is not saturable over the therapeutic range, and the rate of elimination is directly proportional to concentration. For an intravenous dose the amount of drug eliminated from the body is equal to the administered dose. For other dosing routes the amount eliminated is equal to the available dose, i.e. the portion of the administered dose which reaches the systemic circulation intact (*see* under bioavailability):

$$C = \frac{\text{amount eliminated}}{\text{AUC}} = \frac{\text{dose}^{\text{i.v.}}}{\text{AUC}} = \frac{F\ \text{dose}}{\text{AUC}}$$

where F is the systemic availability of the drug and AUC the area under the concentration–time curve from time zero to infinity.

For drugs that exhibit saturable or dose-dependent (non-linear) elimination, the rate of elimination and clearance varies with concentration. The sum of the clearances by all such pathways can be expressed in the classical Michaelis-Menten kinetic terms of V_{max}, K_m and concentration:

$$C = \sum_{i=1}^{n} \frac{V_{max,i}}{K_{m,i} + c}$$

where there are n such different pathways and $K_{m,i}$ and $V_{max,i}$ represent, respectively, the plasma concentration at which half of the maximal rate of elimination is reached (in units of mass per volume) and the maximal rate of elimination (in units of mass per time) for the ith pathway. For a compound to exhibit linear kinetics, K_m values for all routes of elimination must be substantially greater than the encountered concentrations. Clearance is approximated by the ratio of $V_{\max}$ to K_m summed over all routes, and is constant. A more detailed consideration of concentration-dependent elimination is given by Wilkinson (1984).

Elimination of a compound by an individual body organ can also be defined in terms of blood flow through the organ and its concentration in the blood. The rate of presentation of compound to the organ is the product of blood flow (Q) and entering concentration (c_{in}), whereas the rate of exit of drug from the organ is the product of blood flow and exiting drug concentration (c_{out}). The difference between these rates at steady state is the rate of drug elimination. Dividing this by the concentration of drug entering the eliminating organ (c_{in}), we obtain an expression for organ clearance of drug:

$$C_{\text{organ}} = \frac{Qc_{\text{in}} - Qc_{\text{out}}}{c_{\text{in}}} = Q\frac{c_{\text{in}} - c_{\text{out}}}{c_{\text{in}}} = QE$$

The ratio $(c_{\text{in}} - c_{\text{out}})/c_{\text{in}}$ is referred to as the extraction ratio (E) of the drug.

These simple concepts have important implications for compounds eliminated by the liver. For a drug that is efficiently removed by hepatic processes, the extraction ratio will approach unity and clearance will be limited only by blood flow. The rate of elimination of these highly extracted drugs is restricted not by intrahepatic processes but by the rate at which they can be transported in the blood to hepatic sites of elimination. It is apparent that this simple equation for clearance does not take into account the complexity of hepatic or renal drug elimination. It does not account for protein binding of drug to blood and tissue components, nor does it permit an estimation of the intrinsic ability of the liver or kidney to eliminate a drug in the absence of limitations imposed by blood flow.

It is desirable to include expressions for protein binding and intrinsic clearance (C_{int}) in any formula describing organ elimination of drugs. The simplest variant is the so-called venous equilibration or 'well-stirred model' (Rowland, Benet and Graham, 1973; Wilkinson and Shand, 1975), which assumes that the unbound concentration leaving the organ is equal to that inside the organ. The blood clearance for the eliminating organ then becomes

$$C = Q\frac{f_{\text{B}}\, C^{\text{u}}_{\text{int}}}{Q + f_{\text{B}}\, C^{\text{u}}_{\text{int}}}$$

where f_{B} is the unbound fraction of drug in blood and $C^{\text{u}}_{\text{int}}$ is the intrinsic ability of the organ to clear unbound drug.

If the capacity of the eliminating organ to metabolize the drug is large in comparison to the rate of drug presentation to the organ ($f_{\text{B}}\, C^{\text{u}}_{\text{int}}$ much greater than Q), the clearance will approximate the organ blood flow; that is, drug elimination will be limited by blood flow rate, and the compound is said to have a high extraction ratio. On the other hand, if metabolic capacity is small in comparison to the rate of drug presentation, clearance will be proportional to the unbound fraction of drug in blood and to the intrinsic clearance. The drug is then said to have a low extraction ratio.

The availability of a drug that is highly extracted by the liver is influenced by even small changes in the hepatic extraction ratio. The systemic availability (F) of a drug that is completely absorbed from the gastrointestinal tract and enters the liver without biotransformation in the portal vein or gut wall is diminished by the extent of hepatic extraction:

$$F = 1 - E$$

Small changes in E for a highly extracted drug can result in dramatic changes in availability. For example, a small decrease in E from 0.95 to 0.90 results in a doubling of F from 0.05 to 0.10 and a doubling of the available dose.

Volume of distribution

Volume of distribution (V) is a measure of the apparent body space available to contain the drug, and relates the amount of drug in the body to its concentration in blood or plasma. The plasma volume of a healthy person (70 kg) is about 3 liters, compared to a blood volume of 5.5 liters, extracellular fluid volume outside plasma of 12 liters, and total body water of approximately 42 liters. Yet many drugs exhibit a distribution volume far in excess of these known body fluid volumes; thus V does not represent a real body volume but the theoretical size of the pool required if the drug were equally distributed throughout all portions of the body.

In a 'one-compartment' model administration occurs directly into the body (or central) compartment, and the drug is instantaneously distributed throughout volume V. Clearance from this compartment shows first-order kinetics. The decline of plasma concentration with time for a drug injected into such a body compartment is defined as:

$$c = \frac{\text{dose}}{V} \, e^{-\lambda t}$$

where λ is the rate constant for elimination of drug from the body compartment, and is inversely related to the half-life ($T½$) of the drug ($\lambda = 0.693/T½$). When T equals 0, the volume of distribution can be calculated as the dose divided by the concentration at time zero; the latter may be estimated by extrapolating the concentration–time curve back to the y-axis.

Unfortunately, the one-compartment model is not adequate for drugs whose distribution changes with time. In these cases, multi-compartment volume terms representing the total body distribution space must be utilized. V_{area} represents the apparent distribution space during the terminal logarithmic linear phase of drug elimination.

$$V_{area} = \frac{C}{\lambda_z} = \frac{F \text{ dose}}{\lambda_z \text{ AUC}}$$

where λ_z is determined by dividing 0.693 by the terminal half-life of elimination.

Although V_{area} is a convenient and easily calculated parameter, it varies when the rate constant for elimination changes, even when there has been no change in distribution space.

Since we may wish to know whether a disease state or coadministered drug influences either clearance or distribution independently, a volume which is theoretically independent of changes in the rate of elimination may be more useful.

The volume of distribution at steady state (V_{ss}) represents the volume in which a drug would appear to be distributed at steady state if its concentration throughout that volume was the same as that in the analyzed fluid (blood or plasma). This volume can be determined using areas as described by Benet and Galeazzi (1979).

For drugs discussed in this chapter, volumes of distribution are preferentially expressed as V_{ss} when this value is available. However, since it is necessary to administer a drug intravenously to obtain a valid estimate of V_{ss}, many oral dose studies yield only V_{area}/F. In these cases, the volume parameter cannot be calculated independent of bioavailability.

Half-life

Half-life is an expression of the relationship between volume and clearance. For the simple one-compartment model

$$T\frac{1}{2} = 0.693\ V/C$$

For many drugs the plasma concentration–time course appears to follow a multiexponential pattern, which would require more than one compartment and more than one half-life value. In these cases, $T\frac{1}{2}$ as calculated above reflects an overall half-life for the compound in the body.

Half-life is a useful kinetic parameter, in that it indicates the time required to attain, and decay from, steady-state conditions, and the maximum and minimum concentrations for a particular dosage regimen. However, as an indicator of whether disease results in changes in either elimination or distribution, half-life is of little value. Disease can affect either or both of the physiologically related parameters volume and clearance. The derived parameter, $T\frac{1}{2}$, will not necessarily reflect the changes in drug disposition.

Bioavailability

Drug input efficiency is described as bioavailability, a term used to measure both the amount of an administered dose that reaches the general circulation and the rate at which this occurs. In most clinical situations the extent of drug availability (F) is considered to be more important than the rate of availability, and may be defined as the fraction of unchanged drug reaching the systemic circulation after administration by any route. For intravenously given drugs, bioavailability is considered to be unity. However, drugs administered orally may be incompletely absorbed, undergo enterohepatic cycling with incomplete reabsorption, or be metabolized in the gut, gut wall, portal blood or liver before entering the systemic circulation. Loss of absorbed drug before it reaches the systemic circulation is known as the 'first-pass' effect and may be substantial. It can be largely avoided by using sublingual tablets or vaginal or topical preparations because the blood supply from these areas does not pass through the liver before reaching the systemic circulation. Rectal administration partially avoids first-pass metabolism, since about 50% of the dose can be expected to bypass the liver (de Boer *et al.*, 1982).

The extent of availability of a particular drug in a particular dosage form may be calculated as either absolute or relative bioavailability. Absolute bioavailability (F_{abs}) may be determined if the drug can be administered intravenously:

$$F_{abs} = \frac{AUC_{test}}{AUC_{i.v.}} \times \frac{dose_{i.v.}}{dose_{test}}$$

If no intravenous dosage form exists, bioavailability must be determined relative to some other dosage form which will serve as the standard (std).

$$F_{rel} = \frac{AUC_{test}}{AUC_{std}} \times \frac{dose_{std}}{dose_{test}}$$

For an oral tablet test dose the standard might be an oral solution or suspension, or a tablet prepared by another manufacturer.

Although only the extent of availability is important in the design of multiple-dose regimens, differences in rate of availability may be important for drugs given as a single dose. If a measure of the rate of availability is desired, it is necessary to fit the plasma concentration-time curve to a pharmacokinetic model to obtain a rate constant for absorption. Comparisons of peak time and peak concentration are sufficient for most clinical situations.

Macromolecular binding

Macromolecular substances capable of binding drugs include albumin, α_1-acid glycoprotein, lipoprotein and a group of specialized binding globulins such as corticosteroid-binding globulin (CBG or transcortin). Non-linear (saturable) binding may occur, depending on the total number of available binding sites, the number of molecules competing for the binding sites and the dissociation constant for the interaction(s) involved. The specialized binding globulins and α_1-acid glycoprotein are most often subject to saturation because of their lower concentrations and lower number of available binding sites. The unbound (free) fraction of a drug is the portion considered to be biologically active, and so saturable protein binding is of great significance when it occurs. Increases in the dose or total concentration can lead to a disproportionate increase in the unbound concentration, which may be reflected in an unexpected increase in pharmacological and/or toxicological effects.

Changes in protein binding may alter not only the clearance but also the volume of distribution of a drug (Øie and Tozer, 1979). If a drug is more strongly bound to plasma proteins than tissue proteins, the volume of distribution will be small, while if the converse holds it will be large. Independent of protein binding, the lipophilicity of a compound also affects the volume of distribution, since highly lipid-soluble drugs partition preferentially out of the blood into peripheral fatty tissues. Finally, when plasma protein binding becomes saturated, increased free drug in the plasma may bind more readily to tissue proteins or distribute out of the blood into other body tissues.

As discussed above, clearance of highly extracted drugs may be independent of protein binding. Therefore, saturable protein binding could lead to an increased volume of distribution but no change in clearance. However, half-life would increase since less drug would be located in the blood and available for clearance by the organs of elimination.

PHARMACOKINETICS OF GLUCOCORTICOIDS

Early pharmacokinetic studies were compromised by problems with the determination of glucocorticoid levels in body fluids. Prednisone assays, for example, in some cases also measured cortisone; lack of assay sensitivity would often result in an inability to follow drug levels for sufficient time to characterize the pharmacokinetics. A second problem is that sometimes half-life was the only pharmacokinetic parameter reported. Nevertheless, some of these early studies are included here for comparative purposes. This section will deal primarily with values derived from healthy adult volunteers, while later sections will discuss the effects of disease states and age.

Natural glucocorticoids: cortisol and cortisone

The principal natural glucocorticoid in humans is cortisol (or hydrocortisone), which is systemically interconvertible with cortisone, an inactive metabolite (Hollander *et al.*, 1951; Robinson and Robinson, 1956; Jenkins, Lowe and Titterington, 1964). Because of this interconversion, either can be administered to achieve the same pharmacological effects.

The measurement of cortisol and cortisone levels in body fluids for pharmacokinetic analysis is complicated by the endogenous presence of these same compounds, and their circadian rhythm of production. In most recent pharmacokinetic studies, endogenous glucocorticoids have been suppressed by pretreatment with dexamethasone. A major drawback of dexamethasone suppression is the lack of information regarding its effect on the pharmacokinetics of the natural glucocorticoids. Other studies have used radiolabeled (hydrogen-3 or carbon-14) cortisol or cortisone.

Clearance, volume of distribution and half-life

Brown *et al.* (1954) administered cortisol 1 mg/kg intravenously to 11 normal non-suppressed subjects and found a mean half-life of 1.91 hours for disappearance of 17-hydroxycorticosteroids. Peterson *et al.* (1955) used intravenous doses of 50 to 500 mg of cortisol as 20–30 min infusions in 20 non-suppressed subjects; the mean half-life for all subjects was 1.9 hours. A further ten subjects received tracer doses of 4[^{14}C] cortisol; half-life for disappearance of ^{14}C ranged from 1.0 to 1.5 hours. Fell (1972) used hydrocortisone hemisuccinate 10 mg in eight healthy dexamethasone-suppressed volunteers; the mean elimination half-life was 1.65 hours.

Using tracer doses of ^{14}C- or ^{3}H-labeled cortisol in four normal subjects, Kehlet and Binder (1973) found a mean volume of distribution (V) of 17.2 liters, and a mean half-life of 1.25 hours. The mean metabolic clearance was 160 ml/min (*Table 11.1*).

Dazord, Saez and Bertrand (1972) infused tracer quantities of [^{3}H] cortisol and [^{14}C] cortisone in nine normal subjects and found a mean clearance of 542 ml/min per 1.73 m^2 for cortisone, and 186 ml/min per 1.73 m^2 for cortisol (*Table 11.1*).

de Lacerda, Kowarski and Migeon (1973) studied the diurnal variation of cortisol clearance by giving [^{14}C] cortisol at 2000 hours and [^{3}H] cortisol nine hours later. The reported mean clearance was 201 ml/min per 1.73 m^2 over the period between

Table 11.1 Summary of pharmacokinetic parameters for cortisol (hydrocortisone) determined after intravenous doses of cortisol in healthy volunteers

Reference	*Subjects*	*Dose (mg)*	*Clearance (ml/min)*	*Half-life (hours)*	*Volume (l)*
Kehlet and Binder (1973)	4	*	160 ± 57	1.25 ± 0.22	17.2 ± 5.0
Dazord, Saez and Bertrand (1972)	9	*	186 ± 48		
de Lacerda, Kowarski and Migeon (1973)	11	*	201 ± 45 (0500–1100 hours)		
			120 ± 27 (2000–0200 hours)		
Scavo, Cugini and DiLascio (1978)	10	*	197 ± 29	1.30 ± 0.19	19.3 ± 4.3
Lima and Jusko (1980)	12	37	398 ± 87	1.71 ± 0.50	56.8 ± 11.5
Toothaker and Welling (1982)	6	5	362 ± 73	1.3 ± 0.3	20.7 ± 7.3
	6	10	377 ± 40	1.3 ± 0.2	20.8 ± 4.3
	6	20	413 ± 76	1.7 ± 0.2	26.0 ± 4.1
	6	40	509 ± 59	1.9 ± 0.1	37.5 ± 5.8

* Tracer dose.

0500 and 1100 hours, and 120 ml/ ... er 1.73 m^2 over the period from 2000–0200 hours (*Table 11.1*).

Scavo, Cugini and DiLascio (1978) administered tracer doses of 1,2-[^{3}H] cortisol to ten normal volunteers as an intravenous bolus, and obtained similar results to those of Kehlet and Binder (1973) (*Table 11.1*).

Colburn *et al.* (1980) studied the pharmacokinetics of oral cortisone acetate in 23 dexamethasone-suppressed healthy volunteers. The protocol included single tablet doses of 25, 10 and 5 mg, and a 25 mg dose consisting of five 5 mg tablets. For cortisone, the ratio *C/F* was 2500, 1350, 783 and 2580 ml/min, respectively, while for cortisol it was 265, 191, 109 and 283 ml/min. These data clearly demonstrate dose dependence which can be attributed to changes in clearance, availability or both.

Lima and Jusko (1980) administered hydrocortisone hemisuccinate by i.v. infusion over two hours to 12 normal healthy subjects suppressed by dexamethasone 1 mg taken 8–12 hours before dosing. The infused dose of hydrocortisone hemisuccinate was equivalent to 37.0 ± 6.2 mg cortisol. The pharmacokinetic parameters are shown in *Table 11.1*. Toothaker and Welling (1982) used single i.v. bolus doses of hydrocortisone sodium succinate equivalent to 5, 10, 20 and 40 mg cortisol in six healthy volunteers suppressed with 2 mg dexamethasone. Plasma clearance was increased significantly only at the highest dose (*Table 11.1*). Similarly, the steady-state volume of distribution (V_{ss}) determined from the 40-mg dose was significantly increased, as was the overall elimination half-life. The authors concluded that the elimination and distribution of cortisol are independent of dose size up to 20 mg, but dose-dependent at higher doses, perhaps due to the saturable protein binding of cortisol. Toothaker *et al.* (1982) also studied three healthy dexamethasone suppressed volunteers after single oral doses of 10, 30 and 50 mg cortisol as tablets, and found a half-life of 1.50 ± 0.17, 1.29 ± 0.09 and 1.62 ± 0.09 hours respectively; the three means were not statistically different. The ratio *C/F* was 352, 522 and 535 ml/min for the 10, 30 and 50 mg doses, while V_{area}/F was 44.9, 58.0 and 74.5 liters. Dose-dependence of *F* and/or *C* and V_{area} may be responsible for the trends observed in these values. The authors believe that since the cortisol concentrations found in this study are in the range found for the lower doses (5, 10 and 20 mg) in their intravenous study (Toothaker and Welling, 1982), for which no dose-dependence of clearance or volume was found, the oral route is associated with a dose-dependent decrease in bioavailability.

In another study by this group (Toothaker, Craig and Welling, 1982), oral doses of 5, 10, 20 and 40 mg cortisol suspension were given to eight healthy dexamethasone-suppressed volunteers. The half-life at the different doses was 1.2 ± 0.1, 1.3 ± 0.2, 1.5 ± 0.1 and 1.7 ± 0.3 hours respectively. *C/F* was 284, 373, 399 and 498 ml/min, and V_{area}/F was 28.4, 39.2, 52.1 and 71.1 liters for the 5, 10, 20 and 40 mg doses respectively. The authors suggest that, as the dose increases, transient high drug levels in the splanchnic circulation saturate the binding capacity of plasma proteins and lead to increased local clearance, an increased first-pass effect, and decreased bioavailability.

The pharmacokinetic parameters described above for intravenous cortisol doses are summarized in *Table 11.1*. Results with oral doses, where *C/F* and *V/F* must be determined, appear consistent with these findings. The values for clearance and volume should be considered 'apparent' parameters since in each case calculations were made assuming linear kinetics for the drug. Overall, terminal half-lives for cortisol are reported to be 1.3 to 1.9 hours. Early studies yielded clearance values approximately half of those obtained with more specific and sensitive analytical procedures. Clearance is dose-dependent, increasing from about 190 ml/min for tracer doses to about 300–500 ml/min over the 5–50 mg dose range. Volume terms also appear to be dose-dependent, increasing from approximately 20 liters at low doses to 40–70 liters at doses of 40–50 mg. So far, no kinetic results for unbound cortisol have been reported. Such results would be useful in deciding whether dose-dependent changes result from changes in protein binding, as might be expected.

Bioavailability

Absorption of cortisol has been shown to occur from the small intestine by diffusion, with a distally decreasing gradient (Schedl and Clifton, 1963; Schedl, 1965). While studies alluding to the bioavailability of cortisol from various oral and intramuscular routes will be discussed here, rigorous studies of the absolute bioavailability of cortisol have been performed only for the rectal route of administration.

Plumpton, Besser and Cole (1969) studied the effect of steroid cover on 15 patients with compromised adrenal function receiving long-term steroid therapy as they underwent various surgical procedures. Five patients received intramuscular cortisone acetate 100 mg for three days (beginning the evening before the operation), followed by four days of oral cortisone acetate 50 mg. The other ten received intramuscular hydrocortisone hemisuccinate 100 mg at the time of surgical premedication and for three days thereafter. In the five patients receiving cortisone acetate, maximum plasma cortisol levels after the i.m. dose averaged 190 ng/ml; when oral cortisone acetate was started, this increased to 470 ng/ml. Those receiving i.m. hydrocortisone hemisuccinate had mean maximum plasma cortisol levels of 960 ng/ml, suggesting that cortisone acetate is either poorly absorbed or poorly converted to cortisol after i.m. administration. Its bioavailability is improved by oral administration. The authors conclude that intramuscular administration of hydrocortisone hemisuccinate is the only satisfactory presurgical treatment for patients with inadequate adrenocortical response to stress. Similar results for intramuscular administration were reported by Kehlet, Nistrup-Masden and Binder (1974) and Fariss *et al.* (1978).

Kehlet, Binder and Blichert-Toft (1976) compared the results of oral cortisol and cortisone acetate in seven adrenalectomized patients. On the first day, each patient received a total dose of cortisol of 25 mg/g urine creatinine per 24 hours, divided into two doses: two-thirds at 0800 hours and one-third at 1800 hours. The next day, cortisone acetate 33 mg/g urine creatinine per 24 hours was administered in similarly divided doses. Following cortisol administration, mean peak concentrations of 420 ng/ml (range 270–610) and 190 ng/ml (range 120–280) were achieved 1.02 hours (range 0.67–1.5) after the morning and evening dose respectively. When cortisone acetate was given, mean cortisol levels of 270 ng/ml (range 220–400) and 180 ng/ml (range 120–220) were achieved 1.5 hours (range 1.0–2.0) after the morning and evening dose respectively. Similar results have been reported by Fariss *et al.* (1978).

Colburn *et al.* (1980) determined AUC (*see earlier*) values for cortisone and cortisol following four single doses of cortisone acetate (three single tablets of 25, 10 and 5 mg, and one dose of 25 mg as five 5 mg tablets). Mean AUC measurements for cortisone were 163.2, 123.4, 106.4 and 161.7 ng · hour · ml^{-1} respectively. Equivalent measurements for cortisol were 1570, 873, 761 and 1474 ng · hour · ml^{-1}. There was no difference in bioavailability between the single 25 mg tablet and the five 5 mg tablets. Dose-corrected AUCs were 6.53, 12.3 and 21.3 ng · hour · ml^{-1} · mg^{-1} for cortisone and 62.8, 87.3 and 152.2 ng · hour · ml^{-1} · mg^{-1} for cortisol for the 25, 10 and 5 mg doses respectively. Since non-linear kinetics are involved, it is incorrect to calculate relative bioavailability for the different doses. Thus, one may only conclude that increasing dose leads to decreasing dose-corrected area under the curve. This could be explained by a decreased extent of bioavailability, an increased clearance (which is known from the intravenous studies) or a combination of both.

The absolute extent of cortisol bioavailability (F_{abs}) can be estimated from the results of studies carried out by Toothaker and colleagues (Toothaker and Welling, 1982; Toothaker *et al.*, 1982; Toothaker, Craig and Welling, 1982). The mean AUCs for cortisol were 410, 790, 1480 and 2290 ng · hour · ml^{-1} for intravenous administration of cortisol in doses of 5, 10, 20 and 40 mg respectively, compared to 293, 447, 835 and 1340 ng · hour · ml^{-1} following the same doses of an oral hydrocortisone suspension. These values gave an F_{abs} of 0.71, 0.56, 0.56 and 0.58 respectively, indicating again that bioavailability may be dose-dependent. Only one dose (10 mg) was common to both the intravenous and oral administration studies. The mean AUC for the 10 mg tablet was 474, yielding an F_{abs} of 0.60, which is nearly identical to that (0.56) determined from a 10 mg dose given as an oral suspension.

Technically, for a drug showing non-linear kinetics, a measure of availability may only be made using two doses which yield the same plasma concentration–time curves, since the higher the concentration the greater the clearance of cortisol. Thus, the accuracy of the above estimates may be tested by calculating F_{abs} for the 10 mg oral suspension and tablet doses relative to the 5 mg i.v. dose (0.55 for the 10 mg oral suspension and 0.58 for the 10 mg tablet), the 20 mg oral suspension relative to the 10 mg i.v. dose (F_{abs} = 0.53), and the 40 mg oral suspension relative to the 20 mg i.v. dose (F_{abs} = 0.45). The calculated F_{abs} of the 10 and 20 mg oral doses is comparable to the values calculated using identical i.v. and oral doses. However, for the 40 mg oral dose a marked increase is noted in the calculated value when comparable areas are used rather than comparable doses (0.45 compared to 0.58). This difference emphasizes the significant increase in cortisol clearance at higher doses, and the difficulty in obtaining good measures of bioavailability.

A careful and detailed study of the rectal bioavailability of cortisol from retention enemas has been performed by Lima and Jusko (1980) and Lima *et al.* (1980). Because of suspected concentration-dependent changes in hydrocortisone clearance, they chose to determine bioavailability by using an intravenous infusion (hydrocortisone hemisuccinate 40 mg over two hours) designed to give serum cortisol concentrations equal to those obtained from a rectal hydrocortisone 100 mg enema. The mean bioavailability calculated for the two different types of enema was 0.63 and 0.53 respectively. Serum cortisol concentrations were fitted to a one-compartment model employing a first-order absorption rate constant (k_a). The k_a estimates, which fell into two groups, ranged widely (0.208–1.30/hour). Group 1 were slow absorbers with a mean k_a of 0.361 $hour^{-1}$, equivalent to an absorption half-life of about two hours. Group 2 absorbed hydrocortisone significantly more rapidly, with a mean k_a of 1.05 $hour^{-1}$ (absorption half-life about 40 min). The mean extent of bioavailability for group 1 was significantly higher (F_{abs} 0.818) than for group 2 (F_{abs} 0.502). The authors suggest that there may be two sites of absorption, a rectal site from which slow absorption occurs directly into the systemic circulation, and a colonic site from which more rapid absorption occurs into the portal circulation.

In conclusion, the absolute bioavailability of cortisol ranges from approximately 45 to 80% following oral or rectal administration. Studies using varying oral doses of both cortisol and cortisone indicate a decreasing extent of bioavailability with increasing dose. While cortisone appears to be well absorbed orally, intramuscular injection of cortisone is associated with only slight increases in circulating cortisol levels.

Macromolecular binding

Cortisol in plasma may exist in three states: free (unbound); bound to a specific, high affinity α_1-glycoprotein known as corticosteroid-binding globulin; and bound to albumin. The free portion is generally considered to be the pharmacologically active form (Slaunwhite *et al.*, 1962; Beisel, DiRaimondo and Forsham, 1964). However, Siiteri *et al.* (1982) have suggested that CBG-bound hormone is not inactive but is the form in which steroids are taken up by target cells, a contention that deserves further examination.

The molecular weight of CBG is approximately 52 000 (Seal and Doe, 1962; Muldoon and Westphal, 1967). Its affinity for cortisol is very high, and at physiological levels more than 90% is bound. The glucuronide conjugates of cortisol are not significantly bound to CBG (Sandberg and Slaunwhite, 1959). Early studies reported a K_a of about 3×10^8 l/mol at 37 °C, with a linear increase to about double this value at 4 °C (Daughaday and Mariz, 1961). Increasing the temperature to 41 °C leads to a slight decrease in affinity (Few and Haspineall, 1977), while a further increase to 45 °C leads to a several-fold decrease in K_a (Westphal, 1967) and temperatures of 60 °C or more cause irreversible denaturation (Daughaday and Mariz, 1961). More recent work has determined the affinity of cortisol for CBG in the presence of prednisolone (Rocci *et al.*, 1982b; Legler and Benet, unpublished observations). The K_a (37 °C) was found to vary between 1.0 and 2.4×10^7 l/mol, a value less than one-tenth of that reported above. These differences may be due either to the presence of prednisolone or to significant differences in methodology.

The capacity of CBG is very low, of the order of 1×10^{-1} mol/l (Slaunwhite and Sandberg, 1959), so that at pharmacological concentrations a significant portion of cortisol is bound to albumin (mol. wt. approximately 65 000), which is present in plasma at a much higher concentration (about 10^{-4} mol/l), although its affinity for cortisol is lower.

Smith, Nolan and Jubiz (1980) studied the relationship between total and unbound endogenous cortisol levels in 24 normal subjects. The percentage of unbound cortisol was 9.6 ± 1.3 in the morning and 6.9 ± 1.1 in the afternoon, suggesting that the higher morning cortisol secretion may begin to saturate CBG and increase the free fraction. Unbound cortisol concentrations obtained by multiplying the total concentration by the percentage of unbound cortisol were 12.0 ± 4.0 and 4.0 ± 1.0 ng/ml for the morning and afternoon samples respectively.

Addition *in vitro* of CBG to human liver microsome incubations increased the ratio of cortisol to reduced metabolites from 0.35 to 1.0 (Sandberg and Slaunwhite,

1963). CBG apparently competes efficiently with the enzyme systems responsible for reduction of the steroid in the liver.

Few and Haspineall (1977) examined the protein binding of cortisone and cortisol simultaneously in plasma samples from five subjects and found that cortisone is less strongly bound than cortisol. The mean free percentages were 26.0 and 8.4 respectively. Unfortunately, there is no indication as to which plasma protein(s) bind cortisone or if cortisone competes with cortisol for binding sites.

Synthetic glucocorticoids: prednisolone and prednisone

Attempts to separate the mineralocorticoid and glucocorticoid characteristics of the natural adrenal steroids resulted in the synthesis of many new steroids, including prednisolone (Δ^1-hydrocortisone), which became available for clinical use in 1955. Prednisolone is systemically interconvertible with its metabolite prednisone, which, like cortisone, is assumed to be inactive. Two comprehensive reviews of the pharmacokinetics of prednisolone and prednisone have been published by Pickup (1979) and Gambertoglio, Amend and Benet (1980).

Clearance, volume of distribution, and half-life

Ely, Done and Kelley (1956) determined the 17-hydroxycorticosteroid concentration after oral administration of prednisolone 45 mg as tablets in nine healthy subjects. The mean half-life in eight of the nine subjects was 3.2 hours. Similar results were obtained by Nugent, Eik-Nes and Tyler (1959), who infused prednisolone hemisuccinate 1 mg/kg over 30 min. The reported half-life was 3.5 hours in males and 3.4 hours in females. Coburg *et al.* (1970) administered prednisolone phosphate 1 g by intravenous infusion over 60 min. The disappearance half-life of the mean 17-hydroxycorticosteroid concentration–time curve between 4 and 16 hours was approximately 3.2 hours, similar to that found with smaller doses.

Kozower, Veatch and Kaplan (1974) used intravenous bolus doses of prednisolone phosphate 0.3 mg/kg followed by a tracer dose of [^{3}H]prednisolone in seven normal volunteers. Their results are shown in *Table 11.2.*

DiSanto and DeSante (1975), using oral doses of prednisone 50 mg, reported a mean serum half-life of 3.25 ± 1.47 hours. C/F and V_{area}/F can be calculated from the mean AUCs reported in this study. They are 117 ml/min and 33.3 liters respectively.

Meikle, Weed and Tyler (1975) studied two doses (10 and 40 mg) of intravenous prednisolone in two groups of normal volunteers. All three parameters (clearance, volume of distribution, and half-life) increased with increasing dose (*Table 11.2*), but only the increase in volume was significant.

Pickup *et al.* (1977) administered three different doses of prednisolone phosphate to four normal volunteers by slow intravenous injection. The doses consisted of a tracer quantity of [^{3}H]prednisolone alone, or with 0.15 or 0.3 mg/kg prednisolone phosphate (total dose approximately 10 and 20 mg respectively). All values showed significant increases from the tracer to the 0.15 and 0.3 mg/kg doses, except for the half-life, which increased only slightly from the 0.15 to the 0.3 mg/kg dose (*see Table 11.2*). Similar data have been reported by Loo *et al.* (1978). The mean C/F

Table 11.2 Summary of prednisolone pharmacokinetic parameters for total drug measurements following intravenous administration of prednisolone in healthy volunteers

Reference	*Number of subjects*	*Dose (mg)*	*Clearance (ml/min)*	*Half-life (hours)*	*Volume (l)*
Kozower, Veatch and Kaplan (1974)	7	20	97 ± 16	4.2 ± 1.1	37 ± 9
Meikle, Weed and Tyler (1975)	8	10	98 ± 24	4.0 ± 0.9	12 ± 5
	4	40	120 ± 14	5.0 ± 0.7	31 ± 14
Pickup *et al.* (1977)	4	*	71 ± 10	2.6 ± 0.6	16 ± 3
	4	10	101 ± 13	3.3 ± 0.6	30 ± 7
	4	20	140 ± 31	3.8 ± 0.9	45 ± 6
Hsueh, Paz-Guevara and Bledsoe (1979)	5	30	210 ± 53		
Tanner *et al.* (1979a)	3	20	188 ± 37	3.6 ± 0.4	58 ± 10
	3	100	290 ± 22	3.6 ± 0.4	90 ± 14
Al-Habet and Rogers (1980)	7	16	114 ± 26	4.5 ± 1.2	43 ± 7
	7	32	130 ± 50	4.5 ± 1.3	44 ± 10
	7	48	135 ± 56	3.7 ± 0.4	42 ± 15
	7	64	125 ± 58	3.6 ± 0.7	37 ± 11
Rose, Yurchak and Jusko (1981)	6	5	111 ± 5	2.7 ± 0.4	24 ± 3
	6	20	157 ± 25	2.6 ± 0.5	34 ± 13
	6	40	194 ± 29	3.5 ± 0.5	52 ± 8
Frey and Frey (1982)	11	56	175 ± 24		45 ± 5
Legler, Frey and Benet (1982)	9	5†	74 ± 17		
	10	60†	178 ± 35		
Bergrem, Grøttum and Rugstad (1983)	6	10	61 ± 9	2.7 ± 0.6	15 ± 2
	6	20	80 ± 8	2.7 ± 0.6	20 ± 3
Boekenoogen, Szefler and Jusko (1983)	13	40	198 ± 38	3.2 ± 0.6	57 ± 14
Legler and Benet (unpublished observations)	6	10	113 ± 21	2.0 ± 0.3	19 ± 2
	6	35	240 ± 77	2.1 ± 0.2	48 ± 9

* Tracer dose.
† 7-Hour infusion.

values for prednisolone after oral doses of 10 and 20 mg were 1.67 ± 0.20 and 2.17 ± 0.28 ml/min per kg respectively. Neither Pickup *et al.* nor Loo *et al.* measured the protein binding of prednisolone, but both groups attribute the observed dose-dependent changes in prednisolone pharmacokinetics to changes in binding.

Hsueh, Paz-Guevara and Bledsoe (1979) compared the clearance of 30 mg doses of intravenous prednisolone and oral prednisone in five normal subjects. The intravenous clearance was calculated using both one- and two-compartment models. These yielded equivalent values of 2.98 ± 0.75 and 3.03 ± 0.75 ml/min per kg (about 210 ml/min per 70 kg) respectively (*see Table 11.2*). From the oral data, mean values for *C*/*F* of 2.61 ml/min per kg, for V_{area}/F of 58.6 liters, and $T½$ of 4.1 hours were calculated. The authors suggest that the greater value of intravenous *C* than oral *C*/*F* may be due to initial high plasma concentrations following rapid intravenous injection.

Tanner *et al.* (1979a) gave two intravenous doses (20 and 100 mg) of prednisolone to three healthy volunteers, two of whom also received identical doses of oral prednisolone. The clearance, volume and half-life for the oral and intravenous routes were equivalent for each dose. Clearance and V_{area} were significantly increased after the 100 mg dose, while half-life was unchanged (*see Table 11.2*). Al-Habet and Rogers (1980), however, could not confirm dose-dependence of prednisolone pharmacokinetics (*see Table 11.2*).

Rose, Yurchak and Jusko (1981) studied pharmacokinetic changes after 5, 20 and 50 mg doses of oral prednisone, and 5, 20 and 40 mg equivalents of prednisolone phosphate given intravenously. Clearance and steady-state volume increased significantly with increasing i.v. dose (*see Table 11.2*). In contrast to these total prednisolone values, clearance and volume of unbound prednisolone did not change significantly with increasing dose (*Table 11.3*). The half-life for unbound prednisolone was also not significantly different (*Table 11.3*). These results support the hypothesis that increases in clearance and volume of total prednisolone are due to an increasing fraction of unbound prednisolone with increasing total prednisolone concentrations.

Following oral administration of prednisone, mean *C*/*F* values of 572, 1034 and 2271 ml/min per 1.73 m^2 were calculated for prednisone for the 5, 20 and 50 mg doses respectively, showing a significant increase with increasing dose. Half-life, on the other hand, remained constant for prednisone, with mean values of 3.50, 3.34 and 3.36 hours over the dosage range. Prednisolone mean half-life increased from 2.27 to 2.88 and 3.41 hours for the 5, 20 and 50 mg doses respectively. Renal clearance was determined in some of the subjects and also appeared to increase with increasing dose.

Legler, Frey and Benet (1982), in their study of ten healthy volunteers, approached the question of dose dependence in a different manner. In order to achieve a steady-state concentration, prednisolone (as the sodium phosphate) was administered as an intravenous bolus and a 7-hour infusion at two different doses. The results of this study, summarized in *Table 11.4*, prove conclusively that prednisolone kinetics are non-linear. For example, a 12-fold increase in infusion rate resulted in only a five-fold increase in total prednisolone steady-state concentrations, which is reflected in the increase in total prednisolone clearance with the higher dose (*Table 11.4*). Most of the non-linearity may be explained by changes in protein binding, since unbound prednisolone concentrations increase almost twice as much as total levels. However, there was also a slight but significant increase in unbound prednisolone clearance in this cross-over study. Furthermore,

Table 11.3 Summary of prednisolone pharmacokinetic parameters for unbound drug measurements in healthy volunteers

Reference	*Number of subjects*	*Dose (mg)*	*Clearance (ml/min per 70 kg or 1.73 m^2)*	*Half-life (hours)*	*Volume (l/70 kg or 1.73 m^2)*
Rose, Yurchak and Jusko (1981)	6	5*	1043 ± 251	1.7 ± 0.2	123 ± 60
	6	20*	828 ± 418	1.9 ± 0.3	117 ± 46
	6	40*	1050 ± 231	2.2 ± 0.3	168 ± 24
Legler, Frey and Benet (1982)	9	5†	608 ± 111		
	10	60†	763 ± 168		
Frey and Frey (1982)	11	56*	679 ± 112		106 ± 15
Boekenoogen, Szefler and Jusko (1983)	10	40*	1022 ± 227	2.2 ± 0.5	194 ± 52
Bergrem, Grøttum and Rugstad (1983)	6	10*	379 ± 62		
	6	20*	343 ± 83		
English, Dunne and Marks (1983)	6	15§			
		0600 hours	2120 ± 193	1.8 ± 0.2	
		1200 hours	2330 ± 233	1.3 ± 0.1	
		1800 hours	3330 ± 333	2.3 ± 0.5	
		2400 hours	2590 ± 174	2.3 ± 0.4	
Legler and Benet (unpublished observations)	6	10*	610 ± 146	1.4 ± 0.2	83 ± 11
	6	35*	713 ± 105	1.7 ± 0.2	105 ± 11

* Intravenous bolus.
† 7-Hour infusion.
§ Oral dose.

since total prednisone steady-state concentrations did not increase as much as prednisolone levels, the reversible conversion between prednisolone and prednisone may also exhibit non-linear kinetics.

Frey and Frey (1982), using intravenous bolus doses of prednisolone 0.8 mg/kg found a mean total and unbound prednisolone clearance of 2.5 and 9.7 ml/min per kg respectively. V_{ss} for total and unbound prednisolone was 0.64 and 1.52 l/kg respectively.

Bergrem, Grøttum and Rugstad (1983) reported significant increases in total clearance (61.0 ± 8.5 to 80.6 ± 8.4 ml/min) with increasing i.v. doses of prednisolone (10 to 20 mg) (*see Table 11.2*), while mean unbound clearance remained virtually unchanged at 379 and 343 (*see Table 11.3*). As *Table 11.2* shows, V_{ss} for total prednisolone increased significantly with the higher dose. Half-lives following oral or intravenous administration of 10 or 20 mg of prednisolone were equivalent, with an overall mean of 2.72 hours.

Lowe and Dixon (1983) attempted to correlate total or unbound prednisolone concentrations with salivary prednisolone concentrations after slow intravenous administration of 0.15 and 0.3 mg/kg, but the relationships varied both with dose and between individuals, and no simple correlation was found.

English, Dunne and Marks (1983) studied the diurnal variation of prednisolone pharmacokinetics. Significant variation was found for both total and unbound pharmacokinetic parameters. Total *Cl/F*, *V/F* and half-life were highest at 1800 hours, and decreased through the night to reach their lowest values at noon. Mean values are 4.54, 3.85, 3.51 and 3.30 ml/min per kg for *Cl/F*; 0.88, 0.84, 0.59 and 0.40 l/kg for *V/F*; and 5.20, 5.12, 3.40 and 2.00 hours for the half-life, all at 1800 hours, midnight, 0600 hours and noon respectively. Half-life values of unbound drug followed a similar pattern (*see Table 11.3*). Unbound *Cl/F* was also highest at 1800 hours, but followed a slightly different pattern, decreasing up to 0600 hours and beginning to rise again by noon. Starting at 1800 hours, the respective mean values were 47.6, 37.0, 30.3 and 33.3 ml/min per kg. These results conflict with the earlier data of McAllister, Mitchell and Collins (1981), who at a 20 mg dose found mean *Cl/F* values of 189 ml/min at 0800 hours and 203 ml/min at 2000 hours, and V_{area}/F values of 49.3 and 54.4 liters respectively, although the differences are not statistically significant. The clearances determined by English, Dunne and Marks from doses of approximately 14 mg are two to three times higher than those from other authors using 10–20 mg doses.

In summary, it appears that the clearance of total prednisolone is non-linear. Most of the intravenous data (summarized in *Table 11.2*) and oral single-dose data suggest dose-dependent pharmacokinetics for total drug. Clearance increases from approximately 70 to nearly 300 ml/min for doses ranging from trace quantities to 100 mg. Volume estimates also increase, from about 15 to 90 liters over this dosage range. However, the studies of Al-Habet and Rogers (1980) show no dose-dependent changes in either clearance or volume. It is difficult to explain these results considering the consistent changes observed in other studies.

Half-life may also increase with dose but a good deal of disagreement still exists on this point. Half-life values range from two to five hours. It is important to remember that half-life is a poor indicator of changes in drug elimination. In the case of prednisolone, where both clearance and volume increase with increasing dose, changes in half-life will necessarily be less than clearance.

Pharmacokinetic parameters for unbound prednisolone are reviewed in *Table 11.3*. No dose dependence was found in the single-dose studies by Rose, Yurchak

Table 11.4 Prednisolone pharmacokinetic parameters (mean ± s.d.) determined at steady state following low- and high-dose intravenous infusions over 7 hours in ten healthy volunteers. (From Legler, Frey and Benet, 1982, courtesy of the Editor and Publishers, *Journal of Clinical Endocrinology and Metabolism*)

Parameter	*Low dose*	*High dose*	*Ratio of high dose to low dose*
Infusion rate (ng/min per kg)	91.8 ± 8.3	1067 ± 115	11.7 ± 1.5
Total prednisolone steady-state concentration (ng/ml)	91 ± 25	437 ± 116	4.9 ± 0.3
Unbound prednisolone steady-state concentration (ng/ml)	11 ± 2	103 ± 27	9.5 ± 2.0
Total prednisolone apparent clearance (ml/min per kg)	1.06 ± 0.24	2.54 ± 0.50	2.47 ± 0.29
Unbound prednisolone apparent clearance (ml/min per kg)	8.69 ± 1.58	10.9 ± 2.4	1.29 ± 0.24
Total prednisone steady-state concentration (ng/ml)	13 ± 2	37 ± 4	2.8 ± 0.4

and Jusko (1981) and Bergrem, Grøttum and Rugstad (1983). However, the more sensitive steady-steady study of Legler, Frey and Benet (1982) did show a small but significant increase in clearance. The unbound clearance values of Bergrem, Grøttum and Rugstad (1983) are surprisingly low, while both unbound and total clearance values of English, Dunne and Marks (1983) are so high as to question the validity of the measurements. Note that the unbound clearance and volume values of Benet and his collaborators (Frey and Frey, 1982; Legler, Frey and Benet, 1982; Legler and Benet, unpublished observations) are about 30% lower than those of Jusko (Rose, Yurchak and Jusko, 1981; Boekenoogen, Szefler and Jusko, 1982), although within the values from each center there is good consistency from study to study. This probably reflects differences in assay, dialysis techniques and free drug calculations.

The limited work performed on the pharmacokinetics of prednisone indicates that its clearance may also be dose-dependent. Both Rose *et al.* (1981a) and Legler, Frey and Benet (1982) show non-linear changes in the prednisolone to prednisone ratio with increasing dose. Lack of an intravenous dosage form of prednisone makes separation of changes in clearance or volume from changes in bioavailability difficult. The clearance, volume and half-life of prednisolone, however, seem independent of whether prednisolone or prednisone is administered.

Finally, the traditional equations for clearance and volume do not hold for compounds which undergo metabolic interconversion, and it is therefore important to realize that they are apparent values and have limitations. This problem is discussed in depth by Wagner *et al.* (1981).

Bioavailability

Cases of therapeutic inequivalency of various oral preparations of prednisone and prednisolone (Campagna *et al.*, 1963; Levy, Hall and Nelson, 1964) led to concern with the gastrointestinal absorption of these drugs. The *Ad Hoc* Committee on Drug Product Selection of the Academy of General Practice of Pharmacy and the Academy of Pharmaceutical Sciences has placed both prednisone and prednisolone on their list of drugs with high risk potential for therapeutic inequivalence because of differences in bioavailability (Sugita and Niebergall, 1975; Thiessen, 1976).

Because peptic ulceration is a suspected side-effect of oral glucocorticoid administration, the use of enteric-coated tablets has been proposed by West (1959). Leclercq and Copinschi (1974) studied the relative bioavailability of prednisolone 20 mg as conventional and enteric-coated tablets. The pattern of plasma prednisolone concentrations was similar after both forms of administration but peak levels were delayed by about 1.5 hours with the enteric-coated form. Mean peak concentrations were 185 ng/ml (535 nmol/l and 168 ng/ml (485 nmol/l) for the conventional and enteric-coated tablets respectively. The time to peak was significantly longer with the enteric-coated tablets (3.8 ± 0.3 hours compared to 2.1 ± 0.9 hours). The relative bioavailability of the two dosage forms, defining the conventional tablets as the standard, is 0.95. Similar results are reported by Wilson, May and Paterson (1977) and Mant (1979). However, in a similar study by Hulme, James and Rault (1975), much lower peak plasma prednisolone concentrations were found in addition to the delay in peak time.

DiSanto and DeSante (1975) administered prednisone 50 mg either as a single tablet or as ten 5 mg tablets, and found no significant difference in peak prednisolone concentration or time to peak. Mean values were 926 ng/ml

(2680 nmol/l) and 894 ng/ml (2590 nmol/l), and 2.07 and 1.73 hours for the single and ten-tablet regimen respectively. The relative bioavailability is 0.94 when the ten-tablet dose is defined as the standard.

Tembo *et al.* (1976) studied the effect of food on the bioavailability of two brands of prednisone tablets. For one of the tablets, mean peak prednisolone concentrations were 186 ng/ml (540 nmol/l) and 165 ng/ml (480 nmol/l), and peak times 3.25 and 3.00 hours following the fasting and non-fasting regimens respectively. For the other, the respective values were 226 ng/ml (655 nmol/l) and 201 ng/ml (580 nmol/l), and 1.63 and 1.50 hours. Interestingly, there was more variation between tablets than between the fasting and non-fasting treatments for the same tablet. Lee *et al.* (1979a) also found that food has no effect on plasma prednisolone levels after ingestion of enteric-coated tablets.

Sullivan *et al.* (1976) compared the bioavailability of eight commercial prednisone tablets. Significant differences were found among the values for prednisone peak concentration, time to peak and AUC. However, for the active moiety, prednisolone, AUCs were not significantly different in any of the products tested. In only two of the eight products were significant differences in peak concentration and time found. Mean peak prednisolone concentrations ranged from 198 to 266 ng/ml (575 to 770 nmol/l) and peak times from 1.17 to 2.58 hours. The mean AUC range of 1116 to 1342 ng · hour · ml^{-1} leads to a minimum F_{rel} of 0.83 for the least available product compared to the most available.

A similar comparison of different commercially available prednisolone tablets (Tembo *et al.*, 1977) also showed no significant differences in peak prednisolone concentration, peak time or AUC, and values were very similar to those reported by Sullivan *et al.* (1976).

Tse and Welling (1979) studies the bioavailability of oral doses of prednisone and prednisolone relative to measures plasma concentrations of prednisolone. F_{rel} for prednisolone from prednisone tablets when prednisolone tablets are defined as the standard is 0.77.

In another study, Lee *et al.* (1979a) compared two different prednisolone tablets in six healthy volunteers. In one, the enteric coating was based on cellulose acetate phthalate (CAP), in the other on shellac. Peak mean plasma prednisolone concentration was significantly higher with the tablets containing the CAP-based coating (456 compared to 143 ng/ml or 1320 compared to 415 nmol/l). Peak times were not significantly different (mean 4.33 compared to 5.42 hours). These results help to explain the conflicting results of Leclerq and Copinschi (1974) and Hulme, James and Rault (1975) since the former used tablets with a CAP-based coating, and the latter tablets with a shellac-based coating.

Rose, Yurchak and Jusko (1980) compared the bioavailability of two 50 mg tablets of prednisone and found no significant differences between a film-coated and a reference tablet in peak concentration, peak time or AUC of either prednisone or prednisolone. The dose-corrected mean absolute bioavailability was calculated as 0.77 and 0.80 respectively. This value does not distinguish between incomplete absorption and incomplete conversion of prednisone to prednisolone.

They subsequently studied the dose dependence of prednisone bioavailability. Comparison of the AUC following oral doses of 5, 20 and 50 mg to AUC following intravenous doses of 5, 20 and 40 mg (and correcting for dose when necessary) yielded mean F_{abs} values of 0.99, 0.85 and 0.86 for the 5, 20 and 50 mg doses respectively. There were no significant differences among these values (Rose, Yurchak and Jusko, 1981).

Georgitis, Flesher and Szefler (1982) compared a liquid prednisone preparation to both a standard prednisone tablet and an intravenous infusion of prednisolone sodium phosphate. Mean peak plasma prednisolone levels were not significantly different (393 and 377 ng/ml [1140 and 1090 nmol/l] for the liquid and tablet formulations respectively). Peak concentration was reached significantly earlier with the liquid than with the tablet (0.54 compared to 1.23 hours). The AUC of prednisolone is significantly greater for the tablet than for the liquid, yielding a mean F_{rel} of 0.90. The F_{abs} when compared to an intravenous infusion was 0.88 for the liquid and 0.97 for the tablet. These two values are not significantly different.

In summary, the bioavailability of both prednisone and prednisolone tablets is generally measured in terms of prednisolone plasma levels. Although clinical inequivalency has been reported (Campagna *et al.*, 1963; Levy, Hall and Nelson, 1964), subsequent testing of a variety of prednisone and prednisolone tablets has failed to demonstrate significant differences in their bioavailability. Nevertheless, once a patient is stabilized on a particular preparation, switching products remains unwise. The absolute bioavailability of prednisolone from all forms of oral prednisone ranges from 0.77 to 0.99 and appears to depend more on formulation than dose size. The peak time ranges from approximately 1–3 hours for conventional tablets, increasing to 4 hours or more when enteric-coated preparations are administered. Despite the increased peak time, the extent of bioavailability of enteric-coated tablets can be equivalent to that of conventional tablets. Coatings based on cellulose acetate phtalate are more reliable.

Macromolecular binding

Like cortisol, prednisolone exists in plasma in unbound, CBG-bound and albumin-bound states, and competes with cortisol for the same binding sites (DeMoor, Deckx and Steeno, 1963). Prednisolone protein binding has been extensively studied because of the report by Lewis *et al.* (1971) correlating serum albumin concentrations, mean daily prednisone dose and frequency of side-effects. In a series of 240 prednisone-treated patients, 37% of those with serum albumin levels below 25 g/l experienced steroid-related side-effects, compared to only 15% of those with serum albumin levels above 25 g/l. Side-effects were also more common in patients receiving higher prednisone doses, increasing in frequency from 12% to 28% as the mean daily dose increased from less than 25 mg to more than 50 mg.

Angeli *et al.* (1978) studied the diurnal variation of prednisolone binding and found that CBG binding capacity for prednisolone was highest at midnight ($4.87 \pm 0.72 \times 10^{-7}$ mol/l, falling rapidly to the lowest value at 0800 hours ($3.02 \pm 0.66 \times 10^{-7}$ mol/l) before slowly rising again throughout the day and evening. However, English, Dunne and Marks (1983) found highest values for clearance and volume of distribution at 1800 hours and lowest values at noon.

Concentration-dependent binding of prednisolone was demonstrated in the plasma from a single normal subject by Agabeyoglu *et al.* (1979). An increase in prednisolone concentration from 29.5 to 428 ng/ml (85 to 1240 nmol/l) resulted in a decrease of the proportion of prednisolone bound from 89.6 to 63.0%. Association constants for CBG and albumin were 5.6×10^7 l/mol and 1.8×10^3 l/mol respectively.

Rose, Yurchak and Jusko (1981) determined the binding capacities and affinity constants for the interaction of prednisolone with CBG and albumin in six healthy volunteers. For CBG they found a mean binding capacity of 5.69×10^{-7} mol/l and an affinity constant of 3.01×10^{7} l/mol. The respective values for albumin were 6.23×10^{-4} mol/l and 2.05×10^{3} l/mol.

Rocci *et al.* (1982b) made a detailed study of the macromolecular binding interaction of prednisolone and cortisol in 5% albumin solution and pooled human serum. In the albumin solution, addition of cortisol caused an increase in the mean affinity constant from 4.32×10^{2} to 9.43×10^{2} l/mol and a decrease in the apparent number of binding sites from four to two. The authors suggest that allosteric or conformational changes may occur in the structure of albumin in the presence of cortisol. In pooled human serum, affinity constants for prednisolone alone were 1.50×10^{7} and 4.34×10^{2} l/mol for CBG and albumin respectively. In the presence of cortisol, these values increased to 2.47×10^{7} and 9.43×10^{2} l/mol. Affinity constants for cortisol in the presence of prednisolone were 1.01×10^{7} l/mol for CBG and 2.79 l/mol for albumin. Thus, prednisolone binding is approximately 2.5 times stronger to transcortin and more than 300 times stronger to albumin than simultaneous cortisol binding.

Milsap and Jusko (1983) have studied the contribution of α_1-acid glycoprotein (AGP or orosomucoid) to the overall binding of prednisolone. Although prednisolone did bind to AGP, because of the low affinity ($K_a = 5 \times 10^{3}$ l/mol) for the interaction and the low concentration of AGP in plasma (about 900 mg/l), this binding in the presence of normal CBG and albumin levels was less than 3%. The increased levels of AGP associated with inflammatory diseases and some cancers (Schmid, 1975) are therefore probably of little consequence to the unbound concentration of prednisolone.

Legler and Benet (unpublished observations) have also investigated the effect of prednisone and cortisol on plasma protein binding of prednisolone in two healthy subjects (one male and one female). Prednisone appeared to have no effect on prednisolone binding even in concentrations more than 35 times those of prednisolone. In contrast to the lack of competition by prednisone, cortisol was clearly shown to compete for CBG binding. The affinity constant for cortisol in the presence of prednisolone was 1.17×10^{7} l/mol in the male and 2.40×10^{7} l/mol in the female, values quite close to that (1.01×10^{-7} l/mol) reported by Rocci *et al.* (1982b). They also reported CBG binding capacities of 5.83 and 7.94×10^{-7} mol/l, CBG–prednisolone affinities of 1.51 and 1.38×10^{7} l/mol, and albumin-prednisolone affinities of 2.36 and 1.58×10^{3} l/mol for the male and female plasma respectively.

Measurements of free prednisolone concentration and reported binding parameters also vary with the methods employed. Firstly, for a drug showing non-linear protein binding, transfer of substantial amounts of that drug from the plasma to the buffer side of an equilibrium dialysis membrane will result in changes in the free fraction during dialysis (Behm and Wagner, 1979); secondly, osmotic effects can cause substantial volume shifts during all protein binding measurements. Tozer *et al.* (1983) reported an average volume shift of 31% during prednisolone equilibrium dialysis procedures. Finally, for prednisolone, the competition for binding by endogenous glucocorticoids is generally ignored in the calculations so that 'effective' prednisolone binding parameters are reported. All of these errors can be avoided by following the method proposed by Tozer *et al.* (1983) and incorporating corrections for volume shifts and drug transfer.

Other synthetic glucocorticoids

The other glucocorticoids available for systemic administration include methylprednisolone, dexamethasone, betamethasone, triamcinolone, paramethasone, fluprednisolone and meprednisone. Pharmacokinetic information on these compounds is extremely limited, and in the case of the latter three there is none. Except for increased potency (both therapeutic and toxic) and generally less mineralocorticoid activity, these other synthetic glucocorticoids have no apparent advantages over the natural glucocorticoids and prednisolone/prednisone.

Methylprednisolone

Methylprednisolone (6α-methyl-Δ^1-cortisol) is another widely used glucocorticoid. Methylprednisone has been identified as a metabolite of methylprednisolone following intravenous administration (Szefler *et al.*, 1980) and like cortisone and prednisone, is assumed to be inactive. Methylprednisone itself has not been administered to humans and therefore the possible reversible metabolism has not been substantiated.

Jenkins and Schemmel (1958) gave methylprednisolone 50 mg to five healthy volunteers by intravenous infusion over 10 min. The mean half-life was 3.1 hours. However, Slaunwhite and Sandberg (1961) found a much shorter half-life (1.3 hours) following administration of trace quantities of [^{14}C]-methylprednisolone as a 2–3 min intravenous injection.

Oral administration of methylprednisolone 40 mg resulted in maximum mean plasma concentrations of 329 ng/ml at a peak time of two hours in all subjects. The half-life determined from the mean plasma concentration–time curve was 2.5 hours (Colburn and Buller, 1973).

In a study of 13 healthy volunteers (Stjernholm and Katz, 1975) intravenous injection of trace quantities of tritium-labeled methylprednisolone and subsequent plasma analysis revealed a mean clearance, volume of distribution and half-life of 266 ml/min, 61 liters and 2.8 hours respectively (*see Table 11.5*).

Garg *et al.* (1978) determined the hydrolysis rate of methylprednisolone acetate *in vitro*. When incubated at a concentration of 500 ng/ml with fresh whole blood, the half-life at 37 °C for disappearance of the acetate ester was approximately 19 min. However, when methylprednisolone acetate was administered orally as a suspension (one volunteer), no methylprednisolone acetate could be detected in any of the blood samples. The authors concluded that the ester must be hydrolyzed during absorption or first-pass through the liver. The C/F ratio for methylprednisolone following administration of the acetate was 892 ml/min and the half-life 1.82 hours. The clearance value is much greater than that found for an intravenous tracer dose by Stjernholm and Katz (*see above*). Dose-dependent clearance or particularly low bioavailability could be involved.

Rectal absorption of [^{14}C]methylprednisolone was first studied by Spencer, Kirsner and Palmer (1960). The mean proportion of administered radio-label in the urine of these subjects was 37%. Rectal and oral absorption of methylprednisolone acetate were examined by Garg *et al.* (1979). Comparison of methylprednisolone AUCs following three different treatments yielded a mean relative bioavailability of 0.24 for the rectal methylprednisolone acetate and 0.90 for the oral methylprednisolone, defining the oral methylprednisolone acetate as the standard

Table 11.5 Representative pharmacokinetic parameters for natural and synthetic glucocorticoids in healthy adult volunteers

	Dose (mg)	*Clearance (ml/min)*	*Volume (l)*	*Half-life (hours)*	F_{abs}	f*	*Relative glucocorticoid potency*†
Cortisol (hydrocortisone)	5	362§	21§	1.3	0.58	0.90§	1.0
Prednisolone	5	111§	24§	2.7	0.90	0.90§	5.2
Methylprednisolone	**	266	61	2.8	ND	0.50	13.7
Dexamethasone	6.66	247	63	3.5	0.78	0.23	95.3
Betamethasone	0.5	148††	65††	5.1	ND	0.38	62.1
Triamcinolone	ND	ND	ND	ND	ND	0.60	17.9

* Unbound fraction in plasma.
† From Lerner *et al.* (1964), based on liver glycogen deposition in male rats.
§ Concentration dependence proven or suspected.
** Tracer dose.
†† Data derived from oral dose, i.e. values are C/F and V/F.
ND = no data available.
F_{abs} = absolute bioavailability.

in each case. Pharmacokinetic parameters for methylprednisolone calculated for the two oral treatments were not statistically different. Mean *C/F* was 553 and 498 ml/min, mean V_{area}/F was 115 and 103 liters, and mean half-life 2.39 and 2.38 hours following administration of methylprednisolone and the acetate ester respectively.

Albert *et al.* (1979) compared the bioavailability of three lots of methylprednisolone tablets to each other and to an oral methylprednisolone suspension. There were no differences between the three tablet lots in peak serum concentration, peak time, AUC or half-life. The bioavailability of the tablet formulations relative to the aqueous suspension was 0.99.

Intramuscular injection of methylprednisolone acetate 200 mg in two subjects (Ferrari and Fantoli, 1964) was followed by peak plasma levels of 17-hydroxycorticosteroids of 450 to 550 ng/ml (1290 to 1580 nmol/l) on day 5, with a slow decline to pretreatment levels by day 21.

Little is known about the protein binding of methylprednisolone. Florini and Buyske (1961) found that high concentrations of methylprednisolone could displace some cortisol from its binding sites in plasma. It has been stated that methylprednisolone exhibits little concentration-dependent binding to plasma proteins and that 40–60% binding is observed in human plasma (Assael *et al.*, 1982).

Overall, methylprednisolone pharmacokinetics are not well characterized, and so far no studies using specific and sensitive analytical procedures in the determination of absolute bioavailability or pharmacokinetics after intravenous dosing have been published. The half-life appears to be in the range of 2–3 hours.

Dexamethasone

Dexamethasone (9α-fluoro-16α-methyl-Δ^1-hydrocortisone) was first synthesized in 1958. The first pharmacokinetic study was reported by Peterson (1959), who, using infusions of 'pharmacological' quantities of dexamethasone, found a half-life of 3.3 hours in normal subjects. Haque *et al.* (1972) found a dexamethasone mean clearance, volume of distribution and half-life of 201 ml/min, 43.6 liters and 4.2 hours respectively after intravenous administration of 1.5 mg dexamethasone.

Meikle, Lagerquist and Tyler (1973) administered dexamethasone 1 mg by mouth to two subjects. Peak plasma levels occurred at 30 min in one subject and at 60 min in the other. The half-life was 4.0 and 4.5 hours respectively. A similar half-life (4.1 ± 0.9 hours) was found after intravenous injection of dexamethasone 2 mg/kg in five normal subjects (Hichens and Hogans, 1974). English *et al.* (1975) studied the disposition of dexamethasone 2 mg orally in five healthy volunteers. Peak plasma concentrations of dexamethasone varied widely, from 8.5 to 27.0 ng/ml (23 to 72 nmol/l). Peak time was 1 hour in four of the five subjects and 3 hours in the fifth. The mean plasma half-life was 3.8 hours.

The bioavailability of dexamethasone tablets and elixir was determined by comparison to an intravenous infusion by Duggan *et al.* (1975). Plasma clearance averaged 209 ml/min, and mean renal clearance of unchanged drug was 21 ml/min. Comparison of AUCs after oral and intravenous administration yielded an absolute bioavailability of 0.86 for the elixir and 0.78 for the tablets. The difference is not statistically significant.

Hare *et al.* (1975) found that, following administration of dexamethasone phosphate by intravenous infusion, unchanged ester could be measured in plasma. About 90% of the administered dexamethasone phosphate was converted to dexamethasone *in vivo*. The half-life of the conversion was approximately 10 min; this is about 25 times faster than the *in vitro* hydrolysis in whole blood.

The difference in disposition of intravenously administered dexamethasone 6.66 mg between healthy females and males was studied by Tsuei *et al.* (1979; *see also Table 11.5*). Clearance was not significantly different between the two groups. Mean plasma and renal clearances respectively were 243 and 6.1 ml/min in women and 248 and 6.9 ml/min in men. V_{ss} was significantly larger in men (mean 63.1 compared to 41.3 liters in women). There was, however, no difference when the volume was corrected for body weight (0.71 l/kg for females and 0.82 l/kg for males), as the men were significantly heavier. The half-life of 2.38 ± 0.17 hours in women was significantly shorter than that in men (3.55 ± 0.87 hours).

Miyabo *et al.* (1981) investigated the effect of two dexamethasone esters, the sulfate and the phosphate, on ACTH suppression. Following intravenous injection of dexamethasone phosphate 20 mg, no intact ester was found in the urine, but about 9% of the dose appeared in the urine as free dexamethasone. Plasma cortisol and ACTH were suppressed for more than 24 hours. In contrast, dexamethasone sulfate caused no suppression of hydrocortisone or ACTH levels and nearly 60% of the sulfate ester was excreted intact in the urine, suggesting that dexamethasone sulfate is ineffective.

Studies of macromolecular binding of dexamethasone (Peets, Staub and Symchowicz, 1979) have shown that over a plasma concentration range of 40 ng/ml to 4.63 μg/ml the proportion of bound dexamethasone did not change (77.4 ± 1.1%). Dexamethasone did not displace cortisol from its binding sites in plasma. In a solution of 4% human plasma albumin, 83.7 ± 0.6% of dexamethasone is bound, 6.5% of this to macromolecules other than albumin.

Only limited studies of the pharmacokinetics of dexamethasone are available. Clearances of approximately 200–250 ml/min and half-lives of between 2.5 and 4.5 hours have been reported. Volume of distribution ranges from about 40 to 60 liters, with some indication that it increases with increasing body size.

Betamethasone

The synthetic glucocorticoid betamethasone (9α-fluoro-16β-methyl-Δ^1-hydrocortisone) was synthesized in 1958. Fell (1972) studied the pharmacokinetics following rapid intravenous injection of 10 mg ^{14}C-labeled betamethasone in two healthy volunteers and found a half-life of 5.6 hours in one subject and 2.2 hours in the other.

In a comparison of three different formulations of betamethasone (two conventional and one effervescent 0.5 mg tablet), Loo *et al.* (1981) found no significant differences in any of the pharmacokinetic parameters determined except in the absorption rate constant (k_a), which was significantly higher for the effervescent tablet (mean 1.76/hour compared to 1.02 and 1.22/hour for the conventional tablets). Overall means for other pharmacokinetic parameters include a *Cl/F* of 2.12 ml/min per kg; *V/F* of 0.93 l/kg; half-life of 5.12 hours (range 3.7–7.7 hours); peak plasma concentration of 24.5 ng/ml (65 nmol/l) and peak time of 2.15 hours.

Macromolecular plasma binding of betamethasone has been studied in man by Peets, Staub and Symchowicz (1969). Over a concentration range spanning more than 150-fold increases, betamethasone binding remained constant at 62.5 ± 1.3% bound. Even at 300-fold excess, betamethasone did not compete with cortisol for plasma binding sites, a finding similar to that of DeMoor, Deckx and Steeno (1963). Betamethasone is 60.0 ± 1.9% bound in a solution of 4% human plasma albumin. Approximately 8.5% of the betamethasone bound in plasma is associated with macromolecules other than albumin.

Triamcinolone

Only limited information is available on the macromolecular binding of triamcinolone (9α-fluoro-16α-hydroxy-Δ^1-hydrocortisone). Florini and Buyske (1961) found it to be approximately 40% bound (mainly to albumin) and unaffected by changes in concentration from 0.1 to 10.0 μg/ml (265 to 26 500 nmol/l). A 383-fold excess of triamcinolone displaced only 9.5% of cortisol in human plasma, a result confirmed by DeMoor, Deckx and Steeno (1963).

Summary

As can be seen from the summary presented in *Table 11.5*, there is no clear overall relationship between the pharmacokinetic and pharmacodynamic variables of any of the glucocorticoids. For cortisol and prednisolone, there may be a relationship between clearance and relative potency, as the more potent glucocorticoid prednisolone is cleared more slowly. For the other synthetic glucocorticoids, this relationship apparently does not hold, but so little is known about the pharmacokinetics of these compounds that it is difficult to be sure.

The only clear relationship is between the unbound fraction and volume of distribution. As mentioned previously, compounds that are less strongly bound to plasma proteins are more likely to distribute into the tissues and exhibit a larger volume of distribution. Both cortisol and prednisolone are strongly bound to CBG, and especially at low concentrations this binding keeps their volumes of distribution fairly small. The other synthetic glucocorticoids are not bound to CBG to any significant extent and their weak binding to albumin allows larger distribution volumes.

EFFECTS OF DISEASE ON GLUCOCORTICOID PHARMACOKINETICS

Pharmacokinetic information from healthy individuals is often invaluable in determining how the human body deals with a particular drug, and yet it is the diseased rather than the healthy individual to whom most drugs are administered.

Effects of liver disease

The liver is a major site of steroid elimination (Samuels and West, 1952). Since glucocorticoids are widely used in the treatment of liver disease, it is important to understand if and how glucocorticoid pharmacokinetics change in the presence of impaired liver function. Two problems might be anticipated: (1) the overall elimination of glucocorticoids might be decreased, and (2) an inactive compound, e.g. prednisone, might not be converted to the active moiety, in this case prednisolone, in the manner expected. Corticosteroid pharmacokinetics in liver disease have been reviewed by Uribe and Go (1979).

Brown *et al.* (1954) found that after intravenous doses of cortisol 1 mg/kg the mean half-life of plasma 17-hydroxycorticosteroids in patients with liver disease was nearly double that in normal controls (3.7 compared to 1.9 hours). However, in a similar study, Nugent, Eik-Nes and Tyler (1959) found no significant differencc in the half-life of plasma 17-hydroxycorticosteroids between patients with liver disease (4.0 hours) and normal controls (3.5 hours). Jenkins and Sampson (1967) gave single oral doses of cortisone or prednisone to four patients with liver disease and nine normal subjects. The reduction of both to their active metabolites was well maintained since similar levels of the active species were found in the patients and the controls. In contrast, Powell and Axelsen (1972) found impaired conversion of prednisone to prednisolone in 15 patients with acute hepatitis or active chronic liver disease. Significantly lower peak prednisolone levels were found after a single oral dose of prednisone 20 mg than after an equivalent dose of oral prednisolone. Normal subjects had approximately equal peak prednisolone concentrations after administration of prednisone or prednisolone. Furthermore, increased levels of unbound prednisolone were found in those with liver disease, most likely secondary to the decreased serum albumin concentrations in these patients.

Uribe, Go and Summerskill (1976) found insignificant differences in the pharmacokinetics (peak concentration, peak time, half-life and serum binding) of prednisolone after oral administration of prednisone or prednisolone 30 mg to six healthy volunteers and five patients with chronic active liver disease (CALD).

Schalm, Summerskill and Go (1977) studied the effects of intravenous prednisone in 29 subjects. Group 1 consisted of ten healthy volunteers; group 2 of six untreated patients with severe CALD; group 3 of ten patients with prednisone-induced remission of CALD; and group 4 of three patients with CALD which continued to deteriorate despite prednisone treatment. Most of the pharmacokinetic parameters for both prednisone and prednisolone were not significantly different among the four groups. However, the volume of distribution of prednisone was significantly smaller in the two groups exposed to previous prednisone therapy (groups 3 and 4). In addition, the free fraction of serum prednisolone was increased in groups 2 and 4, the two most seriously ill groups, who had decreased serum albumin levels.

Uribe, Summerskill and Go (1977) attempted to correlate the steroid side-effects observed in some patients treated for CALD to hypoalbuminemia and hyperbilirubinemia. *In vivo*, unbound serum prednisolone concentrations after oral administration of prednisone 30 mg were significantly higher in the five patients with CALD who had severe side-effects than in the seven who had no side-effects or the six healthy controls. *In vitro*, both decreasing serum albumin and increasing serum bilirubin levels were associated with increases in free serum prednisolone. The authors concluded that steroid side-effects may be related to a decrease in the number of binding sites in some CALD patients due to hypoalbuminemia and/or competition by bilirubin for albumin binding sites leading to increased exposure to unbound prednisolone.

Although the relative bioavailability of prednisolone after oral administration of prednisolone or prednisone 10 mg in ten normal controls and 25 patients with chronic active hepatitis (Davis *et al.*, 1978) varied widely, the area under the plasma concentration–time curve for prednisolone was not significantly different between the two groups.

The absolute bioavailability of prednisone 10 mg in five healthy volunteers and 11 patients with CALD (Uribe *et al.*, 1978) was found to be approximately 1.0 in all subjects, regardless of whether they were healthy, had severe CALD or were in remission, indicating that prednisone is well absorbed and effectively converted to prednisolone both in healthy volunteers and in patients with CALD.

Scavo, Cugini and DiLascio (1978), using tracer doses of ^{3}H-hydrocortisone in nine patients with liver cirrhosis and ten normal volunteers, found a highly significant decrease in clearance (71 ml/min compared to 197 ml/min in volunteers). The volume of distribution was also decreased in cirrhotics (11.5 ± 3.5 liters compared to 19.3 ± 4.3 liters in normals). The half-life was 1.3 ± 0.2 hours in the normal volunteers and 2.0 ± 0.8 hours in those with cirrhosis.

Madsbad *et al.* (1980) compared serum prednisolone concentrations after oral administration of prednisone 0.3 mg/kg and equal doses of prednisolone in seven patients with severely impaired liver function and seven with slightly impaired liver function. The mean prednisolone concentration after oral prednisone was 47% lower in the severely impaired group, but levels after oral prednisolone were independent of the severity of the disease. They concluded that oral prednisolone should be the drug of choice in patients with decreased liver function.

To determine whether treatment failures in CALD could be attributed to differences in prednisone/prednisolone disposition, Uribe, Summerskill and Go (1982) compared concentrations of prednisolone and prednisone (in both serum and urine) in healthy volunteers, CALD patients in prednisone-induced remission, and CALD patients who had failed to respond to prednisone therapy. No significant differences were observed in any of the parameters studied and the authors concluded that pharmacokinetic variables are probably not the cause of treatment failures in these patients.

Ui *et al.* (1982) studied the absorption of 30 mg doses of oral prednisone and prednisolone in five healthy volunteers and seven patients with decompensated liver cirrhosis. Although results after oral prednisolone were similar in both groups, mean peak concentrations after oral prednisone were very different. In the controls, peak levels occurred 60 min after dosing and were 23.7 and 171 ng/ml (69 and 493 nmol/l) for prednisone and prednisolone respectively, while in those with cirrhosis peak levels occurred at 30 min and were 332 ng/ml (960 nmol/l) for prednisone and 113 ng/ml (325 nmol/l) for prednisolone, indicating impaired conversion to the active compound.

In conclusion, there is probably impaired elimination of cortisol in liver disease, with decreased clearance and increased half-life. Conversion of cortisone to cortisol appears to be unaffected (Jenkins and Sampson, 1967). The effect of liver disease on the prednisone/prednisolone interconversion has been studied extensively. While most results indicate that the conversion is similar in health and liver disease, it has been shown to be impaired in some patients. It is probably wise to use prednisolone rather than prednisone in the treatment of liver disease since its disposition remains relatively normal. A final point to remember is the possibility of increased unbound prednisolone in patients with liver disease secondary to hypoalbuminemia. These patients should be carefully monitored for steroid-related side-effects.

Effect of kidney disease

Glucocorticoids are used widely in the treatment of renal disease such as glomerulonephritis and nephrotic syndrome, and in the immunosuppression of kidney transplant recipients. Since the kidney is known to be a site of steroid metabolism (Samuels and West, 1952), the effect of kidney disease on glucocorticoid pharmacokinetics is an important field of study.

Nugent, Eik-Nes and Tyler (1959) compared the half-life of plasma 17-hydroxycorticosteroids following a 30 min infusion of prednisolone 1 mg/kg in ten normal volunteers and five patients with various types of renal disease. The observed difference in half-life (about 3.7 hours compared to 3.5 hours in controls) was not statistically significant.

In a similar study, Coburg *et al.* (1970) infused prednisolone 1 g over 60 min into ten patients and six healthy subjects. Of the ten patients studied, five had functional renal homografts and five were essentially anephric and awaiting transplantation. The half-life of plasma 17-hydroxycorticosteroids was approximately 3 hours in each of these groups, regardless of the level of kidney function.

Turcotte *et al.* (1972) measured the half-life of methylprednisolone in three kidney transplant patients receiving large (30 mg/kg) 'pulses' of intravenous methylprednisolone to treat acute rejection episodes. The half-life in these patients (2.3 ± 0.7 hours) is comparable to that obtained with lower doses in healthy volunteers. Park, Greene and Bacon (1974) found that the same is true for intravenous prednisolone in doses 30 mg/kg. The half-life of 3.3 hours in four patients undergoing rejection of renal grafts is similar to that found in normal subjects by others.

To determine whether differences in the pharmacokinetics of methylprednisolone contribute to the rejection of renal allografts, Sells *et al.* (1978) studied two groups of renal transplant recipients at least six months postoperatively. One group had well-functioning grafts, while the other had rejected renal transplants and was being maintained by dialysis. Following an oral dose of 0.5 mg/kg, peak serum methylprednisolone levels and elimination half-lives were not significantly different between the two groups, indicating that the failure of renal allografts was not due to pharmacokinetic factors in these patients.

Frey *et al.*, (1981a) found no significant differences in any of the pharmacokinetic parameters for prednisolone in 15 stable renal transplant patients after oral prednisone or intravenous prednisolone. Seven of these patients were grossly cushingoid, and eight appeared normal. An interesting finding, which may shed some light on the genesis of cushingoid side-effects, was an increased AUC for cortisol in the cushingoid patients. The higher cortisol levels in the cushingoid group may be either a cause or a result of the side-effects.

Gambertoglio *et al.* (1982) studied the absolute bioavailability of prednisolone from oral prednisone and prednisolone tablets in nine kidney transplant patients, based on both total and unbound levels of prednisolone. The mean absolute bioavailability from oral prednisone and prednisolone was 0.84 and 0.95 respectively, using unbound drug concentrations, and 0.86 and 0.94 respectively, using total levels of prednisolone. None of these differences are statistically significant. With regard to the rate of bioavailability, the peak prednisolone concentrations and times were also not significantly different between oral prednisone and prednisolone in these patients.

Bergrem (1983) compared the pharmacokinetics of prednisolone 20 mg i.v. in six patients with chronic renal disease and six healthy controls. Both total and unbound prednisolone clearance was significantly decreased and the half-life for total prednisolone significantly increased in those with renal disease, indicating that patients with decreased renal function may require lower prednisolone doses to obtain plasma levels similar to those in healthy subjects.

Frey and Frey (1982) studied the impact of low plasma protein levels on the pharmacokinetics of prednisolone and found that following an intravenous bolus of prednisolone 0.8 mg/kg, patients with nephrotic syndrome had higher unbound fractions of prednisolone and lower total prednisolone plasma concentrations than healthy volunteers. The unbound plasma concentrations of prednisolone, however, were similar to those found in the normal volunteers. Thus, while total clearance and V_{ss} are higher in the nephrotic syndrome, the pharmacokinetic parameters for unbound prednisolone are not significantly different.

The effect of hemodialysis on plasma methylprednisolone levels was studied by Sherlock and Letteri (1977). Mean dialysance (dialysis clearance) was 18.4 ml/min and mean half-life 2.5 hours. The dialysance of methylprednisolone is a significant elimination pathway in these patients and dose adjustment may be necessary.

Frey *et al.* (1982) studied ten renal transplant recipients undergoing hemodialysis for acute tubular necrosis and found that clearance of total prednisolone by dialysis is concentration-dependent, leading to a hyperbolic increase in the amount of prednisolone lost to the dialysate with increasing concentration. In contrast, the unbound clearance of prednisolone was constant (76 ml/min) in the patients studied.

In summary, the pharmacokinetics of glucocorticoids appear to be unchanged in most patients with renal disease. This is probably due to the low percentage of hydrocortisone, prednisolone and methylprednisolone excreted unchanged by the kidney (less than 20%). Although the kidney is a site of steroid metabolism, a significant decrease in clearance and/or increase in half-life was found only by Bergrem (1983). The most significant effects on clearance were due to hemodialysis, and plasma glucocorticoid levels should be monitored in these patients.

Effect of bowel disease

Inflammatory bowel disease (e.g. ulcerative colitis, Crohn's disease) is often treated with glucocorticoids. Intravenous, oral and rectal routes have all been used with apparently good clinical results. However, because of the changes in intestinal mucosa and motility associated with these diseases, questions arose regarding the bioavailability of the oral and rectal dosage forms. Schwartz *et al.* (1958) examined the effects of cortisol 200 mg as a 20-min rectal infusion in five patients with ulcerative colitis. While clinical improvement was noted in all five patients, plasma cortisol levels were not statistically different from those at zero time at any of the sampling times, indicating that the effect is mainly local. Similar results were reported with cortisol 10 mg as a suppository in three patients with ulcerative colitis (Patterson, 1958).

Spencer, Kirsner and Palmer (1960) measured rectal absorption of ^{14}C-labeled methylprednisolone in three patients (two with ulcerative colitis and one with ileocolitis) and one normal subject. A dose of 40 mg was administered as a retention enema. In the three patients with bowel disease, 18–64% of the administered radioactivity was excreted in the urine within 48 hours. In the normal subject, the value was 37%. It therefore appears that methylprednisolone is absorbed from the rectum in both normal and diseased colons and that some of the beneficial effects may be systemic.

Sanbar and West (1961) compared the absorption of oral and rectal doses of methylprednisolone 8.5 mg in three patients with ulcerative colitis and six normal volunteers. After the oral doses, 24–30% of the administered radioactivity was recovered from the urine in the first 24 hours; rectal absorption was found to be only 22% of the oral absorption. Since the dose size in this study is only about 20% of that used by Spencer, Kirsner and Palmer (1960), dose dependence of rectal methylprednisolone absorption might be involved.

Rectal administration of cortisol enemas in six patients with ulcerative colitis and five normal subjects (Sampson and Brooke, 1963) led to mean peak plasma concentrations of 220 ng/ml two to four hours after administration. The mean values for the normal and ulcerative colitis groups were not significantly different, indicating that cortisol can be absorbed from both normal and diseased colons, in contrast to the earlier finding of Schwartz *et al.* (1958).

Farmer and Schumacher (1970) reported comparable clinical results with 21-day series of either cortisol or cortisol acetate administered rectally in 15 patients with ulcerative colitis. Absorption of cortisol was moderate, while the cortisol acetate was less well absorbed. Clinical results in this study were not related to the amount of glucocorticoid absorbed, implying that a local response is involved in the beneficial effects of hydrocortisone enemas.

Powell-Tuck *et al.* (1976) administered prednisolone 20 mg as a retention enema to seven patients with colitis. Prednisolone was detectable in the plasma of all patients following the enema, with peak prednisolone concentrations of 154 ± 72 ng/ml attained within the first three hours, indicating the possibility of significant systemic effects by rectally administered prednisolone.

Lee *et al.* (1980) compared plasma prednisolone levels after administration of retention enemas containing 20 mg of prednisolone as either the phosphate or metasulfobenzoate ester in healthy subjects and patients with proctocolitis. Plasma levels of prednisolone were significantly lower after the administration of the metasulfobenzoate enemas. However, both preparations appeared to be equally effective in the treatment of proctocolitis, supporting the view that the beneficial effect of these enemas is predominantly local.

Pickup *et al.* (1979) studied oral absorption of glucocorticoids in 15 patients with celiac disease and ten normal subjects. Peak prednisolone plasma concentrations, time to peak and AUC were not significantly different, irrespective of the presence or severity of disease. Similar results were reported by Tanner, Halliday and Powell (1981) for patients with celiac and Crohn's disease. In contrast, Elliott *et al.* (1980) reported reduced peak plasma prednisolone levels and increased peak times in six patients with acute colitis.

Milsap *et al.* (1983) examined the bioavailability and pharmacokinetics of prednisolone after oral doses of prednisone and intravenous infusions of prednisolone in seven patients with inflammatory bowel disease during active disease and again during remission. Conversion of prednisone to prednisolone was complete, and pharmacokinetic parameters for total and unbound prednisolone were not significantly different in the two phases of the disease.

In conclusion, the site of action of rectal glucocorticoids remains a point of debate. Some absorption through the rectal mucosa is reported by most investigators, but the significance of this absorption is questioned by some. Final analysis may show that both systemic and local actions are involved. Absorption of oral prednisolone, and presumably other glucocorticoids, is similar in persons with diseased and healthy colons.

Effect of respiratory disease

Glucocorticoids are important drugs in the treatment of bronchial asthma and other respiratory diseases. Boye, Djøseland and Haugen (1974) studied the disappearance rate of intravenous cortisol in seven asthmatic patients who had never received steroids, seven who had been treated with steroids, and six healthy subjects. The overall mean half-life was approximately two hours and was not related to disease or previous steroid treatment. Collins *et al.* (1970) found no significant differences between the plasma 11-hydroxycorticosteroid levels in patients with acute bronchial asthma or chronic airway obstruction compared to controls following injection of hydrocortisone hemisuccinate. Wilson *et al.* (1975) found peak prednisolone plasma concentrations and plasma half-lives in patients with various respiratory diseases that agreed well with their earlier findings in healthy volunteers. Rose *et al.* (1980) found no significant differences in the distribution, clearance or macromolecular binding of intravenous prednisolone between severe, steroid-dependent asthmatics and healthy volunteers. McAllister, Mitchell and Collins (1981) examined diurnal variation in the pharmacokinetics of prednisolone in normal volunteers and chronic, stable asthmatics. They found no significant differences in distribution or elimination parameters for doses given at 0800 or 2000 hours in either group and no significant differences between the two groups.

There is no evidence that the pharmacokinetics of glucocorticoids are altered in patients with respiratory diseases. A recent study of 17 patients with lung disease (Braude and Rebuck, 1983) indicates, however, that all glucocorticoids are not equally able to penetrate lung tissue. Following administration of intravenous methylprednisolone or oral prednisone, the ratio of bronchoalveolar fluid to serum concentrations for methylprednisolone was 0.5, while the ratio for prednisone was 0.3. Unfortunately, levels for the active drug prednisolone were not reported, but these findings may support the clinical use of methylprednisolone in preference to prednisone in respiratory disease patients.

Effect of other disease states

Glucocorticoids are often used to reduce the inflammation of rheumatoid arthritis. Agabeyoglu *et al.* (1979) found increased *in vitro* plasma protein binding of prednisolone in 20 patients with rheumatoid arthritis (74.5 ± 2.1%) compared to 20 healthy volunteers (72.4 ± 2.2%). While this difference is statistically significant, its therapeutic significance is questionable.

Reeback *et al.* (1980) and Armstrong *et al.* (1981) have shown that significant absorption of prednisolone and methylprednisolone occurs after intra-articular injection of each steroid acetate, and that serum cortisol levels can be suppressed for up to one week following such injections.

The effect of altered thyroid function on the half-life of total and unbound cortisol was studied by Beisel *et al.* (1964). In normal subjects, the mean half-lives of total and unbound drug were 134 and 81 min respectively. In hyperthyroid patients these values decreased to 53 and 40 min, while in hypothyroid patients they increased to 288 and 160 min respectively. This is potentially important in hypothyroid patients with primary or secondary adrenal insufficiency, in whom thyroxine therapy may enhance cortisol metabolism and precipitate a crisis of acute adrenal insufficiency.

Brumback (1980) reported a patient with acute inflammatory polyradiculoneuropathy (Guillain-Barré syndrome) in whom neither oral prednisone nor methylprednisolone could prevent deterioration while intravenous methylprednisolone led to therapeutic improvement. Gastrointestinal motility is disturbed in these patients and may lead to impaired absorption of orally administered drugs.

The pharmacokinetics of intravenous dexamethasone were studied in 14 patients undergoing craniotomy for various neurological diseases (McCafferty *et al.*, 1981). Both mean clearance (7.2 ml/min per kg) and volume of distribution (1.2 l/kg) were significantly increased compared to the values found by Tsuei *et al.* (1979) in normal volunteers (3.6 ml/min per kg and 0.76 l/kg). It would appear that some aspect of either neurological disease or the surgical procedure alters dexamethasone pharmacokinetics.

EFFECTS OF SIMULTANEOUSLY ADMINISTERED DRUGS ON GLUCOCORTICOID PHARMACOKINETICS

It is very common for patients to receive other drugs along with glucocorticoids to treat a variety of diseases. The possibilities are nearly endless and the consequences (e.g. decreased pharmacological and/or increased toxic effects) can be serious.

Effect of hepatic enzyme induction

Several drugs are known to increase the level of microsomal oxidation in the liver. Among them are phenytoin (diphenylhydantoin), carbamazepine, phenobarbital (phenobarbitone) and rifampin (rifampicin). Werk, MacGee and Sholiton (1964) noted that patients receiving phenytoin excreted increased amounts of urinary 6β-hydroxycortisol, a metabolite of cortisol known to be a product of hepatic microsomal oxidation. Adverse effects on renal allograft function have been reported in patients receiving rifampin (Buffington *et al.*, 1976), phenobarbital and phenytoin (Wassner *et al.*, 1976, 1977). Jubiz *et al.* (1970) described a lack of plasma and urinary corticosteroid response to a low-dose dexamethasone suppression test in patients receiving phenytoin.

Choi *et al.* (1971) studied the pharmacokinetics of dexamethasone before and after six weeks of phenytoin therapy. Clearance increased from 220 ± 71 to 275 ± 87 ml/min and half-life decreased from 1.00 ± 0.34 to 0.93 ± 0.17 hours. Volume of distribution was not significantly altered by phenytoin administration. Similar results were reported by Haque *et al.* (1972). Concurrent phenobarbital administration results in qualitatively similar pharmacokinetic changes for dexamethasone (Brooks *et al.*, 1972). The metabolism of methylprednisolone is similarly affected by the concurrent administration of phenytoin or phenobarbital

(Stjernholm and Katz, 1975). Addition of phenobarbital to the therapeutic regimen of prednisolone-dependent rheumatoid arthritis patients has been shown to lead to significant clinical deterioration; simultaneous pharmacokinetic determinations in nine such patients revealed a decrease in prednisolone half-life from 2.2 ± 0.8 to 1.7 ± 0.5 hours after 14 days of treatment (Brooks *et al.*, 1976). Petereit and Meikle (1977) studied the pharmacokinetics of intravenous prednisolone in five normal subjects before and after three weeks of phenytoin administration. The clearance of prednisolone increased from 1.16 ± 0.29 to 2.00 ± 0.41 ml/min per kg, while the half-life decreased from 3.17 ± 0.38 to 1.74 ± 0.26 hours. Bioavailability of oral prednisolone before and after the phenytoin administration was not altered significantly (mean 0.82 before and 0.87 after phenytoin dosing), suggesting that absorption and first-pass clearance of prednisolone are not significantly affected by phenytoin. A single patient with tuberculous Addison's disease was studied by Edwards *et al.* (1974). Clinically, the patient required increased cortisol supplementation while receiving rifampin. The half-life at that time was 0.97 hours. One month after termination of rifampin therapy, the half-life was found to be 1.37 hours, an increase of 41%.

Hendrickse *et al.* (1979) reported similar changes in a child with nephrotic syndrome who failed to respond to prednisolone in doses up to 3 mg/kg per day during treatment with rifampin. The clearance, volume of distribution and half-life of prednisolone were 5.67 ml/min per kg, 15.7 liters and 1.29 hours respectively. Four weeks after withdrawal of rifampin, the respective values were 3.33 ml/min per kg, 15.9 liters and 2.17 hours, showing the expected decrease in clearance and increase in half-life in the absence of the enzyme inducer. Similar results were reported in adolescents and adults by McAllister *et al.* (1983).

Bergrem and Refvem (1982) studied the effect of rifampin on both total and unbound prednisolone. Before rifampin therapy, clearances of total and unbound prednisolone were 73.5 ± 14.6 and 339 ± 74.8 ml/min respectively. After rifampin therapy, these values rose to 142.7 ± 35.8 and 766 ± 285 ml/min. The half-life of total prednisolone decreased from 3.72 ± 0.37 to 2.11 ± 0.23 hours. Volume of distribution was not significantly altered.

Clearly, enzyme-inducing drugs have a great effect on the elimination of glucocorticoids, increasing clearance by as much as twofold. Since distribution is not altered, half-life must decrease. Concurrent administration of these drugs in patients taking glucocorticoids certainly requires careful monitoring and may necessitate an increase in the glucocorticoid dose. In patients with unrecognized Addison's disease or hypopituitarism, these drugs may precipitate adrenal insufficiency.

Effect of estrogen and oral contraceptives

Sandberg and Slaunwhite (1959) reported increased binding of cortisol by CBG in pregnancy, and in males and females receiving exogenous estrogens. This was subsequently attributed to increased CBG concentrations (Sandberg, Slaunwhite and Carter, 1960). This finding, together with the observation of clinical relief of the symptoms of rheumatoid arthritis during pregnancy, first reported by Hench in 1938, suggests that changes in the pharmacokinetics of cortisol and prednisolone might be expected in patients taking estrogens as a result of increased levels of CBG.

Mills *et al.* (1960) studied the effect of intramuscular estradiol benzoate and oral ethinylestradiol on cortisol half-life and macromolecular binding. Half-life increased from 1.3 ± 0.4 to 2.4 ± 0.7 hours when estrogen was administered, total cortisol concentration increased, and the unbound fraction fell. The unbound concentration, however, was unchanged. Similar findings have been reported by Brien (1975). In contrast, Plager, Schmidt and Staubitz (1964) found relatively higher levels of unbound cortisol in patients treated with diethylstilbestrol for 1 to 3 weeks. Burke (1969) reported similar findings in 13 women taking oral contraceptives compared to 67 normal subjects. The total plasma cortisol concentrations increased threefold in those taking contraceptives and, while the unbound cortisol concentrations were within the normal range, the median value was significantly increased. A decrease in clearance and volume of distribution and an increase in half-life of prednisolone in women taking oral contraceptives was reported by Kozower, Veatch and Kaplan (1974). Hsueh, Paz-Guevara and Bledsoe (1979) studied a single subject before and after two months of ethinylestradiol therapy. Oral doses of prednisone 30 mg results in C/F values of 3.68 ml/min per kg initially and 1.20 ml/min per kg during the estrogen therapy.

Recently, Boekenoogen, Szefler and Jusko (1983) studied total and unbound prednisolone pharmacokinetics in women receiving oral contraceptives. Total clearance and volume were decreased and half-life was increased, and the mean CBG concentration (13.9×10^{-7} mol/l) was increased compared to controls (7.74×10^{-7} mol/l). Unbound clearance and volume were also decreased, and the unbound half-life increased. Legler and Benet (unpublished observations) have recently shown a qualitatively similar but quantitatively larger change in prednisolone pharmacokinetics. Unbound prednisolone clearance in those taking oral contraceptives was 3.75 ± 0.66 ml/min per kg compared to 10.2 ± 1.5 ml/min per kg in the controls, a decrease of nearly two-thirds. Unbound V_{ss} was reduced by one-third (0.96 ± 0.19 l/kg compared to 1.50 ± 0.16 l/kg in controls), and unbound half-life was increased from 1.72 ± 0.17 to 2.98 ± 0.35 hours. Possible reasons for these changes in the pharmacokinetics of unbound prednisolone may include destruction of cytochrome P450 by the synthetic ethinyl steroids in oral contraceptives (Ortiz de Montallano *et al.*, 1979), and estrogen-induced intrahepatic cholestasis (Meyers *et al.*, 1980). Preliminary results from our laboratory indicate that natural estrogen (i.e. no ethinyl function) also results in a significant decrease in total and unbound clearance of prednisolone.

In summary, women taking oral contraceptives and patients receiving estrogen therapy may be at greater risk of glucocorticoid side-effects. The decreased clearance and volume of distribution and increased half-life reported for cortisol and prednisolone expose these groups to more persistent levels of both total and unbound glucocorticoid. Careful monitoring of such patients is indicated. Reduced doses may be required.

Effect of other drugs

Several other drugs have been reported to influence the pharmacokinetics of glucocorticoids. The effects of the macrolide antibiotic troleandomycin (TAO) have been studied by Szefler *et al.* (1980) since earlier clinical studies had indicated that this agent may have a 'steroid-sparing' effect in patients with severe steroid-dependent asthma (Spector, Katz and Farr, 1974). Szefler *et al.* were able to show that TAO significantly decreased the elimination of methylprednisolone in asthmatic patients, from 406 to 146 ml/min per 1.73 m^2. Mean half-life was concomitantly increased from 2.46 to 4.63 hours, and volume of distribution decreased from 78.4 to 56.4 liters/1.73 m^2. Similar results were reported for erythromycin, another macrolide antibiotic (LaForce *et al.*, 1983).

Non-steroidal anti-inflammatory drugs have also been reported to have a 'steroid-sparing' effect (Flores and Rojas, 1975). When Rae *et al.* (1982) studied the effect of indomethacin and naproxen on prednisolone pharmacokinetics, they found that while total prednisolone levels were unchanged, unbound prednisolone concentrations had increased significantly after two weeks of therapy with either of these agents. Both indomethacin and naproxen are highly protein-bound and may displace prednisolone from its binding sites.

Chulski and Forist (1958) studied the effects of several antacid agents on prednisolone *in vitro* and found that prednisolone was adsorbed by magnesium trisilicate and degraded by magnesium oxide, whereas aluminum hydroxide, calcium carbonate and magnesium carbonate had no effect. However, Tanner *et al.* (1979b) found no *in vivo* effect on serum prednisolone levels of an antacid mixture containing aluminum hydroxide, magnesium trisilicate and magnesium hydroxide. Lee *et al.* (1979b) found a slight but insignificant decrease in prednisolone plasma AUC when oral prednisone was given together with either magnesium trisilicate or aluminum hydroxide. Bergrem *et al.* (1981) found similar absorption parameters whether oral prednisolone was given alone or with an antacid mixture containing aluminum hydroxide, magnesium hydroxide and magnesium carbonate. Thus it appears that antacids may be given with prednisolone and presumably other glucocorticoids without concern for decreased availability or altered pharmacokinetics.

The bioavailability of prednisolone was not impaired by concurrent administration of the anion exchange resin cholestyramine (Audetat and Bircher, 1976), and cimetidine did not alter its bioavailability or clearance (Morrison *et al.*, 1980); neither did smoking have any effect on the pharmacokinetics of prednisolone or dexamethasone (Rose *et al.*, 1981b). A single dose of azathioprine did not alter the pharmacokinetics of prednisolone (Frey *et al.*, 1981b), and absorption of prednisolone was unaffected by simultaneous administration of melphalan (Taha *et al.*, 1982).

EFFECTS OF AGE ON GLUCOCORTICOID PHARMACOKINETICS

Though little is known about the effect of age on glucocorticoid metabolism, pharmacokinetic parameters may be altered during growth in childhood and with advancing age.

Glucocorticoid pharmacokinetics in pregnancy and in the neonate

Glucocorticoids are often given to pregnant women with premature labor in order to stimulate fetal lung surfactant and prevent respiratory distress syndrome in the neonate. Johnson, Youssefnejadian and Craft (1976) used betamethasone 2 mg/hour as an 8-hour infusion in a single patient and found a peak betamethasone plasma concentration of approximately 35 ng/ml at the end of the infusion. The elimination half-life was approximately 4.4 hours. Osathanondh *et al.* (1977) administered dexamethasone 8 mg by mouth to five pregnant women 8–11 hours before elective cesarean section. Plasma samples were collected from a maternal peripheral vein, the umbilical vein and the umbilical artery, along with a sample of amniotic fluid. The mean dexamethasone concentrations in these samples were 22, 29, 28 and 25 ng/ml (59, 77, 75 and 67 nmol/l) respectively, and were not significantly different in the maternal and fetal compartments. Cortisol concentrations were suppressed in both compartments. A similar study was reported by Petersen *et al.* (1980) for betamethasone. The mean plasma concentration ratio of the umbilical vein versus a maternal peripheral vein was 0.28, indicating that betamethasone does not penetrate the fetal compartment as well as dexamethasone. Anderson, Rotchell and Kaiser (1981) found a mean concentration ratio for umbilical versus maternal vein plasma of 0.48 for methylprednisolone. In a recent study of dexamethasone in ten pregnant women (Kream *et al.*, 1983), the mean plasma dexamethasone concentration two hours after injections of 5 mg i.m. was 37 ng/ml (99 nmol/l). After the last of four injections the plasma concentration–time curve decayed, with a mean half-life of 3.6 hours. Dexamethasone could not be detected in the plasma of the neonates 24 hours after delivery, and normal cortisol levels suggest that there was insignificant adrenal suppression in the neonates.

Concern about the exposure of infants to glucocorticoids via breast milk prompted the study by McKenzie, Selley and Agnew (1975) in which seven lactating women were given [^{3}H]-prednisolone orally. Only 0.14% (range 0.07–0.23) of the radioactivity was recovered per liter of breast milk. Similar results were reported by Katz and Duncan (1975) for oral prednisone.

Glucocorticoid pharmacokinetics in children

Green *et al.* (1978) determined the half-life of prednisolone following oral administration of prednisone in children with various diseases. The mean half-life was 2.2 hours (somewhat shorter than in adults), but significantly longer (5.0 hours) in four children. There was no correlation with sex, size, disease, dose or duration of therapy. Subsequent analysis of their data by Rose, Jusko and Nickelsen (1979) revealed a curvilinear relationship between dose and peak plasma prednisolone, showing the expected dose dependence.

Rose *et al.* (1981a) administered prednisolone 40 mg i.v. to ten steroid-dependent asthmatic children. The mean clearance, volume of distribution and half-life were 246 ml/min per 1.73 m^2, 52.8 l/1.73 m^2 and 2.5 hours respectively. These values are similar to those found in healthy and asthmatic adults.

Rocci *et al.* (1982a) studied the pharmacokinetics of intravenous prednisolone and oral prednisone in children with nephrotic syndrome. The total clearance was greater than that reported in asthmatic children (*see above*), probably due to the loss of albumin in nephrotic syndrome and the increased unbound concentrations of plasma prednisolone. The disposition of methylprednisolone in adults and children with nephrotic syndrome was compared by Assael *et al.* (1982). Mean clearance per kg body weight was greater in children than in adults. However, when mean clearances were normalized for body surface area, no significant difference was found between children and adults. The mean volume of distribution was 1.27 l/kg for all subjects. The half-life was slightly, but not significantly, shorter in the children.

Glucocorticoid pharmacokinetics in older subjects

The total clearance of albumin-bound glucocorticoids might be expected to increase in older patients because of the decreased albumin concentrations (Chesrow *et al.,* 1958). Preliminary results from our laboratory suggest that the unbound clearance of intravenously administered prednisolone may be decreased in older women. More investigation is needed before the effects of age on glucocorticoid pharmacokinetics can be clearly delineated. From the available data it appears that clearance is similar in children and adults, but may decrease with advancing age. Volume of distribution seems best correlated with body weight so that children generally have smaller absolute volumes. The relationship between clearance and volume described earlier in this chapter dictates a shorter half-life in children.

References

AGABEYOGLU, I. T., BERGSTROM, R. F., GILLESPIE, W. R., WAGNER, J. G. and KAY, D. R. (1979) Plasma protein binding of prednisolone in normal volunteers and arthritic patients. *European Journal of Clinical Pharmacology,* **16,** 399–404

ALBERT, K. S., BROWN, S. W., DESANTE, K. A., DISANTO, A. R., STEWART, R. D. and CHEN, T. T. (1979) Double Latin square study to determine variability and relative bioavailability of methylprednisolone. *Journal of Pharmaceutical Sciences,* **68,** 1312–1316

AL-HABET, S. and ROGERS, H. J. (1980) Pharmacokinetics of intravenous and oral prednisolone. *British Journal of Clinical Pharmacology,* **10,** 503–508

ANDERSON, G. G., ROTCHELL, Y. and KAISER, D. G. (1981) Placental transfer of methylprednisolone following maternal intravenous administration. *American Journal of Obstetrics and Gynecology,* **140,** 699–701

ANGELI, A., FRAJRIA, R., DEPAOLI, R., FONZO, D. and CERESA, F. (1978) Diurnal variation of prednisolone binding to serum corticosteroid-binding globulin in man. *Clinical Pharmacology and Therapeutics,* **23,** 47–53

ARMSTRONG, R. D., ENGLISH, J., GIBSON, T., CHAKRABORTY, J. and MARKS, V. (1981) Serum methylprednisolone levels following intra-articular injection of methylprednisolone acetate. *Annals of the Rheumatic Diseases,* **40,** 571–574

ASSAEL, B. M., BANFI, G., APPIANI, A. C., EDEFONTI, A. and JUSKO, W. J. (1982) Disposition of pulse dose methylprednisolone in adult and paediatric patients with the nephrotic syndrome. *European Journal of Clinical Pharmacology,* **23,** 429–433

AUDETAT, V. and BIRCHER, J. (1976) Bioavailability of prednisolone during simultaneous treatment with cholestyramine. *Gastroenterology,* **71,** 1110–1111

BEHM, H. L. and WAGNER, J. G. (1979) Errors in interpretation of data from equilibrium dialysis protein binding experiments. *Research Communications in Chemical Pathology and Pharmacology,* **26,** 145–160

BEISEL, W. R., DIRAIMONDO, V. C., CHAO, P. Y., ROSNER, J. M. and FORSHAM, P. H. (1964) The influence of plasma protein binding on the extra-adrenal metabolism of cortisol in normal hyperthyroid and hypothyroid subjects. *Metabolism,* **13,** 942–951

BEISEL, W. R., DIRAIMONDO, V. C. and FORSHAM, P. H. (1964) Cortisol transport and disappearance. *Annals of Internal Medicine,* **60,** 641–652

BENET, L. Z. and GALEAZZI, R. L. (1979) Noncompartmental determination of the steady-state volume of distribution. *Journal of Pharmaceutical Sciences,* **68,** 1071–1074

BENET, L. Z., MASSOUD, N. and GAMBERTOGLIO, J. G. (Eds) (1984) *Pharmacokinetic Basis for Drug Treatment.* New York: Raven Press

BERGREM, H. (1983) The influence of uremia on pharmacokinetics and protein binding of prednisolone. *Acta Medica Scandinavica,* **213,** 333–337

BERGREM, H., DJØSELAND, O., JERVELL, J. and RUGSTAD, H. E. (1981) Absorption of prednisolone. I. The effect of fasting, food, and food combined with antacids. *Scandinavian Journal of Urology and Nephrology* (Suppl.), **64,** 167–173

BERGREM, H., GRØTTUM, P. and RUGSTAD, H. E. (1983) Pharmacokinetics and protein binding of prednisolone after oral and intravenous administration. *European Journal of Clinical Pharmacology,* **24,** 415–419

BERGREM, H. and REFVEM, O. K. (1982) Changes in prednisolone pharmacokinetics and protein binding during treatment with rifampicin. *Proceedings of the European Dialysis and Transplant Association,* **19,** 552–557

BOEKENOOGEN, S. J., SZEFLER, S. J. and JUSKO, W. J. (1983) Prednisolone disposition and protein binding in oral contraceptive users. *Journal of Clinical Endocrinology and Metabolism,* **56,** 702–709

BOYE, N. P., DJØSELAND, O. and HAUGEN, H. N. (1974) Protein binding and metabolic clearance of cortisol in asthmatic patients. *Scandinavian Journal of Respiratory Disease,* **55,** 200–206

BRAUDE, A. C. and REBUCK, A. S. (1983) Prednisone and methylprednisolone disposition in the lung. *Lancet,* **2,** 995–997

BRIEN, T. G. (1975) Cortisol metabolism after oral contraceptives: total plasma cortisol and the free cortisol index. *British Journal of Obstetrics and Gynaecology,* **82,** 987–991

BROOKS, P. M., BUCHANAN, W. W., GROVE, M. and DOWNIE, W. W. (1976) Effect of enzyme induction on metabolism of prednisolone. Clinical and laboratory study. *Annals of the Rheumatic Diseases,* **35,** 339–343

BROOKS, S. M., WERK, E. E., ACKERMAN, S. J., SULLIVAN, I. and THRASHER, K. (1972) Adverse effects of phenobarbital on corticosteroid metabolism in patients with bronchial asthma. *New England Journal of Medicine,* **286,** 1125–1128

BROWN, H., WILLARDSON, D. G., SAMUELS, L. T. and TYLER, F. H. (1954) 17-Hydroxycorticosteroid metabolism in liver disease. *Journal of Clinical Investigation,* **33,** 524–532

BRUMBACK, R. A. (1980) Failure of oral versus parenteral corticosteroids in a case of acute inflammatory polyradiculoneuropathy (Guillain-Barré syndrome). *Australian and New Zealand Journal of Medicine,* **10,** 224–226

BUFFINGTON, G. A., DOMINGUEZ, J. H., PIERING, W. F., HEBERT, L. A., KAUFFMAN, H. M. and LEMANN, J. (1976) Interaction of rifampin and glucocorticoids. Adverse effect on renal allograft function. *Journal of the American Medical Association,* **236,** 1958–1960

BURKE, C. W. (1969) Biologically active cortisol in plasma of oestrogen-treated and normal subjects. *British Medical Journal,* **2,** 798–800

CAMPAGNA, F. A., CURETON, G., MIRIGIAN, R. A. and NELSON, E. (1963) Inactive prednisone tablets USP XVI. *Journal of Pharmaceutical Sciences,* **52,** 605–606

CHESROW, E. J., BRONSKY, D., ORFEI, E., DYNIEWICZ, H., DUBIN, A. and MUSCI, J. (1958) Serum proteins in the aged. *Geriatrics,* **13,** 20–24

CHOI, Y., THRASHER, K., WERK, E. E., SHOLITON, L. J. and OLINGER, C. (1971) Effect of diphenylhydantoin on cortisol kinetics in humans. *Journal of Pharmacology and Experimental Therapeutics,* **176,** 27–34

CHULSKI, T. and FORIST, A. A. (1958) The effects of some solid buffering agents in aqueous suspension on prednisolone. *Journal of the American Pharmaceutical Association,* **47,** 553–555

COBURG, A. J., GRAY, S. H., KATZ, F. H., PENN, I., HALGRIMSON, C. and STARZL, T. E. (1970) Disappearance rates and immunosuppression of intermittent intravenously administered prednisolone in rabbits and human beings. *Surgery, Gynecology and Obstetrics,* **131,** 933–942

COLBURN, W. A. and BULLER, R. H. (1973) Radioimmunoassay for methylprednisolone (Medrol). *Steroids,* **22,** 687–697

COLBURN, W. A., DISANTO, A. R., STUBBS, S. S., MONOVICH, R. E. and DESANTE, K. A. (1980) Pharmacokinetic interpretation of plasma cortisol and cortisone concentrations following a single oral administration of cortisone acetate to human subjects. *Journal of Clinical Pharmacology,* **20,** 428–436

COLLINS, J. V., HARRIS, P. W. R., CLARK, T. J. H. and TOWNSEND, J. (1970) Intravenous corticosteroids in treatment of acute bronchial asthma. *Lancet,* **2,** 1047–1050

DAUGHADAY, W. H. and MARIZ, I. K. (1961) Corticosteroid binding globulin: its properties and quantification. *Metabolism,* **10,** 936–950

DAVIS, M., WILLIAMS, R., CHAKRABORTY, J., ENGLISH, J., MARKS, V., IDEO, G. and TEMPINI, S. (1978) Prednisone or prednisolone for the treatment of chronic active hepatitis? A comparison of plasma availability. *British Journal of Clinical Pharmacology,* **5,** 501–505

DAZORD, A., SAEZ, J. and BERTRAND, J. (1972) Metabolic clearance rates and interconversion of cortisol and cortisone. *Journal of Clinical Endocrinology and Metabolism,* **35,** 24–34

DEBOER, A. G., MOOLENAAR, F., DELEEDE, L. G. J. and BREIMER, D. D. (1982) Rectal drug administration: clinical pharmacokinetic considerations. *Clinical Pharmacokinetics,* **7,** 285–311

DE LACERDA, L., KOWARSKI, A. and MIGEON, C. J. (1973) Diurnal variation of the metabolic clearance rate of cortisol. *Journal of Clinical Endocrinology and Metabolism,* **36,** 1043–1049

DEMOOR, P. M., DECKX, R. and STEENO, O. (1963) Influence of various steroids on the specific binding of cortisol. *Journal of Endocrinology,* **27,** 355–356

DISANTO, A. R. and DESANTE, K. A. (1975) Bioavailability and pharmacokinetics of prednisone in humans. *Journal of Pharmaceutical Sciences,* **64,** 109–112

DUGGAN, D. E., YEH, K. C., MATALIA, N., DITZLER, C. A. and MC MAHON, F. G. (1975) Bioavailability of oral dexamethasone. *Clinical Pharmacology and Therapeutics,* **18,** 205–209

EDWARDS, O. M., COURTENAY-EVANS, R. J., GALLEY, J. M., HUNTER, J. and TAIT, A. D. (1974) Changes in cortisol metabolism following rifampicin therapy. *Lancet,* **2,** 549–551

ELLIOTT, P. R., POWELL-TUCK, J., GILLESPIE, P. E., LAIDLOW, J. M., LENNARD-JONES, J. E., ENGLISH, J., CHAKRABORTY, J. and MARKS, V. (1980) Prednisone absorption in acute colitis. *Gut,* **21,** 49–51

ELY, R. S., DONE, A. K. and KELLEY, V. C. (1956) Δ^1-Hydrocortisone: plasma 17-hydroxycorticosteroid concentrations following oral and i.v. administration. *Proceedings of the Society for Experimental Biology and Medicine,* **91,** 503–506

ENGLISH, J., CHAKRABORTY, J., MARKS, V. and PARKE, A. (1975) A radioimmunoassay procedure for dexamethasone. *European Journal of Clinical Pharmacology,* **9,** 239–244

ENGLISH, J., DUNNE, M. and MARKS, V. (1983) Diurnal variation in prednisolone kinetics. *Clinical Pharmacology and Therapeutics,* **33,** 381–385

FARMER, R. G. and SCHUMACHER, O. P. (1970) Treatment of ulcerative colitis with hydrocortisone enemas. *American Journal of Gastroenterology,* **54,** 229–236

FARISS, B. L., HANE, S., SHINSAKO, J. and FORSHAM, P. H. (1978) Comparison of absorption of cortisone acetate and hydrocortisone hemisuccinate. *Journal of Clinical Endocrinology and Metabolism,* **47,** 1137–1140

FELL, P. J. (1972) Kinetic studies of cortisol and synthetic corticosteroids in man. *Clinical Endocrinology,* **1,** 65–72

FERRARI, S. and FANTOLI, U. (1964) Duration of blood and urinary levels of corticosteroid following intramuscular injection of methylprednisolone acetate. *Clinical Medicine,* **71,** 706–709

FEW, J. D. and HASPINEALL, J. R. (1977) Some physiological factors affecting the binding of cortisol by human plasma proteins. *Annals of Clinical Biochemistry,* **14,** 35–38

FLORES, J. J. B. and ROJAS, S. V. (1975) Naproxen: corticosteroid-sparing effect in rheumatoid arthritis. *Journal of Clinical Pharmacology,* **15,** 373–377

FLORINI, J. R. and BUYSKE, D. A. (1961) Plasma protein binding of triamcinolone-H^3 and hydrocortisone-C^{14}. *Journal of Biological Chemistry,* **236,** 247–251

FREY, F. J., AMEND, W. J. C., LOZADA, F., FREY, B. M., HOLFORD, N. H. G. and BENET, L. Z. (1981a) Pharmacokinetics of prednisolone and endogenous hydrocortisone levels in cushingoid and non-cushingoid patients. *European Journal of Clinical Pharmacology,* **21,** 235–242

FREY, F. J., LOZADA, F., GUENTERT, T. and FREY, B. M. (1981b) A single dose of azathioprine does not affect the pharmacokinetics of prednisolone following oral prednisone. *European Journal of Clinical Pharmacology,* **19,** 209–212

FREY, F. J. and FREY, B. M. (1982) Altered prednisolone kinetics in patients with the nephrotic syndrome. *Nephron,* **32,** 45–48

FREY, F. J., GAMBERTOGLIO, J. G., FREY, B. M., BENET, L. Z. and AMEND, W. J. C. (1982) Nonlinear plasma protein binding and haemodialysis clearance of prednisolone. *European Journal of Clinical Pharmacology,* **23,** 65–74

GAMBERTOGLIO, J. G., AMEND, JR, W. J. C. and BENET, L. Z. (1980) Pharmacokinetics and bioavailability of prednisone and prednisolone in healthy volunteers and patients: a review. *Journal of Pharmacokinetics and Biopharmaceutics,* **8,** 1–52

GAMBERTOGLIO, J. G., FREY, F. J., HOLFORD, N. H. G., BIRNBAUM, J. L., LIZAK, P. S., VINCENTI, F., FEDUSKA, N. J., SALVATIERRA, JR, O. and AMEND, JR, W. J. C. (1982) Prednisone and prednisolone bioavailability in renal transplant patients. *Kidney International,* **21,** 621–626

GARG, D. C., NG, P., WEIDLER, D. J., SAKMAR, E. and WAGNER, J. G. (1978) Preliminary *in vitro* and *in vivo* investigations on methylprednisolone and its acetate. *Research Communications in Chemical Pathology and Pharmacology,* **22,** 37–48

GARG, D. C., WAGNER, J. G., SAKMAR, E., WEIDLER, D. J. and ALBERT, K. S. (1979) Rectal and oral absorption of methylprednisolone acetate. *Clinical Pharmacology and Therapeutics,* **26,** 232–239

GEORGITIS, J. W., FLESHER, K. A. and SZEFLER, S. J. (1982) Bioavailability assessment of a liquid prednisolone preparation. *Journal of Allergy and Clinical Immunology,* **70,** 243–247

GIBALDI, M. and PERRIER, D. (1982) *Pharmacokinetics,* 2nd Edn. New York: Marcel Dekker

GREEN, O. C., WINTER, R. J., KAWAHARA, F. S., PHILLIPS, L. S., LEWY, P. R., HART, R. L. and PACHMAN, L. M. (1978) Pharmacokinetic studies of prednisolone in children. *Journal of Pediatrics,* **93,** 299–303

HAQUE, N., THRASHER, K., WERK, E. E., KNOWLES, H. C. and SHOLITON, L. J. (1972) Studies on dexamethasone metabolism in man: effect of diphenylhydantoin. *Journal of Clinical Endocrinology and Metabolism,* **34,** 44–50

HARE, L. E., YEH, K. C., DITZLER, C. A., MCMAHON, F. G. and DUGGAN, D. E. (1975) Bioavailability of dexamethasone. II. Dexamethasone phosphate. *Clinical Pharmacology and Therapeutics,* **18,** 330–337

HENCH, P. S. (1938) The ameliorating effect of pregnancy on chronic atrophic (infectious rheumatoid) arthritis, fibrositis, and intermittent hydroarthrosis. *Proceedings of the Staff Meetings of the Mayo Clinics,* **13,** 161–167

HENDRICKSE, W., MCKIERNAN, J., PICKUP, M. and LOWE, J. (1979) Rifampicin-induced non-responsiveness to corticosteroid treatment in nephrotic syndrome. *British Medical Journal,* **1,** 306

HICHENS, M. and HOGANS, A. F. (1974) Radioimmunoassay for dexamethasone in plasma. *Clinical Chemistry,* **20,** 266–271

HOLLANDER, J. L., BROWN, E. M., JESSAR, R. A. and BROWN, C. Y. (1951) Hydrocortisone and cortisone injected into arthritic joints. *Journal of the American Medical Association,* **147,** 1629–1635

HSUEH, W. A., PAZ-GUEVARA, A. and BLEDSOE, T. (1979) Studies comparing the metabolic clearance rate of 11β,17,21-trihydroxypregn-1,4-diene-3,20-dione (prednisolone) after oral 17,21-dihydroxypregn-1,4-diene-3,11,20-trione and intravenous prednisolone. *Journal of Clinical Endocrinology and Metabolism,* **48,** 748–752

HULME, B., JAMES, V. H. T. and RAULT, R. (1975) Absorption of enteric and non-enteric coated prednisolone tablets. *British Journal of Clinical Pharmacology,* **2,** 317–320

JENKINS, J. S., LOWE, R. D. and TITTERINGTON, E. (1964) Effect of adrenocortical hormones on release of free fatty acids and uptake of glucose in human peripheral tissues. *Clinical Science,* **26,** 421–427

JENKINS, J. S. and SAMPSON, P. A. (1967) Conversion of cortisone to cortisol and prednisone to prednisolone. *British Medical Journal,* **2,** 205–207

JENKINS, D. and SCHEMMEL, J. E. (1958) Metabolic effects of 6-methylprednisolone. *Metabolism,* **7,** 416–424

JOHNSON, M. W., YOUSSEFNEJADIAN, E. and CRAFT, I. (1976) A radioimmunoassay for betamethasone. Preparation of a specific antiserum to betamethasone-3(O-carboxymethyl)oxime-bovine serum albumin and evaluation of method. *Journal of Steroid Biochemistry,* **7,** 795–799

JUBIZ, W., MEIKLE, A. W., LEVINSON, R. A., MIZUTANI, S., WEST, C. D. and TYLER, F. H. (1970) Effect of diphenylhydantoin on the metabolism of dexamethasone. Mechanism of the abnormal dexamethasone suppression in humans. *New England Journal of Medicine,* **283,** 11–14

KATZ, F. H. and DUNCAN, B. R. (1975) Entry of prednisolone into human milk. *New England Journal of Medicine,* **293,** 1154

KEHLET, H. and BINDER, C. H. R. (1973) Alterations in distribution volume and biological half-life of cortisol during major surgery. *Journal of Clinical Endocrinology and Metabolism,* **36,** 330–333

KEHLET, H., BINDER, C. and BLICHERT-TOFT, M. (1976) Glucocorticoid maintenance therapy following adrenalectomy: assessment of dosage and preparation. *Clinical Endocrinology,* **5,** 37–41

KEHLET, H., NISTRUP-MASDEN, S. and BINDER, C. (1974) Cortisol and cortisone acetate in parenteral glucocorticoid therapy. *Acta Medica Scandinavica,* **195,** 421–423

KOZOWER, M., VEATCH, L. and KAPLAN, M. M. (1974) Decreased clearance of prednisolone, a factor in the development of corticosteroid side-effects. *Journal of Clinical Endocrinology and Metabolism,* **38,** 407–412

KREAM, J., MULAY, S., FUKUSHIMA, D. K. and SOLOMON, S. (1983) Determination of plasma dexamethasone in the mother and the newborn after administration of the hormone in a clinical trial. *Journal of Clinical Endocrinology and Metabolism,* **56,** 127–133

LAFORCE, C. F., SZEFLER, S. J., MILLER, M. F., EBLING, W. and BRENNER, M. (1983) Inhibition of methylprednisolone elimination in the presence of erythromycin therapy. *Journal of Allergy and Clinical Immunology,* **72,** 34–39

LECLERQ, R. and COPINSCHI, G. (1974) Patterns of plasma levels of prednisolone after oral administration in man. *Journal of Pharmacokinetics and Biopharmaceutics,* **2,** 175–187

LEE, D. A. H., TAYLOR, M., JAMES, V. H. T. and WALKER, G. (1980) Rectally administered prednisolone – evidence for a predominantly local action. *Gut,* **21,** 215–218

LEE, D. A. H., TAYLOR, G. M., WALKER, J. G. and JAMES, V. H. T. (1979a) The effect of food and tablet formulation on plasma prednisolone levels following administration of enteric-coated tablets. *British Journal of Clinical Pharmacology,* **7,** 523–528

LEE, D. A. H., TAYLOR, G. M., WALKER, J. G. and JAMES, V. H. T. (1979b) The effect of concurrent administration of antacids on prednisolone absorption. *British Journal of Clinical Pharmacology,* **8,** 92–94

LEGLER, U. F., FREY, F. J. and BENET, L. Z. (1982) Prednisolone clearance at steady state in man. *Journal of Clinical Endocrinology and Metabolism,* **55,** 762–767

LERNER, L. J., TURKHEIMER, A. R., BIANCHI, A., SINGER, F. M. and BORMAN, A. (1964) Comparison of anti-granuloma, thymolytic and glucocorticoid activities of anti-inflammatory steroids. *Proceedings of the Society for Experimental Biology and Medicine,* **116,** 385–388

LEVY, G., HALL, N. A. and NELSON, E. (1964) Studies on active prednisone tablets USP XVI. *American Journal of Hospital Pharmacy,* **21,** 402

LEWIS, G. P., JUSKO, W. J., BURKE, C. W. and GRAVES, L. (1971) Prednisone side-effects and serum protein levels. *Lancet,* **2,** 778–781

LIMA, J. J., GILLER, J., MACKICHAN, J. J. and JUSKO, W. J. (1980) Bioavailability of hydrocortisone retention enemas in normal subjects. *American Journal of Gastroenterology,* **73,** 232–237

LIMA, J. J. and JUSKO, W. J. (1980) Bioavailability of hydrocortisone retention enemas in relation to absorption kinetics. *Clinical Pharmacology and Therapeutics,* **28,** 262–269

LOO, J. C. K., MCGILVERAY, I. J., JORDAN, N. and BRIEN, R. (1981) Pharmacokinetic evaluation of betamethasone and its water-soluble phosphate ester in humans. *Biopharmaceutics and Drug Disposition,* **2,** 265–272

LOO, J. C. K., MCGILVERAY, I. J., JORDAN, N., MOFFAT, J. and BRIEN, R. (1978) Dose-dependent pharmacokinetics of prednisone and prednisolone in man. *Journal of Pharmacy and Pharmacology,* **30,** 78

LOWE, J. R. and DIXON, J. S. (1983) Salivary kinetics of prednisolone in man. *Journal of Pharmacy and Pharmacology,* **35,** 390–391

MANT, T. G. K. (1979) Investigation of a case of prednisolone malabsorption. *Postgraduate Medical Journal,* **55,** 421–422

MADSBAD, S., BJERREGAARD, B., HENRIKSEN, J. H., JUHL, E. and KEHLET, H. (1980) Impaired conversion of prednisone to prednisolone in patients with liver cirrhosis. *Gut,* **21,** 52–56

MCALLISTER, W. A. C., MITCHELL, D. M. and COLLINS, J. V. (1981) Prednisolone pharmacokinetics compared between night and day in asthmatic and normal subjects. *British Journal of Clinical Pharmacology,* **11,** 303–304

MCALLISTER, W. A. C., THOMPSON, P. J., AL-HABET, S. M. and ROGERS, H. J. (1983) Rifampicin reduces effectiveness and bioavailability of prednisolone. *British Medical Journal,* **286,** 923–925

MCCAFFERTY, J., BROPHY, T. R. O'R., YELLAND, J. D.., CHAM, B. E., BOCHNER, F. and EADIE, M. J. (1981) Intraoperative pharmacokinetics of dexamethasone. *British Journal of Clinical Pharmacology,* **12,** 434–436

MCKENZIE, S. A., SELLEY, J. A. and AGNEW, J. E. (1975) Secretion of prednisolone into breast milk. *Archives of Disease in Childhood,* **50,** 894–896

MEIKLE, A. W., LAGERQUIST, L. G. and TYLER, F. H. (1973) A plasma dexamethasone radioimmunoassay. *Steroids,* **22,** 193–202

MEIKLE, A. W., WEED, J. A. and TYLER, F. H. (1975) Kinetics and interconversion of prednisolone and prednisone studied with new radioimmunoassays. *Journal of Clinical Endocrinology and Metabolism,* **41,** 717–721

MEYERS, M., SLIKKER, W., PASCOE, G. and VORE, M. (1980) Characterization of cholestasis induced by estradiol-17β-D-glucuronide in the rat. *Journal of Pharmacology and Experimental Therapeutics,* **214,** 87–93

MILLS, I. H., SCHEDL, H. P., CHEN, P. S. and BARTTER, F. C. (1960) The effect of estrogen administration on the metabolism and protein binding of hydrocortisone. *Journal of Clinical Endocrinology and Metabolism,* **20,** 515–528

MILSAP, R. L., GEORGE, D. E., SZEFLER, S. J., MURRAY, K. A., LEBENTHAL, E. and JUSKO, W. J. (1983) Effect of inflammatory bowel diseases on absorption and disposition of prednisolone. *Digestive Diseases and Sciences,* **28,** 161–168

MILSAP, R. L. and JUSKO, W. J. (1983) Binding of prednisolone to α_1-acid glycoprotein. *Journal of Steroid Biochemistry,* **18,** 191–194

MIYABO, S., NAKAMURA, T., KUWAZIMA, S. and KISHIDA, S. (1981) A comparison of the bioavailability and potency of dexamethasone phosphate and sulphate in man. *European Journal of Clinical Pharmacology,* **20,** 277–282

MORRISON, P. J., ROGERS, H. J., BRADBROOK, I. D. and PARSONS, C. (1980) Concurrent administration of cimetidine and enteric-coated prednisolone: effect on plasma levels of prednisolone. *British Journal of Clinical Pharmacology,* **10,** 87–89

MULDOON, T. G. and WESTPHAL, U. (1967) Steroid protein interactions. XV. Isolation and characterization of corticosteroid-binding globulin from human plasma. *Journal of Biological Chemistry,* **242,** 5636–5643

NUGENT, C. A., EIK-NES, K. and TYLER, F. H. (1959) A comparative study of the metabolism of hydrocortisone and prednisolone. *Journal of Clinical Endocrinology and Metabolism,* **19,** 526–534

ØIE, S. and TOZER, T. N. (1979) Effect of altered plasma protein binding on apparent volume of distribution. *Journal of Pharmaceutical Sciences,* **68,** 1203–1205

ORTIZ, DE MONTELLANO, P. R., KUNZE, K. L., YOST, G. S. and MICO, B. A. (1979) Self-catalyzed destruction of cytochrome P450: covalent binding of ethynyl sterols to prosthetic heme. *Proceedings of the National Academy of Sciences, USA,* **76,** 746–749

OSTHANONDH, R., TULCHINSKY, D., KAMALI, H., FENCL, M.DEM. and TAEUSCH, JR, H. W. (1977) Dexamethasone levels in treated pregnant women and newborn infants. *Journal of Pediatrics,* **90,** 617–620

PARK, H. S., GREENE, J. A. and BACON, G. E. (1974) Disappearance of high dose prednisolone from plasma after renal transplantation. *Clinical Nephrology,* **2,** 18–23

PATTERSON, M. (1958) Studies on the absorption of hydrocortisone from the colon of patients with idiopathic ulcerative colitis. *Texas Reports on Biology and Medicine,* **16,** 508–514

PEETS, E. A., STAUB, M. and SYMCHOWICZ, S. (1969) Plasma binding of betamethasone-^{3}H, dexamethasone-^{3}H and cortisol-^{14}C – a comparative study. *Biochemical Pharmacology,* **18,** 1655–1663

PETEREIT, L. B. and MEIKLE, A. W. (1977) Effectiveness of prednisolone during phenytoin therapy. *Clinical Pharmacology and Therapeutics,* **22,** 912–916

PETERSEN, M. C., NATION, R. L., ASHLEY, J. J. and MCBRIDE, W. G. (1980) The placental transfer of betamethasone. *European Journal of Clinical Pharmacology,* **18,** 245–247

PETERSON, R. E. (1959) Metabolism of adrenocorticosteroids in man. *Annals of the New York Academy of Sciences,* **82,** 846–853

PETERSON, R. E., WYNGAARDEN, J. B., GUERRA, S. L., BRODIE, B. B. and BUNIM, J. J. (1955) The physiological disposition and metabolic fate of hydrocortisone in man. *Journal of Clinical Investigation,* **34,** 1779–1794

PICKUP, M. E. (1979) Clinical pharmacokinetics of prednisone and prednisolone. *Clinical Pharmacokinetics,* **4,** 111–128

PICKUP, M. E., FARAH, F., LOWE, J. R., DIXON, J. S. and RECORD, C. O. (1979) Prednisolone absorption in coeliac disease. *European Journal of Drug Metabolism and Pharmacokinetics,* **4,** 87–89

PICKUP, M. E., LOWE, J. R., LEATHAM, P. A., RHIND, V. M., WRIGHT, V. and DOWNIE, W. W. (1977) Dose dependent pharmacokinetics of prednisolone. *European Journal of Clinical Pharmacology,* **12,** 213–219

PLAGER, J. E., SCHMIDT, K. G. and STAUBITZ, W. J. (1964) Increased unbound cortisol in the plasma of estrogen-treated subjects. *Journal of Clinical Investigation,* **43,** 1066–1072

PLUMPTON, F. S., BESSER, G. M. and COLE, P. V. (1969) Corticosteroid treatment and surgery. 2. The management of steroid cover. *Anaesthesia,* **24,** 12–18

POWELL, L. W. and AXELSEN, E. (1972) Corticosteroids in liver disease: studies on the biological conversion of prednisone to prednisolone and plasma protein binding. *Gut,* **13,** 690–696

POWELL-TUCK, J., LENNARD-JONES, J. E., MAY, C. S., WILSON, C. G. and PATERSON, J. W. (1976) Plasma prednisolone levels after administration of prednisolone-21-phosphate as a retention enema in colitis. *British Medical Journal,* **1,** 193–195

RAE, S. A., WILLIAMS, I. A., ENGLISH, J. and BAYLIS, E. M. (1982) Alteration of plasma prednisolone levels by indomethacin and naproxen. *British Journal of Clinical Pharmacology,* **14,** 459–461

REEBACK, J. S., CHAKRABORTY, J., ENGLISH, J., GIBSON, T. and MARKS, V. (1980) Plasma steroid levels after intra-articular injection of prednisolone acetate in patients with rheumatoid arthritis. *Annals of the Rheumatic Diseases,* **39,** 22–24

ROBINSON, R. C. V. and ROBINSON, H. M. (1956) Topical treatment of dermatoses with steroids. *Southern Medical Journal,* **49,** 260–266

ROCCI, M. L., ASSAEL, B. M., APPIANI, A. C., EDEFONTI, A. and JUSKO, W. J. (1982a) Effect on nephrotic syndrome on absorption and disposition of prednisolone in children. *International Journal of Pediatric Nephrology,* **3,** 161–166

ROCCI, M. L., D'AMBROSIO, R., JOHNSON, N. F. and JUSKO, W. J. (1982b) Prednisolone binding to albumin and transcortin in the presence of cortisol. *Biochemical Pharmacology,* **31,** 289–292

ROSE, J. Q., JUSKO, W. J. and NICKELSEN, J. A. (1979) Prednisolone pharmacokinetics in relation to dose. *Journal of Pediatrics,* **94,** 1014–1015

ROSE, J. Q., NICKELSEN, J. A., ELLIS, E. F., MIDDLETON, E. and JUSKO, W. J. (1981a) Prednisolone disposition in steroid-dependent asthmatic children. *Journal of Allergy and Clinical Immunology,* **67,** 188–193

ROSE, J. Q., YURCHAK, A. M., MEIKLE, A. W. and JUSKO, W. J. (1981b) Effect of smoking on prednisone, prednisolone and dexamethasone pharmacokinetics. *Journal of Pharmacokinetics and Biopharmaceutics,* **9,** 1–14

ROSE, J. Q., NICKELSEN, J. A., MIDDLETON, E., YURCHAK, A. M., PARK, B. H. and JUSKO, W. J. (1980) Prednisolone disposition in steroid-dependent asthmatics. *Journal of Allergy and Clinical Immunology,* **66,** 366–373

ROSE, J. Q., YURCHAK, A. M. and JUSKO, W. J. (1980) Bioavailability and disposition of prednisone and prednisolone from prednisone tablets. *Biopharmaceutics and Drug Disposition,* **1,** 247–258

ROSE, J. Q., YURCHAK, A. M. and JUSKO, W. J. (1981) Dose dependent pharmacokinetics of prednisone and prednisolone in man. *Journal of Pharmacokinetics and Biopharmaceutics,* **9,** 389–417

ROWLAND, M., BENET, L. Z. and GRAHAM, G. G. (1973) Clearance concepts in pharmacokinetics. *Journal of Pharmacokinetics and Biopharmaceutics,* **1,** 123–136

ROWLAND, M. and TOZER, T. N. (1980) *Clinical Pharmacokinetics.* Philadelphia: Lea and Febiger

SAMPSON, P. A. and BROOKE, B. N. (1963) Absorption of hydrocortisone from the large bowel. *Lancet,* **1,** 701–702

SAMUELS, L. T. and WEST, C. D. (1952) The intermediary metabolism of the nonbenzenoid steroid hormones. *Vitamins and Hormones,* **10,** 251–265

SANBAR, S. S. and WEST, K. M. (1961) Rectal absorption of radioactive 6-alpha-methyl prednisolone in ulcerative colitis. *Journal Medical Libanais,* **14,** 380–386

SANDBERG, A. A. and SLAUNWHITE, W. R. (1959) Transcortin: a corticosteroid-binding protein of plasma. II. Levels in various conditions and the effects of estrogens. *Journal of Clinical Investigation,* **38,** 1290–1297

SANDBERG, A. A. and SLAUNWHITE, W. R. (1963) Transcortin: a corticosteroid-binding protein of plasma. V. *In vitro* inhibition of cortisol metabolism. *Journal of Clinical Investigation,* **42,** 51–54

SANDBERG, A. A., SLAUNWHITE, W. R. and CARTER, A. C. (1960) Transcortin: a corticosteroid-binding protein of plasma. III. The effects of various steroids. *Journal of Clinical Investigation,* **39,** 1914–1926

SCAVO, D., CUGINI, P. and DILASCIO, G. (1978) The metabolic clearance rate of cortisol in normal subjects and in various diseases states. *Journal of Nuclear Medicine and Allied Sciences,* **22,** 113–123

SCHALM, S. W., SUMMERSKILL, W. H. J. and GO, V. L. W. (1977) Prednisone for chronic active liver disease: pharmacokinetics, including conversion to prednisolone. *Gastroenterology,* **72,** 910–913

SCHEDL, H. P. (1965) Absorption of steroid hormones from the human small intestine. *Journal of Clinical Endocrinology and Metabolism,* **25,** 1309–1316

SCHMID, K. (1975) α_1-Acid glycoprotein. In *The Plasma Proteins,* vol. 1, edited by F. W. Putnam, pp. 184–228. New York: Academic Press, Inc

SCHWARTZ, R. D., COHN, G. L., BONDY, P. K., BRODOFF, M. P., UPTON, G. V. and SPIRO, H. P. (1958) Absorption of cortisol from the colon in ulcerative colitis. *Proceedings of the Society for Experimental Biology and Medicine,* **97,** 648–650

SEAL, U. S. and DOE, R. P. (1962) Corticosteroid-binding globulin. I. Isolation from plasma of diethylstilbestrol-treated men. *Journal of Biological Chemistry,* **237,** 3136–3140

SELLS, R. A., BROOKES, L., BASU, P. and WHITEMORE, D. (1978) Methylprednisolone blood levels in cadaveric renal allograft recipients. *Transplantation Proceedings,* **10,** 651–653

SHERLOCK, J. E. and LETTERI, J. M. (1977) Effect of hemodialysis on methylprednisolone plasma levels. *Nephron,* **18,** 208–211

SIITERI, P. K., MURAI, J. T., HAMMOND, G. L., NISKER, J. A., RAYMOURE, W. J. and KUHN, R. W. (1982) The serum transport of steroid hormones. *Recent Progress in Hormone Research,* **38,** 457–510

SLAUNWHITE, W. R., LOCKIE, G. N., BACK, N. and SANDBERG, A. A. (1962) Inactivity *in vivo* of transcortin-bound cortisol. *Science,* **135,** 1062–1063

SLAUNWHITE, W. R. and SANDBERG, A. A. (1961) Disposition of radioactive 17α-hydroxyprogesterone, 6α-methyl-17α-acetoxyprogesterone and 6α-methylprednisolone in human subjects. *Journal of Clinical Endocrinology and Metabolism,* **21,** 753–764

SMITH, J. B., NOLAN, G. and JUBIZ, W. (1980) The relationship between unbound and total cortisol: its usefulness in detecting CBG abnormalities. *Clinica Chimica Acta,* **108,** 435–444

SPECTOR, S. L., KATZ, F. H. and FARR, R. S. (1974) Troleandomycin: effectiveness in steroid-dependent asthma and bronchitis. *Journal of Allergy and Clinical Immunology,* **54,** 367–379

SPENCER, J. A., KIRSNER, J. B. and PALMER, W. L. (1960) Rectal absorption of 6-alpha-C^{14}-H_3-prednisolone. *Proceedings of the Society for Experimental Biology,* **103,** 74–77

STJERNHOLM, M. R. and KATZ, F. H. (1975) Effects of diphenylhydantoin, phenobarbital, and diazepam on the metabolism of methylprednisolone and its sodium succinate. *Journal of Clinical Endocrinology and Metabolism,* **41,** 887–893

SUGITA, E. T. and NIEBERGALL, P. J. (1975) Prednisone bioavailability monograph. *Journal of the American Pharmaceutical Association,* **15,** 529–532

SULLIVAN, T. J., HALLMARK, M. R., SAKMAR, E., WEIDLER, D. J., EARHART, R. H. and WAGNER, J. G. (1976) Comparative bioavailability: eight commercial prednisone tablets. *Journal of Pharmacokinetics and Biopharmaceutics,* **4,** 157–172

SZEFLER, S. J., ROSE, J. Q., ELLIS, E. F., SPECTOR, S. L., GREEN, A. W. and JUSKO, W. J. (1980) The effect of troleandomycin on methylprednisolone elimination. *Journal of Allergy and Clinical Immunology,* **66,** 447–451

TAHA, A.-K., AHMAD, R. A., GRAY, H., ROBERTS, C. I. and ROGERS, H. J. (1982) Plasma melphalan and prednisolone concentrations during oral therapy for multiple myeloma. *Cancer Chemotherapy and Pharmacology,* **9,** 57–60

TANNER, A., BOCHNER, F., CAFFIN, J., HALLIDAY, J. and POWELL, L. (1979a) Dose-dependent prednisolone kinetics. *Clinical Pharmacology and Therapeutics,* **25,** 571–578

TANNER, A. R., CAFFIN, J. A., HALLIDAY, J. W. and POWELL, L. W. (1979b) Concurrent administration of antacids and prednisone: effect on serum levels of prednisolone. *British Journal of Clinical Pharmacology,* **7,** 397–400

TANNER, A. R., HALLIDAY, J. W. and POWELL, L. W. (1981) Serum prednisolone levels in Crohn's disease and coeliac disease following oral prednisolone administration. *Digestion,* **21,** 310–315

TEMBO, A. V., HALLMARK, M. R., SAKMAR, E., BACHMANN, H. G., WEIDLER, D. J. and WAGNER, J. G. (1977) Bioavailability of prednisolone tablets. *Journal of Pharmacokinetics and Biopharmaceutics,* **5,** 257–270

TEMBO, A. V., SAKMAR, E., HALLMARK, M. R., WEIDLER, D. J. and WAGNER, J. G. (1976) Effect of food on the bioavailability of prednisone. *Journal of Clinical Pharmacology,* **16,** 620–624

THIESSEN, J. J. (1976) Prednisolone bioavailability monograph. *Journal of the American Pharmaceutical Association,* **16,** 143–146

TOOTHAKER, R. D., CRAIG, W. A. and WELLING, P. G. (1982) Effect of dose size on the pharmacokinetics of oral hydrocortisone suspension. *Journal of Pharmaceutical Sciences,* **71,** 1182–1185

TOOTHAKER, R. D., SUNDARESAN, G. M., HUNT, J. P., GOEHL, T. J., ROTENBERG, K. S., PRASAD, V. K., CRAIG, W. A. and WELLING, P. G. (1982) Oral hydrocortisone pharmacokinetics: a comparison of fluorescence and ultraviolet HPLC assays for hydrocortisone in plasma. *Journal of Pharmaceutical Sciences,* **71,** 573–576

TOOTHAKER, R. D. and WELLING, P. G. (1982) Effect of dose size on the pharmacokinetics of intravenous hydrocortisone during endogenous hydrocortisone suppression. *Journal of Pharmacokinetics and Biopharmaceutics,* **10,** 147–156

TOZER, T. N., GAMBERTOGLIO, J. G., FURST, D. E., AVERY, D. S. and HOLFORD, N. H. G. (1983) Volume shifts and protein binding estimates using equilibrium dialysis – application to prednisolone in man. *Journal of Pharmaceutical Sciences,* **72,** 1442–1446

TSE, F. L. S. and WELLING, P. G. (1979) Relative bioavailability of prednisone and prednisolone in man. *Journal of Pharmacy and Pharmacology,* **31,** 492–493

TSUEI, S. E., MOORE, R. G., ASHLEY, J. J. and MCBRIDE, W. G. (1979) Disposition of synthetic glucocorticoids. I. Pharmacokinetics of dexamethasone in healthy adults. *Journal of Pharmacokinetics and Biopharmaceutics,* **7,** 249–262

TURCOTTE, J. G., FEDUSKA, N. J., CARPENTER, E. W., MCDONALD, F. D. and BACON, G. E. (1972) Rejection crises in human renal transplant recipients. Control with high dose methylprednisolone therapy. *Archives of Surgery,* **105,** 230–236

UI, T., MITSUNAGA, M., TANAKA, T. and HORIGUCHI, M. (1982) Determination of prednisone and prednisolone in human serum by high-performance liquid chromatography – especially on impaired conversion of corticosteroids in patients with chronic liver disease. *Journal of Chromatography,* **239,** 711–716

URIBE, M. and GO. V. L. W. (1979) Corticosteroid pharmacokinetics in liver disease. *Clinical Pharmacokinetics,* **4,** 233–240

URIBE, M., GO, V. L. W. and SUMMERSKILL, W. H. J. (1976) Kinetics and interconversion of prednisone and prednisolone compared in chronic active liver disease (CALD) after oral doses. *Gastroenterology,* **71,** 932

URIBE, M., SCHALM, S. W., SUMMERSKILL, W. H. J. and GO, V. L. W. (1978) Oral prednisone for chronic active liver disease: dose responses and bioavailability studies. *Gut,* **19,** 1131–1135

URIBE, M., SUMMERSKILL, W. H. J. and GO, V. L. W. (1977) Why hyperbilirubinemia and hypoalbuminemia predispose to steroid side effects during treatment of chronic active liver disease (CALD). *Gastroenterology,* **72,** 1143

URIBE, M., SUMMERSKILL, W. H. J. and GO, V. L. W. (1982) Comparative serum prednisone and prednisolone concentrations following administration to patients with chronic active liver disease. *Clinical Pharmacokinetics,* **7,** 452–459

WAGNER, J. G. (1975) *Fundamentals of Clinical Pharmacokinetics.* Hamilton, Ill: Drug Intelligence Publications

WAGNER, J. G., DISANTO, A. R., GILLESPIE, W. R. and ALBERT, K. S. (1981) Reversible metabolism and pharmacokinetics: application to prednisone–prednisolone. *Research Communications in Chemical Pathology and Pharmacology,* **32,** 387–406

WASSNER, S. J., MALEKZADEH, M. H., PENNISI, A. J., ETTENGER, R. B., UITTENBOGAART, C. H. and FINE, R. N. (1977) Allograft survival in patients receiving anticonvulsant medications. *Clinical Nephrology,* **8,** 293–297

WASSNER, S. J., PENNISI, A. J., MALEKZADEH, M. H. and FINE, R. N. (1976) The adverse effect of anticonvulsant therapy on renal allograft survival. *Journal of Pediatrics,* **88,** 134–137

WERK, E. E., MACGEE, J. and SHOLITON, L. J. (1964) Effect of diphenylhydantoin on cortisol metabolism in man. *Journal of Clinical Investigation,* **43,** 1824–1835

WEST, H. F. (1959) Prevention of peptic ulceration during corticosteroid therapy. *British Medical Journal,* **2,** 680

WESTPHAL, U. (1967) Steroid protein interactions. XIII. Concentrations and binding affinities of corticosteroid-binding globulins in sera of man, monkey, rat, rabbit and guinea pig. *Archives of Biochemistry and Biophysics,* **118,** 556–567

WILKINSON, G. R. (1984) Pharmacokinetic considerations in toxicology. In *Drug Metabolism and Drug Toxicity*, edited by J. R. Mitchell and M. G. Horning, pp. 213–235. New York: Raven Press

WILKINSON, G. R. and SHAND, D. G. (1975) A physiological approach to hepatic drug clearance. *Clinical Pharmacology and Therapeutics,* **18,** 377–390

WILSON, C. G., MAY, C. S. and PATERSON, J. W. (1977) Plasma prednisolone levels in man following administration in plain and enteric-coated forms. *British Journal of Clinical Pharmacology,* **4,** 351–355

WILSON, C. G., SSENDAGIRE, R., MAY, C. S. and PATERSON, J. W. (1975) Measurement of plasma prednisolone in man. *British Journal of Clinical Pharmacology,* **2,** 321–325

Index